*The Estrogen Elixir*

# The Estrogen Elixir

## A History of Hormone Replacement Therapy in America

ELIZABETH SIEGEL WATKINS

The Johns Hopkins University Press
*Baltimore*

© 2007 The Johns Hopkins University Press
All rights reserved. Published 2007
Printed in the United States of America on acid-free paper
2 4 6 8 9 7 5 3 1

The Johns Hopkins University Press
2715 North Charles Street
Baltimore, Maryland 21218-4363
www.press.jhu.edu

Library of Congress Cataloging-in-Publication Data
Watkins, Elizabeth Siegel.
The estrogen elixir : a history of hormone replacement therapy in America /
Elizabeth Siegel Watkins.
p. cm.
Includes bibliographical references and index.
ISBN-13: 978-0-8018-8602-7 (hardcover : alk. paper)
ISBN-10: 0-8018-8602-3 (hardcover : alk. paper)
1. Estrogen—Therapeutic use—United States—History. I. Title.
RM295.W38 2007
618.1'75061—dc22
2006023196

A catalog record for this book is available from the British Library.

# CONTENTS

The list of institutions and individuals to whom I owe thanks is long. First, I wish to acknowledge the following sources of funding: the National Endowment for the Humanities for a summer stipend in 1999, the American College of Obstetricians and Gynecologists for an ACOG/Ortho-McNeil Fellowship in the History of Obstetrics and Gynecology in 2000, the National Academy of Education and the Spencer Foundation for a postdoctoral fellowship in 2000–2002, the National Library of Medicine for a publication grant in 2003–2006, and the University of California at San Francisco academic senate for a travel grant award in 2005.

My research was enriched by a wide variety of primary sources in formal and informal collections, and I am grateful to the knowledgeable and accommodating individuals who facilitated my access to those collections: Susan Rishworth, Debra Scarborough, and Pamela Van Hine at the American College of Obstetricians and Gynecologists; Cynthia Pearson at the National Women's Health Network; Michelle Bigesby, Suzanne White Junod, and John Swann at the Food and Drug Administration; Kathleen Banks Nutter at the Sophia Smith Collection at Smith College; Ellen M. Shea at the Schlesinger Library at Radcliffe (with special thanks to Abby Bass for sharing with me her inventory of the as yet uncataloged records of the Boston Women's Health Book Collective housed there); Arlene Shaner at the New York Academy of Medicine; Anna Gasner at Pfizer; and Wendy Braun, Georgia Ratliff, and Don Sutton at the Saturday Evening Post editorial offices.

Working in recent history affords the historian the opportunity to speak with many of the actors directly involved in the events being studied. I am especially indebted to the following individuals who generously consented to be interviewed: Deborah Grady, Florence Haseltine, Steve Hulley, Lila Nachtigall, Robert Nachtigall, Rosetta Reitz, Barbara Seaman, and Wulf Utian. Thanks also to Phyllis Greenberger for introducing me to Florence Haseltine, to Gay Becker for introducing me to Robert Nachtigall, and to Robert Nachtigall for introducing me to Lila Nachtigall. I would also like to thank Romano Deghenghi, Lois A. Gaeta, Morris Givner, and Michael Stern for sharing with me their recollections of working for Wyeth-Ayerst.

I was fortunate to present portions of this work on panels at national conferences and in invited presentations (at Georgia Institute of Technology, Johns Hopkins University, University of Toronto, University of Pennsylvania, University of California at Berkeley, UCLA, Yale University, and McGill University). My analysis of the history of HRT was sharpened by the comments of audiences at these meetings and seminars and by the thoughtful prepared commentaries of Adele Clarke, Gwen Kay, Leslie Reagan, and Susan Smith. This book also builds upon my previously published articles: "'Educate Yourself': Women and Information about Hormone Replacement Therapy" in *Medicating Modern America: Prescription Drugs in History,* ed. Andrea Tone and Elizabeth Siegel Watkins (New York: New York University Press, 2007); "'Doctor, are you trying to kill me?': Ambivalence about the Patient Package Insert for Estrogen," *Bulletin of the History of Medicine* 76 (2002), 84–104; and "Dispensing with Aging: Changing Rationales for Long-term Hormone Replacement Therapy, 1960–2000," *Pharmacy in History* 43 (2001), 23–37.

Six colleagues graciously gave their time to read the entire manuscript from beginning to end. I benefited enormously from the careful reviews and insightful suggestions of Rima Apple, Brian Dolan, William Greene, Judith Houck, Wendy Kline, and Dorothy Porter. I am especially grateful to Rima Apple for supporting this project from its earliest days through to the final draft and for introducing me to Judith Houck, to Judith Houck for generously sharing with me her terrific dissertation (now, book) on the cultural history of menopause, and to Dorothy Porter for her insistence that I keep rewriting the introduction until I got it right. Many others read portions of my work, listened to me talk about it, and offered constructive comments; thanks to Charlotte Borst, Ruth Schwartz Cowan, Lara Freidenfelds, Janet Golden, Herb Goldman, Joan Goldman, Jeremy Greene, John Janetos, Gwen Kay, Judith Leavitt, Deborah Levine, Susan Lederer, Barron Lerner, Jack Lesch, Harry Marks, Margaret Marsh, Heather Monroe Prescott, Naomi Rogers, Susan Reverby, Nancy Tomes, Andrea Tone, Elizabeth Toon, Martha Verbrugge, and John Harley Warner. It was a pleasure to work again with the staff at the Johns Hopkins University Press. Bob Brugger nurtured this project from its initial conception and facilitated its completion with his consistent encouragement. Marie Blanchard's expert copyediting skills helped to polish my prose. Thanks also to Carol Zimmerman for her role in the production of the book and to Celeste Newbrough for her work on the index. I also wish to acknowledge the many tangible and intangible contributions made to my academic career over the years by Timothy Haggerty, Everett Mendelsohn, and Steven Schlossman.

Writing can be a lonely process, but I have been lucky to be surrounded by sociable colleagues. Thanks to Jared Day, Laurie Eisenberg, Timothy Haggerty, Rebecca Kluchin, Mary Lindemann, Loretta Lobes, Marie Norman, Richard Pierce, and Stephanie Wallach in the history department at Carnegie Mellon; Brian Dolan, Molly Sutphen, Dorothy Porter, Nancy Rockafellar, Justin Suran, and John Tercier in the history of health sciences division at UCSF; and my academic and administrative colleagues in the department of anthropology, history, and social medicine at UCSF and the department of history at UC Berkeley for engaging me on all sorts of topics related and unrelated to work.

My friends and family also provided much appreciated diversions from this project. Thanks to Jamie D'Angelo, Lisa Keillor, Sarah Mott, and Kristin Clevenger for fun on and off the court. Peter, Karen, Gabriel, and Olivia Siegel housed, fed, and entertained me during my many research trips to Washington. My parents, Edward and Judith Siegel, read the entire manuscript; even more importantly, they gave me the confidence to undertake and complete this book. My children, Emily and Ellen, delighted and continue to delight me with their love, enthusiasm, and good humor. Lastly, I thank Simon for all his support and all that we have shared. This book is dedicated to him.

*The Estrogen Elixir*

Every woman who lives to a certain age reaches menopause. Decreased estrogen production in the years during and after menopause has been blamed for causing everything from hot flashes to heart disease to diminished femininity. One possible remedy has been to replace that lost estrogen with hormones from outside the body. The scientific and commercial development of pharmaceutical estrogen in the 1930s produced the hormone replacement therapy (HRT) that would become the most popular drug in America by the 1990s. Physicians prescribed HRT not only to relieve the temporary symptoms of menopause but also to forestall the diseases of aging and to maintain youthfulness. Opinions on the use of this drug have been sharply divided; it was ballyhooed as "one of medicine's most revolutionary breakthroughs" and condemned as "the greatest experiment ever performed on women."[1]

The story of estrogen is woven from several strands: blind faith in the ability of science and technology to solve a broad range of health and social problems, social and cultural stigmatization of aging, shifting meanings and interpretations of femininity and female identity, and the pitfalls of medical hubris in the twentieth century. Estrogen became much more than a drug in the American pharmacopoeia, as media representations of its roles in both medicine and culture reflected the engagement of the expanding authority of medical science with transformations in perceptions of aging women in society.

Simply put, too much was expected of estrogen. Had it remained a short-term antidote for menopausal symptoms, it would have generated little controversy. Indeed, to this day, estrogen remains the single most effective remedy for the hot flashes of menopause, and few critics dispute its value as a temporary treatment. But because it was promoted as a lifelong therapy, even as an elixir of life, with so many promised benefits, the stakes got higher, and when it failed to live up to the hype, disenchantment and incrimination ensued. The checkered history of this controversial drug therapy is the subject of this book.

Sixty years after the Premarin brand of estrogen received approval from the U.S. Food and Drug Administration (FDA) as a treatment for menopausal symptoms, a government-funded study of its sister product Prempro (a combination of estrogen and progestin) reported that Prempro increased the risk of heart attacks, stroke, blood clots, and breast cancer. But the story of estrogen neither begins in 1942 with FDA approval of Premarin nor ends with the findings of this study in 2002. Its roots date back to the nineteenth century, and its life as a medical therapy continues today. The biography of hormone replacement therapy spans the "long" twentieth century, from its conception in the 1890s, to its infancy in the 1920s and 1930s, through its adolescent growth spurt at midcentury, to its maturity into one the most prescribed drugs in America in the 1980s and 1990s, and, in the past few years, to hints of its senescence. As the times have changed— in terms of the status of science and medicine, the roles of men and women in society, and attitudes toward aging—so too have the rationales for the prescription and use of hormone therapy.

This book tells the story of the rise and fall, and rise again and fall again of estrogen and its promise to stave off the effects of aging. This was not a conspiracy of pharmaceutical manufacturers and physicians to dupe women, although drug makers aggressively promoted their products and many doctors believed in estrogen's healing powers. Women were also active agents, motivated by both personal concerns and cultural forces to use estrogen as one way to take control of the aging process. The added support of research scientists, government regulators, and the popular media helped to build the estrogen empire. Throughout its history, there were dissenting voices from health activists, physicians, and women themselves. But its widespread use—by some 40 percent of the postmenopausal population by the end of the twentieth century—indicates the willingness of many Americans to put their trust in this drug product.[2]

Each rise and each fall in the history of estrogen has its own narrative, because each era raised new questions about the meaning of HRT and its place within particular medical, scientific, social, cultural, and economic contexts. The initial rise covers the longest time period, from the earliest days of endocrinology in the 1890s through the first heyday of HRT in the 1960s. It begins with the dynamic interactions among scientists, pharmaceutical manufacturers, and gynecologists in producing, marketing, and prescribing estrogen in the first decades of the twentieth century, and it locates the significance of this new drug within the historical trend of the medicalization of women's health. The concept of medicalization has been defined by historians and sociologists as "the processes through which aspects of life previously outside the jurisdiction of medicine come to be construed as medical problems."[3] Over time, the medical profession has expanded its sphere of control to include "deviant" conditions, such as madness, alcoholism, and addiction, and "normal" life-course events, such as childbirth, menopause, and aging.

Changes in medical practice in the 1950s and 1960s—from prescribing estrogen for the short-term relief of menopausal symptoms to prescribing it as a long-term therapy for postmenopause—contributed to the increasingly blurred boundaries between normal and pathologized menopause and aging. The medical profession and the pharmaceutical industry based the rationale for long-term hormone replacement therapy on a broad construction of postmenopausal health, one that included not only the maintenance of strong bones but also physical appearance and emotional states, as defined by prevailing cultural standards. Many women took estrogen in their efforts to fight aging and the social stigma of being an older woman in an increasingly youth-centered America.

The first fall of HRT began in 1975, following the clinical research finding that estrogen use caused endometrial cancer (cancer of the uterus), and it lasted for less than a decade. The fallout from the link between estrogen and endometrial cancer in the late 1970s brought two sets of actors to the center of the public debate about the safety of HRT, specifically, and medical intervention in women's lives, more generally. The first group of actors consisted of both the newly organized women's health movement and individual feminists who challenged medical hegemony over women's health and pharmaceutical hubris in intruding on normal life-course events. The second was the Food and Drug Administration, which came into the spotlight in 1976 with its new mandate for patient labeling for estrogen prescribed for menopausal and postmenopausal women. The variety of responses to this action—from women and their husbands, feminist and consumer activists, doctors, pharmacists, and pharmaceutical manufacturers—reflected unresolved tensions between medicine and culture, including informed consent in medicine, government regulation of medical practice, medicalization of menopause and postmenopause, and individual as well as cultural conceptions of aging. Both health activists and government regulators disputed the medical model of menopause and contributed initially to an expansion in critical discourse on and a decline in the popularity of HRT. However, these trends were temporary, which raises interesting questions about the relative power of groups outside the medical community to influence medical practice and its authority in American culture.

The second rise, or resurrection, of HRT took place in the 1980s after two scientific developments: first, confirmation that adding a second hormone (synthetic progesterone, or progestin) to the regimen's cycle seemed to decrease the risk of endometrial cancer, by encouraging the uterine lining to shed, and second, endorsement of the preventive effect of estrogen on bone loss and osteoporosis. Advocates and manufacturers successfully redefined postmenopausal hormone use as a preventive health measure for active senior women. The medical uses of estrogen changed not only in response to scientific evidence but also in accordance with contemporary cultural notions of health and aging and the social position of women. The medical profession and the pharmaceutical industry successfully co-opted both the increased health consciousness among Americans and the feminist battle against negative stereotypes of women—two cultural forces that might have contributed to a decline in the reliance on drug-mediated aging but instead were harnessed to promote and legitimize hormone replacement therapy through the end of the century. The media, whose role is significant as translator of the world of science and medicine to the general public, was quick to popularize the link between osteoporosis and the decreased production of estrogen. Whereas articles in the late 1970s had expressed caution and concern about hormone replacement therapy, those written a decade later touted its benefits for aging women.

Even greater enthusiasm for HRT was generated in the 1990s by clinical reports that posited a decreased risk of heart disease in women who took hormones after menopause. This correlation was not new, but corroboration by epidemiological studies and support from basic science research encouraged government agencies and the media to publicize cardiovascular disease as the primary cause of death among older American women and to associate this deadly disease with the loss of estrogen. Postmenopause became even more medicalized and pathologized, as physicians considered virtually *all* women over the age of 50 as potentially at risk for cardiovascular disease and osteoporosis and therefore as candidates for long-term hormone replacement therapy.

The cultural climate in which this medicalization occurred so rapidly is itself worth investigating, because it was shaped in part by the baby boomers who began to reach menopause in the 1990s. Although they instigated an upsurge in public discourse on the topic, the substance of available information did not incorporate a very wide spectrum of possible approaches to health maintenance in the years during and after menopause. Commentators may have acknowledged the existence of nonhormonal therapies and even the possibility of no therapy at all, but the medical-pharmaceutical paradigm of treating midlife and older women with hormones remained dominant.

Indeed, the second and most recent fall of HRT was brought about not by the critical gaze of feminists or consumers, but by what the media presented as the authoritative word of science. In 2002 the Women's Health Initiative study of estrogen-progestin therapy was abruptly terminated because unexpected results indicated increased risks for heart disease, strokes, blood clots, and breast cancer for those using HRT. This story was widely reported as the definitive answer to questions about the advisability of long-term estrogen use, in spite of heated debates within the medical profession about the design of this epidemiological study and the validity of its results. Feminist health activists hailed the findings from this massive clinical trial as a victory, choosing to accept this particular scientific interpretation over previous scientific accounts as confirmation of their disapproval of estrogen replacement. Although feminists in the 1970s included medicalized menopause and aging as part of a broad critique of the low status of older women in America, by the turn of the twenty-first century, they had jettisoned social and political arguments in favor of what appeared to be objective scientific data, even though the nature of the science behind the Women's Health Initiative was highly disputed.

The contested nature of science is one of the themes guiding this book's explanation of the shifting fortunes of hormone replacement therapy in America. The public face of medical investigation earned an increasing measure of respect over the course of the twentieth century, thanks to positive portraits of medical researchers in literature and film (think of Sinclair Lewis's *Arrowsmith*) and to thrilling accounts of medical success stories, such as Salk's polio vaccine. Americans' faith in science to solve social problems contributed to the initial enthusiasm for estrogen as a long-term therapy at midcentury. Clinical studies became increasingly important in the evaluation of drugs in the postwar years. As drugs such as estrogen earned the scientific seal of approval, many physicians and women accepted their role in the management not just of menopause but also of postmenopause.[4] Findings of clinical studies led to the widespread use of estrogen in the late twentieth century, but ironically it was this belief in the scientific authority of clinical research that brought about the most recent contraction in the popularity of HRT.

Beneath the surface of the cultural authority of medical science lay an unsettled debate within the medical community about the claims of "medicine" to be "scientific." Practicing physicians, clinical researchers, and bench scientists have disagreed on what has constituted acceptable and valid evidence and methods of analysis. The introduction of biostatistics into medical research, the development of the randomized controlled trial, and the clarion call for evidence-based medicine all influenced the design of scientific investigations and the application of results. Changing standards produced tensions between rank-and-file doctors and academic researchers, which tended to play out in debates over the relative value of one physician's empirical observations versus large-scale epidemiological studies, or the merits of qualitative experience (such as symptom reporting) versus quantitative data (such as laboratory analyses and statistical outcomes).

In clinical practice, increasing reliance on standardized measurements to assess a healthy individual's risk of developing disease (for instance, bone density tests) further contributed to the medicalization of menopause and aging. The medicalization and increasing pathologization of the years after menopause has been guided by the reductionist view that female aging is determined by the decreased production of hormones by the ovaries. The nineteenth-century notion of the centrality of the ovary was only made more specific in the twentieth century, namely, that women's essence could be reduced to the hormones their ovaries produce.[5] But while menopause and its alleged aftereffects, along with other aspects of women's health such as childbirth and contraception, increasingly came under the purview of medicine during the 1900s, there was also a parallel trend, particularly in the latter decades of the century, to reclaim menopause as a "natural" life-course event that needed no medical intervention. The boundaries between the two positions were fluid and shifted over time, and tensions between these two views of menopause and aging were reflected in differing positions on the advisability of the use of HRT.

Managing menopause and aging depended to a certain extent on the changing meanings of and frameworks for aging in America.[6] Cultural prescriptions for how women should age influenced the reception of replacement therapy, both as a temporary palliative for hot flashes and as a lifelong preventive for osteoporosis and heart disease. The history of the roles and expectations for older women in America was not a linear progression toward increased participation and acceptance but rather a more tortured path littered with predictable obstacles and unexpected impediments. The status of older women had a significant effect on medical and popular rationales for the prescription and use of HRT.

One last theme to mention deals with information flows and their influences in changing beliefs, values, actions, and behaviors. Where and how women and physicians learned about estrogen for menopausal and postmenopausal use is critical to the biography of HRT.[7] Exchanges—some collaborative, others conflicting—among scientists, doctors, drug companies, government agencies, feminist health activists, the media, and the public affected the production and consumption of information about menopause, aging, and treatments both medical and alternative. As research studies multiplied and consumers demanded greater participation in health care decision-making, the media played an increasingly important role in the dissemination and translation of medical information for midlife and older women.

To say that hormone replacement has been heavily scrutinized since the results of the Women's Health Initiative (WHI) came out in 2002 would be a major understatement. Part of this scrutiny has been a search into the past, for historical clues to the current state of estrogen use. Journalists often included in their news reports and analyses of the WHI short synopses of the history of estrogen, which they usually dated to the 1960s when HRT became popular as a long-term therapy. Similarly, reviews in medical journals sometimes began with a retrospective account of HRT use. In a book published in 2003, Barbara Seaman took a longer view of the history of estrogen, providing a unique perspective from her position as a leading feminist health activist.[8]

As a historian, I take an even longer and wide-angled view, using estrogen as a lens through which to illuminate the complex and changing relationships between menopause and aging, drugs and alternatives, doctors and patients, and providers and consumers of health care services, products, and information. I aim to develop a nuanced interpretation of the process of medicalization, within the broader context of the dominance of health as a cultural preoccupation in recent American society and its implications for multiple approaches to and interpretations of menopause and aging. This requires recognition of the differences among doctors, who too often are considered as one monolithic entity. The same goes for women, whose differences as individuals must also be taken into account.

A few qualifications are in order. First, this book uses *estrogen* as an umbrella term for the various forms of the hormone, both natural and synthetic. Similarly, *progestin* stands for the different synthetic forms of the second female sex hormone, progesterone. The chemical and commercial differences among these products are discussed in chapter 1; thereafter, the umbrella terms are used as shorthand to describe the two families of compounds. *Estrogen replacement therapy* (ERT) refers to treatment with estrogen only and *hormone replacement therapy*

(HRT) refers to a combined regimen of estrogen and progestin; this book will also use HRT as a shorthand generic allusion to any use of exogenous sex hormones (those that come from outside the body).

Second, this study of hormone replacement therapy concentrates on women who reached menopause naturally, as opposed to those who had surgical menopause. Women whose ovaries were removed in their twenties, thirties, and forties lost the ability to produce estrogen; for these women, the use of estrogen (until the age of menopause) was akin to the use of thyroid hormone or insulin in the replacement of a hormone that the body normally made (during the reproductive years). Thus women who had ovariectomies, which were usually accompanied by hysterectomy (removal of the uterus) and referred to by the latter term—especially those operated on before the average age of natural menopause—used estrogen in far greater proportions than women whose reproductive organs remained intact. However, the focus here is on the relationship between HRT and aging in the postmenopausal years. Thus statistics cited for hormone use in women over 50 include both those who have had hysterectomies and those who have not, while much of the discussion in this book concerns the use of estrogen (with or without progestin) by (naturally) menopausal and postmenopausal women. The debate about unnecessary hysterectomies performed on American women is a serious one, but it is beyond the scope of this work.[9]

Third, the women discussed in this book are overwhelmingly white, middle- and upper-class Americans who had the time to seek medical counsel and the money to purchase medical treatments for health issues during menopause and postmenopause. As more women obtained health insurance, either privately or through government-funded programs in the 1960s and 1970s, and as HRT became more popular in the 1980s and 1990s, more women of lower socioeconomic status joined the ranks of hormone users. Women of color took HRT in much lower proportions than white women did; only in the late 1990s did researchers begin to study the reasons for this discrepancy. This is also very much a story about cultural circumstances within the United States. Although women in other developed nations (Australia, New Zealand, South Africa, most countries in Europe, and some in South America) also wrestled with the decision whether or not to take hormones, unique local conditions preclude generalizations or extrapolations from the study of the history of HRT in the United States.[10]

Part of the uniqueness of the American context is this nation's enormous consumption of pharmaceuticals. The U.S. drug bill far exceeds that of any other nation. In the early twenty-first century, one in every two Americans took a prescription medication each day, at a total cost of some $200 billion each year.[11] The history of hormone replacement therapy is bound up not only with the medicalization and pathologization of menopause and postmenopause but also with the commercialization of health care in the twentieth century. As medical consumers became increasingly affluent, midlife and older women were bombarded with a growing array of products and services available for purchase, both from mainstream medicine and alternative healing. The menopause bazaar drew increasing crowds of merchants and shoppers, but for much of its history, hormone replacement therapy dominated the market.

That the small group of molecules known collectively as estrogen would eventually become one of America's best-selling and then most contentious drug products was by no means obvious from its humble origins. Indeed, decades passed between the earliest commercial development of estrogen and its eventual rise to medical fame and infamy. The story of the discovery of estrogen begins in St. Louis in the 1920s, with two biologists, a pile of sow ovaries, and gallons upon gallons of urine.

# CHAPTER ONE

# Beginnings

In 1919, upon their discharge from the United States Army after service in World War I, two young men found themselves in St. Louis. One was a gregarious, heavy-set biologist trained at Brown University; the other was a reserved, slim biochemist trained at Harvard.[1] Both had accepted positions at Washington University School of Medicine. Both were newly wed. Both liked to play baseball. And, within a decade, both would earn credit for their collaborative discovery of the female sex hormone, estrogen.

The biologist, Edgar Allen, was born in Colorado in 1892. Soon thereafter, his family moved east to Rhode Island.[2] The biochemist, Edward Doisy, eighteen months younger than Allen, was born and raised in Illinois.[3] When they met as new instructors in St. Louis, 26-year-old Doisy and 27-year-old Allen formed a faculty baseball team, which beat all of the classes at Washington University except the freshmen.[4] Their friendship secured on the ball field, the two men introduced their wives, and the foursome spent many evenings playing bridge together.[5]

In 1921, the couples moved to neighboring homes in South St. Louis. Doisy had a car, Allen did not. Since public transportation between their houses and their labs was unreliable, Allen and Doisy rode to and from work together in Doisy's noisy Model T Ford. Naturally, the conversation revolved around their scientific research.[6]

For his doctoral dissertation, Allen had studied the estrous cycle of the mouse. His thesis described the cellular changes in the animal's sex organs during its reproductive cycle. In 1922–23 Doisy was involved in the purification of insulin, the hormone secreted by the pancreas and lacking in diabetics. Although insulin had been discovered and isolated by scientists in Toronto the year before, not much of the hormone had yet been produced. When a baby arrived at St. Louis Children's Hospital in a diabetic coma in September 1922, Doisy's boss suggested they make their own batch of insulin to save the child's life. They succeeded, and the work on insulin continued. Six months later, on the way home from work one evening, Doisy was chatting with Allen about the hormonal nature of insulin. As he was describing how the cells of the pancreas gland secrete insulin, which then acts on other body cells to encourage the uptake of glucose sugar from the blood, Allen interrupted to draw a parallel with the female reproductive system. From his work with mice, he had noticed a relationship between the ripening of the follicle cells of the ovaries and the development of the uterus and vagina. Might this relationship be governed by hormones?[7]

To test his hypothesis, Allen decided to inject the liquid from the follicles of sow ovaries (obtained from a nearby meat packing plant) under the skin of female mice and rats whose ovaries had been removed. To obtain the desired fluid, he enlisted the help of his wife, and together they extracted the liquid from the follicle cells at their kitchen table.[8] Back in the laboratory, the substance from the ovaries of the sow caused the vaginas of the mice and rats to become estrous, as evidenced by cellular changes clearly visible under the microscope. Allen then injected the follicular liquid into baby mice and rats, which stimulated the vaginas and uteruses of those animals to rapidly mature to adult size. These results proved the existence of some "estrogenic hormone" (so called because it stimulated estrous growth in the sex organs) in the fluid extracted from the sow ovaries. Allen's work, which continued at the University of Missouri, to which he moved in 1923 to become professor of anatomy and, later, dean of the medical school, led to the development of a quick and easy assay that biochemists could use to test the potency of hormonal extracts.[9]

While Allen was performing these animal experiments, Doisy set to work trying to purify the active substance in the follicular fluid. His work was complicated by the fact that sow ovaries were neither inexpensive nor easy to come by. He negotiated a deal with a pharmaceutical company that had access to a local meat packing factory, but in 1927 a scientific report from Berlin released Doisy from his dependence on pig parts. Two German scientists, Selmar Ascheim and Bernhard Zondek, had discovered that the estrogenic hormone was found in large quantities in the urine of pregnant women. Doisy arranged for a nurse in the obstetrics clinic at the St. Louis University School of Medicine (to which he had moved in 1923 to become professor of biochemistry and, in 1924, director of the department of biochemistry) to supply each pregnant patient with a one-gallon bottle and instructions to return it—filled—at the next visit.[10]

As the work progressed, Doisy and his team needed more urine for processing. Doisy's car (it is unknown whether it was the same Model T Ford that he drove in 1923) was used to transport two-gallon bottles to and from patients' homes. Doisy later recollected, "On one occasion, a graduate student, Jack Curtis, was stopped by a policeman who had seen the bottles of amber liquid. Since this was in the days of prohibition, he thought he had caught a bootlegger. He would not believe Jack's statement that the bottles contained urine, so he was invited to climb in the car and have a drink. He pulled the cork, sniffed the contents and said, 'My God, it is urine! Drive on and quit obstructing traffic.'" The distillation of the hormone from urine continued, and on 13 July 1929, Doisy isolated the first sample of pure crystalline estrogen.[11]

The discovery of estrogen capped eighty years of speculation and research on the secretions of the sex glands.[12] In 1849 the German physician Arnold Berthold was the first to demonstrate that the testes secreted a chemical substance that affected other parts of the male body.[13] His experiments on roosters contributed to work on one of the great problems of nineteenth-century physiology, namely, the secretions of the ductless glands. Contemporary physiologists were also fascinated by the conductive properties of nerve tissue.[14] Interest in electrical (nervous) stimulation and chemical (hormonal) stimulation grew into the twentieth-century fields of neurophysiology and endocrinology, respectively, both of which deal with the integration and control of an organism's activities. Nerve impulses of the nervous system regulate the body via rapid reactions; the hormones of the endocrine system travel more slowly through the bloodstream and thus produce slower responses. In 1855 the eminent French physiologist Claude Bernard first articulated the concept of internal secretion in his description of how the liver acts as a gland by secreting glycogen into circulation.[15] Although glycogen is not a true hormone, or chemical messenger, it is released by a similar mechanism. What Bernard realized was that the internal release of substances must somehow contribute to the maintenance of an organism's internal environment.

In 1896 Emil Knauer, a 29-year-old Austrian working in one of Vienna's gynecological clinics, conducted experiments with rabbit ovaries similar to those performed on rooster testes by Berthold several decades earlier. Knauer excised the ovaries from adult rabbits, cut them into smaller pieces, and then reimplanted the pieces into different locations in the same animals. The animals did not display the usual uterine atrophy following ovarian removal, which led him to postulate that the ovaries must send some sort of substance via the blood, since all nerves to the ovaries had been cut in the surgery.[16] In 1900 Josef Halban, another young Austrian working at Vienna's other gynecological clinic, expanded on Knauer's experiments. Using guinea pigs, Halban implanted pieces of adult ovaries under the skin of infant animals, and he observed that the young pigs' uteruses rapidly developed to maturity. Halban combined his observation with that of his countryman and proposed that ovarian secretions, traveling via the blood, were responsible for the development and maintenance of the female genitals.[17]

This ovarian secretion and the secretions from other ductless, or endocrine, glands became known as hormones in 1905. The term was coined by Ernest Starling, the British scientist who, along with William Bayliss, had discovered secretin, a substance released by cells of the intestine to stimulate the pancreas to release digestive juices. Taken from the Greek meaning "to excite or arouse," the word *hormone* was intended to emphasize the messenger role of these chemical substances in regulating the specific physiological activity of target organs.[18]

By the 1910s, the study of reproductive endocrinology had become a legitimate field of inquiry among research biologists. Moreover, curiosity about the secretions of the testes and ovaries was not limited to gynecologists and physiologists working on animals in laboratories. Clinical doctors were also very interested in the curative or restorative powers of glandular extracts for their patients. In the late nineteenth and early twentieth centuries, doctors had few effective remedies either for treating diseases or for relieving discomforts. "Regular" physicians, so called to distinguish them from sectarian healers (e.g., eclectics, homeopaths), had earned a bad reputation in the nineteenth century because of their "heroic" medicines and procedures, such as mercury and leeches.[19] Often these treatments harmed patients rather than helping them. Sectarian medicine, by contrast, relied on herbals or diluted doses, which may not have had any effect on disease but did no further damage to patients. If regular physicians could harness glandular extracts and use them to replace or augment the body's own production of these substances, then they would have at their disposal a biological and seemingly natural medicine for treating their patients' ills.

Thus it was with keen interest that American physicians followed the news from Paris of the discovery of the "elixir of life." In June 1889, at the advanced age of 72, Charles-Edouard Brown-Séquard, professor of medicine at the Collège de France, reported at a meeting of the Society of Biology that he had experienced rejuvenation after injecting himself with extracts of the crushed testicles of dogs and guinea pigs.

Brown-Séquard was no quack. A respected physician and scientist who made important contributions to neurology, he had been elected to the British Royal Society in 1860, the American Academy of Arts and Sciences in 1867, the American National Academy of Sciences in 1868, and the French Académie des Sciences in 1886. Twenty years before his stunning statement, he had speculated on the physically and intellectually rejuvenating properties of sperm injections for aged men.[20] This suggestion was consistent with the contemporary Victorian belief that men had to conserve their seminal fluid in order to keep their minds and bodies sharp. Physicians and writers of sexual advice manuals blamed excessive ejaculation, through sexual intercourse or, worse, masturbation, for a host of debilitating conditions.[21] Brown-Séquard turned this theory around and reasoned that the weakness observed in old men might be attributed not to seminal overspending but rather to decreased seminal production.[22] Using himself as the proverbial guinea pig, he found, much to his delight, that he felt stronger and more alert after a series of injections of animal testicular extract.

Realizing that he may have been subject to the power of suggestion (this was as far from a controlled clinical study as possible), Brown-Séquard asked his Paris audience to verify his experiments. Physicians in both France and America obliged by testing the effects of testicular extracts on feeble old men. A few months later, Brown-Séquard collected these results and published them in a French medical journal. The article electrified the international medical community, and physicians around the world raced to apply Brown-Séquard's technique to their patients and to identify and exploit other glandular extracts.[23] A book published in Boston called *The "Elixir of Life"—Dr. Brown-Séquard's own account of his Famous Alleged Remedy for Debility and Old Age, Dr. Variot's Experiments, and Contemporaneous Comments of the Profession and the Press* captured and spread the spirit of excitement over this new medical marvel.[24]

Coined "organotherapy," this practice quickly became the latest medical fad. Within a few years, doctors had ground up several different organs in the hopes of extracting some active principle that would cure disease. Thus thyroid gland was used to treat myxoedema (hypothyroidism), brain for neurasthenia (described as exhaustion of the central nervous system), pancreas for diabetes, kidney for uremia, muscle for muscular atrophy, heart for heart disease, and testicles for debility, epilepsy, cancer, cholera, tuberculosis, leprosy, and asthma.[25] By the end of the decade, it was clear that most of these treatments did not work—only thyroid extract was proven to be effective—and organotherapy fell into disrepute.

Two decades later, the male sex organs and their extracts enjoyed a clinical renaissance as part of "rejuvenation" therapy.[26] In the 1920s hundreds of doctors provided thousands of men with putative rejuvenation cures. Some consisted of injections of animal gland extracts, à la Brown-Séquard, but the treatment that received the most attention was testicular transplantation, in which animal sex glands were grafted onto patients' own testicles. This operation was popularized on both sides of the Atlantic.

In Europe, Serge Voronoff, a Russian-born French physician, began to transplant monkey testes into humans in 1920.[27] His reputation as a transplant surgeon had been made six years earlier when he successfully performed bone grafts on wounded soldiers during World War I.[28] A few years later, he undertook testes transplantation experiments in sheep, hoping to reverse the effects of aging in old rams with grafts of the testes of young lambs. He reported the positive results of his work at academic meetings and in medical publications. Although the animals may have appeared healthier simply because of improved diet and care, both the scientific community and the general public accepted Voronoff's claims, and he established himself as an internationally respected medical innovator.[29] *Scientific American* gushed, "Even death, save by accident, may become unknown, if the daring experiments of Dr. Serge Voronoff, brilliant French surgeon, continue to produce results such as have startled the world."[30] By the end of 1926, he estimated that he had performed a thousand testicular transplants.[31]

Voronoff was not the first to perform a testicular transplant in humans. In 1919 Dr. Leo L. Stanley, the chief medical officer at San Quentin Prison in California, transplanted the testicles from an executed murderer into a 60-year-old prison inmate. He continued these operations over the next several years and earned publicity when his revitalized patients bested their younger fellow inmates in the annual Thanksgiving games at the prison in November 1923.[32]

Several other American physicians jumped on the testes transplant bandwagon, but the most enterprising was John R. Brinkley, who ran a private gland-grafting hospital in the tiny hamlet of Milford, Kansas. He used the testicles of goats, which, unlike monkeys, were not susceptible to tuberculosis. Brinkley advertised over the airwaves of his personally owned and operated radio station, and patients came from far and wide to pay $750 for the surgery.[33] An opportunist who made a lot of money (until his medical license was revoked in 1930) from his patients' desires to regain their youthful vitality, Brinkley presented a very different portrait as compared to the reputable Voronoff. The distance between these two very successful practitioners reveals the latitude within the practice of medicine in the 1920s, as well as the willingness of men to undergo dubious treatment in search of rejuvenation.

The rationale for testicular transplantation superimposed a modern interpretation on the older Victorian notion of the conservation of (seminal) energy. Rejuvenators still believed the testes to be the source of a man's virility and manliness, but they did not locate the active principle in the seminal fluid. Instead, they applied the twentieth-century concept of the hormone. The testes produced not only semen (which the individual could control) but also the male hormone (which he could not). A man could not be blamed for the failure of his sex glands to produce enough hormone; modern medicine could help him to replace that which his body could no longer produce, presumably by providing a supplementary supply of hormones via the animal graft. A consequence of this new reasoning meant that men did not have to conserve their semen by avoiding ejaculation; in the 1920s, sexual expression was, if not explicitly encouraged, then at least tacitly understood to be an indication of masculine vigor, consistent with the more relaxed attitudes toward sex evident during that decade.[34]

Rejuvenation therapy, in its various guises, flourished during the early, exciting years of the new science of endocrinology. Laboratory research into the physiology of the endocrine glands and their secretions yielded some impressive results. Perhaps the most dramatic breakthrough was the discovery of insulin by Frederick Grant Banting and Charles Herbert Best at the University of Toronto in 1921. The isolation of insulin from the pancreas provided a life-saving therapy for diabetics. Thyroxine had been isolated from the thyroid gland in 1914, but only tiny amounts of the hormone could be produced (it took three tons of pig thyroids to make 33 grams, or about one ounce, of pure thyroxine) by the initial biochemical methods of purification.[35] Still, researchers were encouraged by these achievements and continued the active pursuit of many different hormones. In the 1920s, the future of this science—and its medical applications—seemed highly promising.

Within this context of high expectations for applied endocrinology, all sorts of claims, ranging from the moderate to the outrageous, were made for rejuvenation therapy. However, the main reason that men pursued it was to improve the quality of the remaining years of their lives.[36] Historian Julia Rechter has identified these men, mostly in their fifties and sixties, as caught between two generations. Too young to abide by the Victorian codes of their fathers and too old to join the new youth culture of the 1920s, they still sought to participate in modern society.[37] Rejuvenation therapy offered them a sense of better health and well-being. Few seriously believed that testicular transplantation could actually turn back the clock, but many hoped that the operation might forestall the negative aspects of aging. In this way, the motivation for rejuvenation therapy for middle-aged men in the years after World War I foreshadowed the rationales for hormone replacement therapy for middle-aged women that would become popular in the decades after World War II.

It is important to point out that, in this earlier period, women were not included in the rejuvenation craze. Ovarian therapy began in the 1890s and continued for the next several decades, but its objective was not to return women to a more youthful state; rather, doctors used ovarian products as medications to treat diseases and ailments. Gynecologists administered a diverse array of preparations to their female patients for conditions assumed to be related to ovarian failure. Women were prescribed solutions of ovarian extract in water, glycerin, or alcohol; tablets of dried ovarian material; even fresh sow or cow ovaries, minced and served in sandwiches.[38] They took these alleged remedies not to restore vitality but to alleviate the symptoms of what their doctors diagnosed as hysteria, chlorosis, menstrual disorders, and menopause.

In the late 1800s, menopause was not just for women over 50. From the 1870s until the end of the century, thousands of younger women were subjected to double ovariotomy (the removal of both ovaries), and, as a result, experienced the symptoms of menopause in their twenties, thirties, and forties. Gynecologists performed this surgery—known as Battey's operation, after its American originator, Robert Battey, who carried out the first of hundreds of these operations in 1872—on women for menstrual disorders, such as painful menstruation or the absence of menstruation, and for other conditions, dubiously attributed to ovarian problems, such as mania and epilepsy. The ovaries themselves were not diseased but were blamed for causing pain and illness, which, according to followers of Battey, justified their excision.[39] Women who survived this surgery (about one in five did not) ceased to menstruate and often suffered from hot flashes. By the early 1900s, ovarian extracts had become a recognized prescription to try to alleviate the discomfort of both surgical and natural menopause.[40] In spite of recurring doubts and concerns about the wisdom and safety of treating menopause with hormones, this therapy continued as part of regular medical practice for the next century.

Some doctors did apply the technique of gland transplantation to the treatment of the menopausal syndrome, by grafting either the patient's own ovary or a donor ovary onto the abdominal wall, presumably to provide a supplemental source of female hormones. John H. Hannan, a physician at the Maternity Center and Hospital for Women in London, rejected this course of treatment for all but young ovariotomized women in his 1927 book, *The Flushings of the Menopause*. He reported that the transplanted ovarian tissue degenerated over time, but this problem was not his main reason for opposing the practice of grafting. Rather, he objected to the possibility of forestalling menopause in older women. "Would we be justified, even if it were possible, in delaying indefinitely, the onset of the menopause? Would the body, as old age advances with generalized degenerative changes, stand the strain of a vigorous sex life?"[41] His answer was no: apparently, menopause signaled the end of women's sex lives, and any further sexual activity after the age of menopause would be perilous to a woman's health.

Hannan also discussed a therapeutic regimen called "pelvic diathermy," in which a ring electrode was inserted into the vagina, a plate electrode was placed over the pubic area, and electrical currents were passed between the two. The goal was to temporarily increase the ovarian hormonal output to relieve symptoms in women with an abrupt onset of menopause. Hannan observed that "the length of the sittings, and their frequency will depend . . . in part upon the tolerance of the patient." He also noted that patients tended to display increased nervous irritability, which could be remedied with increasing doses of sedatives.[42] It is not known how prevalent this seemingly barbaric treatment was in either Europe or America, but it does seem to have been overshadowed by the more popular use of ovarian extracts among those physicians who chose to treat menopausal women with more than just old-fashioned counseling.

In the 1920s laboratory scientists disagreed with gynecologists on the merit of using ovarian extracts in clinical practice. For gynecologists, the fact that patients reported relief from their symptoms gave sufficient justification for the prescription of ovarian preparations; laboratory scientists demanded more stringent proof of the specificity and potency of these extracts. In a special article for the *Journal of the American Medical Association* in 1924, A. J. Carlson (a scientist, not a physician) expressed skepticism about the use of ovarian extracts as substitution therapy, since, at the time, "none of the ovarian hormones have so far been isolated, as determined by reliable biologic or chemical tests."[43] Nevertheless, while scientists labored to identify the precise nature of the hormone allegedly produced by the ovary, some doctors were willing to take it on faith that the hormone was present and active in the supplements they gave to their menopausal patients. And drug companies were happy to cater to this market; in the 1920s, there were several commercial ovarian products available for purchase.[44]

The clinical use of ovarian therapy and the laboratory research on the ovarian hormone did not transpire independently of each other. Scientists relied on gynecologists to gain access to research materials, such as the urine of pregnant women, to carry out their experiments. Gynecologists prescribed the latest ovarian preparations produced by pharmaceutical companies based on developments made by scientists in the laboratory.[45] By the time Edward Doisy isolated estrogen in 1929, researchers, gynecologists, and drug manufacturers had well-established networks in place for the further development and expansion of the medical market for hormone replacement therapy.

Doisy obtained his first crystalline sample of pure estrogen on 13 July 1929, and he reported his results to fellow scientists at the thirteenth meeting of the

International Physiological Congress in Boston in late August. Two months later, the stock market crashed on 29 October, Black Tuesday, and the nation plunged into the Great Depression. While most sectors of the American economy were hit hard in the early 1930s, the pharmaceutical industry appears not to have suffered as badly. Although companies experienced a decline in sales, their research activities continued apace. Parke, Davis and Company began its annual report for 1930 by acknowledging the Depression and resultant "severe contraction in business," but it continued in a more upbeat tone with the assurance that "our scientific research is going ahead in very satisfactory fashion."[46]

Other indices reveal that the expansion in pharmaceutical research that began in the 1920s was sustained through the 1930s. The number of commercial laboratories increased from eleven in 1920 to ninety-six in 1940.[47] A survey of Eli Lilly, Merck, Parke Davis, E. R. Squibb, and Upjohn found that the number of research personnel working for these five companies rose from about one hundred in 1920 to about three hundred in 1940.[48] And even in the depths of the Depression, some companies expended capital to build new facilities for both basic and applied research.[49]

It was not only the drug companies who gave their support to research; the public also put its trust in science. Thanks to the largely positive presentation of science and scientists in mass-circulation magazines in the first half of the twentieth century, Americans came to expect good things from scientific research.[50] By accepting the reliability and accuracy of the scientific method and by reporting the recent achievements of scientists, journalists contributed to the cultural authority and prestige of both the practice and the practitioners of science.[51] The accomplishments of science in the past seemed to portend further successes in the future, and the public readily accepted this optimistic outlook. Surely, in the gloomy years of the Great Depression, Americans preferred stories of a better tomorrow to reports on the miseries of the present.

Readers of the business magazine *Fortune*, many of whom keenly felt the stress of a contracted economy, were probably heartened by the upbeat article on the promise and potential of endocrinology in the November 1933 issue.[52] This "survey of the most important field in medical research" covered fourteen pages, complete with graphic photographs of scientists and doctors in action. It provided a one-page "primer" for those unfamiliar with the pancreas, ovaries, testes, and the pineal, pituitary, thyroid, parathyroid, thymus, and adrenal glands. The article began by quickly dissociating the respectable science of endocrinology from its discredited relative, rejuvenation therapy, and went on to assure its audience that "there are plenty of authentic miracles, however. If the age affords any whisper of magic, you will find it in the endocrine glands."[53]

After characterizing endocrinologists as "sober, patient, cautious, academic gentlemen," the article described recent discoveries as well as the current status of research activities and their potential medical applications for the hormones of the human body. A two-page pictorial spread acclaimed the contributions of twenty-one "heroes" of endocrinology, including René Descartes, who claimed to have located the soul in the pineal gland, Montesquieu, who judged an essay contest on the purpose of the adrenal glands in 1716, Charles-Edouard Brown-Séquard and Serge Voronoff, the suspect rejuvenators, along with more recent investigators, such as Frederick Banting, Edward Doisy, and James Bertram Collip, described as "an amazing Jack-of-all-glands."[54]

Fact and fiction were presented side by side, without any distinction made between the two. On one page, four before-and-after photographs demonstrated the striking changes in the face of a hypothyroid woman treated with thyroxin. On another, an illustration of the Greek poet Sappho was captioned, "Too much adrenal activity makes women masculine in character, may well be the explanation for *Sappho*, first and most famous of all Lesbians."[55] Uncritical readers may have assumed the veracity of both the actual therapy and the speculative hypothesis.

The article also blended scientific findings with contemporary ideas about gender. It claimed that the reason few serious endocrinologists paid attention to the testes in the 1930s was because "the effects of the male hormone are largely limited to the sexual sphere, in contrast to the far-reaching scope of . . . the female sex hormones."[56] Both physiology and poetry—the role of hormones in menstruation, pregnancy, and menopause and Byron's couplet "Man's love is of man's life a thing apart / 'Tis woman's whole existence"—were cited as evidence for the assertion that "women's health and happiness center around her sexual life."[57] Hormones appeared to give a physiological basis to female vagaries. But the diligent men of science could do more than just document hormonal effects; the fruits of their labors in the laboratory had "limitless medical possibilities."

The pharmaceutical preparation of hormones had commercial as well as medical potential. Given this account of the pace and productivity of endocrinological research, astute readers may have interpreted these results as a sign to invest in the drug industry. The article distinguished between ineffective "gland sausages" sold by quacks and purer hormone extracts sold by "ethical" drug firms, and it estimated that the latter probably sold annual totals of between $1 million and $2 million worth of all endocrine products, with the implication that these figures would certainly rise in the future. It also acknowledged the distance between the virtuous scientist, untainted by filthy lucre, and those who profited from his work. "The doctor, the drug manufacturer, both reputable and quack, and the meat packer all reap the golden harvest. But the sower of the seed, the researcher in the laboratory, receives nothing but renown for his labors."[58] Perhaps this observation was the most accurate of all; in the case of estrogen, laboratory scientists wit-nessed significant commercialization of their discoveries during the decade of the 1930s.

Edward Doisy gave up his claim to any financial reward when he assigned his patents for the isolation of estrogen to his employer, the president and board of trustees of St. Louis University.[59] On 3 March 1930, he applied for a patent on the processes and products of obtaining the crystallized hormone from the urine of pregnant women, in a form sufficiently concentrated and free of odor to be used as a therapeutic agent. He applied for a second patent on related processes on 6 October; both patents were awarded on 24 July 1934. St. Louis University then licensed pharmaceutical companies to manufacture and sell the product, known as estrone or theelin. Parke Davis was the first company to take advantage of this relationship, having developed a cooperative relationship with Doisy in the late 1920s. By the end of 1931, the company boasted in its annual report that "the introduction of some of our newer specialties . . . like Theelin . . . are fully meeting our expectations."[60] Later in the decade, two other American firms obtained li-censes to manufacture theelin: Abbott Laboratories and Eli Lilly.

Other companies in both Europe and North America brought estrogen prod-ucts onto the market in the late 1920s and early 1930s. The difference between these medications and earlier estrogen preparations was their greater potency and specificity, attainable through more sophisticated methods of extraction and purification and measurable by biological assay tests; the assays allowed estrogen medications to be standardized. In 1929, E. R. Squibb and Sons began to sell Amniotin, a liquid noncrystalline form of estrone derived from the fetal fluids of cattle.[61] By 1932, two additional estrogen products were available to American physicians: Menformon, produced by the Dutch company Organon, based on the experimental work of Ernst Laqueur, and Progynon, made by the German com-pany Schering, based on the work of Adolf Butenandt.[62] Both Laqueur and Bute-nandt had reported the independent isolation of estrogen just months after Doisy's announcement of his discovery in 1929. In 1934, the Canadian firm of Ayerst, McKenna, and Harrison entered the American market with Emmenin, an estrogen product obtained from placentas, which had been isolated and devel-oped by James Bertram Collip at McGill University in 1930.[63] Emmenin initially offered the advantage of being orally active and thus available in pill form; how-ever, other manufacturers soon came out with their own products in tablets or capsules. Parke Davis, Eli Lilly, and Abbott Laboratories all sold theelin (estrone) in varying concentrations dissolved in oil (vegetable, peanut, or sesame) to be used for injections and theelol (estriol—an estrogenic substance also found in the urine of pregnant women but less active than estrone) in capsules that could be swallowed.[64] Thus, by the middle of the decade, American doctors had a considerable selection of oral and injectable estrogen preparations with which to treat their menopausal patients.

The medical diagnosis of menopause followed from the availability of these new therapeutic agents. One Wisconsin physician, writing in the *American Journal of Obstetrics and Gynecology* in 1932, drew an explicit analogy between diabetes and menopause; he considered both to be deficiency diseases treatable with hormone replacement therapy.[65] He recognized, however, that menopause did not present the same life-threatening risk as did diabetes, because he acknowledged · that "the conversational method" of psychotherapy—"simply explaining to the patient the nature of her difficulty"—often worked as well as hormone therapy in relieving physical and emotional symptoms.[66] On the one hand, this physician-author contributed to the medicalization of menopause with his characterization of it as a deficiency disease. On the other hand, he subscribed to the consensus at the time that menopause was a natural event in the life course, in which some (not all) women experienced symptoms that sometimes necessitated medical intervention.[67]

Three years later, this same author, writing in the *Journal of the American Medical Association,* changed his evaluation of menopause as a deficiency disease. Based on six years of experience with 115 patients (too small a group, he acknowledged, to allow for statistical analysis), he noted that none of his patients had needed more than two and a half years of hormone treatments. Both the treated and untreated seemed able to regain equilibrium after passing through menopause. He now assured his readers that, unlike diabetes, menopause had a limited duration and thus did not require lifelong replacement therapy.[68] The implications of this assessment are important to point out: if menopause was a finite period, then medication might help to smooth its passage but would be completely unnecessary in the postmenopausal years.

Both physicians and women welcomed this new menopause remedy. Schering reported that sales of Progynon tripled in one year, from 30,968 units in 1931 to 92,280 in 1932.[69] But commercially available estrogen preparations remained expensive, which often precluded their continued use by patients.[70] One of the ways companies tried to reduce the cost of estrogen products was to find a more plentiful supply of the raw material, urine. In 1930 Bernhard Zondek, part of the team that found estrogens in pregnant women's urine, suggested an even better source: the urine of pregnant mares. Four years later, he announced the surprising finding that stallions excreted almost twice the quantity of estrogens in their urine as did pregnant mares.[71] Stallions, however, turned out to be impractical sources because of their tendency to kick over the collection buckets. Organon, the Dutch maker of Menformon, arranged to purchase horses' waste product from farmers; by 1937, the company celebrated the milestone of the millionth liter of pregnant mares' urine to be processed.[72] Other companies switched from human to horse urine in the isolation of female sex hormones, but the procedure was still costly, which translated into an expensive medication for consumers.

Cost certainly motivated much of the biochemical research on the structure of estrogen in the 1930s; a synthetic estrogen manufactured in the laboratory would be considerably less expensive than the natural hormones extracted from human or animal products. In August of 1936 both *Time* and *Newsweek* reported the laboratory synthesis of estrone by scientists at the Pennsylvania State College (supported by Parke Davis); three years later, *Newsweek* reported the synthesis of estrone by an alternative method by scientists at the University of Michigan and commented, "If this method of making the female sex factor is developed for commercial use, it will relieve both doctors and chemists of their dependence on natural sources for the hormone and may result in a cheaper product."[73] In 1938 scientists at Schering in Berlin produced another form of estrogen, called ethinyl estradiol. The company marketed this chemically altered urine-derived estradiol compound in both oral and injectable products.[74] Estradiol, estrone, and estriol are the three "natural" forms of this female sex hormone because they come from human or animal sources (extracted from ovaries or excreted in urine); they are "steroidal" because of their chemical structure. Of these three estrogens, estradiol has the greatest biological activity. Estrone is about half as potent as estradiol, and estriol is about one-tenth as active as estradiol. A review of the "Present Status of Commercial Endocrine Preparations" for the *Journal of the American Medical Association* in 1941 noted that there were several natural estradiol-based products, derived from the estrone excreted in horse urine, on the market, but none had yet received the approval of the American Medical Association's Council on Pharmacy and Chemistry (as of that date, only the theelin/estrone, theelol/estriol, and Amniotin products had earned the AMA's seal of approval). The author of this review dismissed the practicality of Schering's synthetic hormone, ethinyl estradiol, because of the frequent occurrence of side effects, namely, nausea and vomiting. (Twenty years later, this compound would be incorporated—in much smaller doses—into oral contraceptives; it remains an important component of many birth control pills today.)

The development of synthetic steroidal estrogens was overshadowed in 1938 by the news from England that scientists had synthesized diethylstilbestrol, more commonly known as DES, the first nonsteroidal estrogen drug. Edward Charles Dodds and his colleagues at the Courtauld Institute of Biochemistry at Middlesex Hospital in London reported that DES (or, as they called it, stilbestrol, because it was derived from a precursor molecule called stilbene) was both highly potent and orally active as an estrogenic substance, despite the fact that its structure was different from both natural and synthetic steroidal estrogens.[75] DES could also be manufactured and sold at a fraction of the cost of the steroidal hormones, thus making it accessible to more women. While some historians have identified the synthesis of DES as a transformative technology that medicalized menopause, it seems more appropriate to consider the development of this drug as part of a much longer trend of medicalization that had already begun in the early years of the twentieth century.[76]

Initial clinical studies of the use of DES to treat menopausal symptoms indicated that women suffered side effects such as nausea, vomiting, dizziness, and headaches, which raised concern about the toxicity of this new drug. Although some investigators reported incidence rates as high as 80 percent, most physicians agreed that 10 to 20 percent of women experienced adverse effects.[77] The AMA's Council on Pharmacy and Chemistry acknowledged that early studies raised the possibility that DES might "produce damage to the various body organs and tissues," but it cited a lack of evidence from further studies to confirm a definite toxic effect in humans. However, the discomforts *produced* by this medication that was supposed to *relieve* unpleasant symptoms deflated the earlier high hopes that DES would revolutionize the treatment of menopause.

It is important to point out that most physicians did not prescribe hormone therapy indiscriminately to their menopausal patients. They recognized that "many a woman passes through the menopause without 'batting an eye' "; only a small proportion of women experienced symptoms severe enough to warrant medical intervention.[78] Furthermore, physicians relied on their patients' subjective experiences to determine the necessity, dosage, and duration of estrogen therapy. Although the "Pap" smear (developed by George Papanicolaou for early detection of cervical cancer) could be used as a diagnostic test to measure the effect of estrogen on vaginal cytology, physicians in the late 1930s and early 1940s preferred to listen to what their patients had to say. They did not consider menopause to be a disease with quantifiable symptoms; rather, they treated individual patients who suffered from temporary disturbances during this normal stage of the female life course.[79]

That DES instigated further discomfort in even a small proportion of the women seeking relief from menopausal disturbances did not sit well with benevolent physicians. After all, replacing hot flashes and night sweats with nausea and vomiting merely traded one set of complaints for another. The Hippocratic oath enjoined physicians to "do no harm," but for a significant proportion of menopausal patients, the administration of DES broke that pledge. A better option would be an estrogen drug as potent and as cost-effective as DES with none of the annoying side effects. When a new estrogen preparation came along promising to do just that, doctors took notice.

This new estrogen product came from north of the border, from the laboratories of the Canadian firm of Ayerst, McKenna and Harrison. Ayerst was already a player in the estrogen market, with its successful product, Emmenin. Sales of Emmenin had suffered from competition with DES because Emmenin, a natural hormone derived from placentas, was much more expensive to produce than synthetic DES. In 1939 scientists at Ayerst revisited the possibility of extracting potent estrogens from horses' urine.[80] They figured out a way to obtain a highly active and very stable water-soluble concentrate of conjugated estrogens (as the sodium salts of estrone sulfate, different from the free or unconjugated estrone of other urine-based products). (Premarin is actually a mixture of at least ten different estrogens, of which estrone sulfate is the largest component—about 50% of the total.) These conjugated estrogens were two to four times more potent than unconjugated estrone (theelin) and nearly as potent as DES.[81] Most importantly, the new product did not cause noxious side effects. Ayerst introduced Premarin (the name came from *pre*gnant *mar*es' ur*in*e) in Canada in 1941 and in the United States in 1942.

Premarin earned wholehearted approval from the physicians who tried the drug on their menopausal patients. Several of these clinicians published reports of their experiences in medical journals, thus spreading the word that Premarin trumped DES. In its February 1943 issue, the *Journal of Clinical Endocrinology* published four separate studies of Premarin in clinical use in Chicago, Louisville, Los Angeles, and Madison, Wisconsin.[82] All four research groups praised Premarin for its ease of administration (via oral tablets), effective relief of symptoms (as gauged by patients' self-evaluations and vaginal smears), lack of side effects, and relative cost-efficiency. The Los Angeles group reported that their overwhelmingly positive results with Premarin "justified discontinuing the use of diethylstilbestrol in routine therapy in the menopause and the substitution of sodium estrone sulfate for all other estrogens."[83]

As important as Premarin's advantage of reduced side effects was the reason given for this superiority: Premarin was a biological product. Physicians specu-lated that Premarin worked better than DES because it came from the natural source of pregnant mares' urine; all the manufacturer did was to concentrate the already existing hormones. DES, by contrast, was an artificial creation made from chemicals unrelated to the hormones as found in nature; it might affect the human body in unknown, possibly detrimental ways. One doctor explained, "The laboratory cannot quite give the finished product the adsorption advantages that nature conjoins in her slow-working elaboration."[84] The fact that Premarin was a natural hormone gave it the edge over synthetic (read: unnatural) DES. Higher potency, oral activity, and lower cost allowed Premarin to compete successfully with the other natural hormones, such as Amniotin and Progynon.

Although physicians preferentially prescribed Premarin to treat the symptoms of menopause, they found another use for DES. Starting in the mid-1940s, DES was prescribed to pregnant women in hopes of reducing the incidence of com-plications, such as miscarriage and toxemia, although no clinical trials ever legiti-mated this practice. An estimated 500,000 to two million pregnant American women took the drug until 1971. In that year a team of doctors led by Arthur Herbst at the Massachusetts General Hospital published a report in the *New England Journal of Medicine* suggesting a causal association between a very rare cancer of the vagina in young women and prenatal exposure to DES; shortly thereafter, the FDA issued a warning that DES should not be used by pregnant women.[85] Subsequent research found additional adverse health effects for DES-exposed daughters and sons, as well as an increased risk of breast cancer in the mothers who took DES. Today, DES is only indicated for palliative treatment of advanced prostate cancer and certain cases of metastatic breast cancer.

Both DES and Premarin were among the first drugs to be approved under the new Federal Food, Drug, and Cosmetic Act of 1938. This law replaced the Pure Food Act of 1906, which prohibited the manufacture of adulterated and mis-branded drugs but did not regulate either effectiveness or safety. Manufacturers were held accountable for the contents but not the therapeutic success or failure of their products. In 1912 Congress passed an amendment to the Pure Food Act which mandated that labels on medicines could not make "false or fraudulent" claims. However, "falseness" and "fraudulence" were difficult to identify. At the time there was no scientific way to verify that a drug did not perform as adver-tised; moreover, it was impossible for the government to prove that a manufac-turer meant to deceive customers about its product.[86] By the early 1930s drug sales had increased by a factor of six, yet the public remained at the mercy of manufacturers.[87] With the blessing of newly elected President Franklin D. Roosevelt, the Food and Drug Administration was charged with rewriting the law.

The law that finally passed Congress in 1938 was considerably longer and more detailed than its predecessor.[88] It addressed weaknesses in the earlier law; for example, government prosecutors no longer had to prove fraudulent intent in order to remove false therapeutic claims from drug labels. Among the many other significant changes were two that had enormous impact on the manufacture and sale of drugs. First, manufacturers had to apply to the government for permission to sell new drugs. Not only did the application have to detail the contents of the product, the process by which it was manufactured, and the uses for which it was intended, but also it had to include scientific evidence of the safety of the product when used as recommended. The government had a period of sixty days (extendable to six months) in which to raise objections to the application; if the government did not respond, the drug was automatically approved and the manufacturer could proceed with the sale of its product.

The second change, which resulted from regulations disseminated by the Food and Drug Administration to clarify the enforcement of the law, created a new class of drugs that could be sold *only* by prescription. It left the decision up to the manufacturers as to which drugs would fit into this category but warned that improper sale of drugs (that is, over-the-counter sale of drugs the FDA considered to be safe only if prescribed by a physician) could lead to the charge of misbranding.[89] While government regulation was designed to protect consumers from unscrupulous drug manufacturers, it entrusted consumers' medical care to physicians, who would, presumably, look out for their patients' best interests. In many cases, patients who might have self-medicated were obliged to visit a doctor to obtain a prescription for treatment. However, for women seeking relief for menopausal symptoms, there was little difference in their medical interactions before and after 1938. The doctor continued to control not only the patient's treatment but also the degree to which she understood the nature and complexities of that treatment, since only the doctor was privy to the estrogen labeling information provided by the manufacturer.

These regulations did change the landscape of the estrogen drug market. Many of the older ovarian preparations had to be taken off the market, because of their dubious ingredients. In December 1939 the Food and Drug Administration issued the following notice to manufacturers:

> There are on the market drug products in liquid form designated as "Ovarian Extract" or by some similar title. In some instances these products have been found not

to contain the known therapeutically and physiologically active constituents of ovary, namely, those having estrogenic . . . activities. The Food and Drug Administration is of the opinion that such inert or essentially inert preparations when sold as "Ovarian Extracts," or under any other designation that such active principles are present, are both adulterated and misbranded as those terms are defined in the Federal Food, Drug, and Cosmetic Act.[90]

The only acceptable biological products were those proven by laboratory assay to contain pure extracts of estrone, estriol, or estradiol. Companies such as Parke Davis, Abbott Laboratories, Eli Lilly, Squibb, Schering, Organon, and Ayerst had little to fear from this warning, since their estrogen products could pass the test; furthermore, these drugs had always been available only by prescription. Other firms, however, quietly withdrew their previously unregulated elixirs from pharmacists' shelves.

The two latest estrogen drugs—synthetic diethylstilbestrol and natural conjugated estrogens—had to go through the new application procedure before they could be marketed and sold. Premarin probably had a simpler passage through the approval process, since its formulation and production were similar to those of other urine-derived estrogen products that had been on the market and in use for at least ten years.[91] DES, however, was a brand-new chemical invention. It presented a further challenge to the newly empowered FDA, because it was the first drug to be reviewed that did not purport to cure a disease yet did have the potential to harm users.[92] Thus regulators took seriously the stipulation that the manufacturers provide sufficient evidence of DES's safety.

The DES case was also unusual in that its creator, Edward Charles Dodds, chose not to patent either the product or the process by which it was made but instead encouraged the widespread dissemination of his invention. Scores of American drug companies explored the possibility of manufacturing and marketing DES; in the end, a consortium of twelve firms pooled their resources and submitted a joint application to the FDA.[93] Representatives from these firms formed a committee in 1941 which designed questionnaires sent out to fifty-four medical experts asking them about their clinical research and experience with DES. The objective was to collect enough data to convince the FDA that the medical profession considered DES safe to use. The manufacturers did not have to demonstrate DES's effectiveness in relieving menopausal symptoms; the FDA would not require proof of efficacy until 1962.[94]

The consensus among the experts polled was that the nausea and vomiting experienced by some users of DES was not an indication of the drug's toxicity.

Rather, the physicians attributed these adverse effects to overdosing by incompetent doctors or to idiosyncratic responses of individual patients.[95] Although they did not yet know the mechanism by which DES caused nausea, they felt that it was unnecessary to understand this etiology before DES could be approved for marketing and sale. One physician commented, "I personally believe that it is safe to prescribe [DES] using the same precautions to avoid toxic effects as in the use of digitalis [a medicine made from the dried leaves of the foxglove plant, still widely used today as a heart stimulant]. The mechanism by which digitalis produces nausea and vomiting was not understood when Withering wrote about it more than 150 years ago and it is fortunate that overanxiety in regard to the toxic symptoms did not lead him to withhold its use then."[96] In other words, it would be foolish, and perhaps even negligent, to deny the drug to women who needed it. In effect, the medical experts agreed that DES was safe enough, so long as it was prescribed by a capable physician. Over time, they hypothesized, greater experience with the drug might yield information as to how and why it produced nausea and vomiting.[97] This decision set an important precedent: patients would take the drug not only as recipients of a therapeutic agent to relieve their individual ailments but also as participants in a larger experiment to better understand its course of action.

The FDA accepted the medical evidence collected and presented by the consortium of pharmaceutical manufacturers and approved the sale of DES in September 1941.[98] Within a year, Ayerst, McKenna and Harrison won approval from the FDA to market their Premarin brand of conjugated estrogens in the United States. The role of the FDA in drug regulation did not end with the initial approval of new drugs; the agency also served as watchdog over the American pharmacopoeia. However, its involvement with menopausal estrogen products was minimal over the next thirty years; every now and then the agency issued a gentle reminder to the trade that "glandular preparations" must be properly labeled. In the 1970s, though, the FDA would come to play a significant part in both the censure of DES and the authorization of estrogen product labeling for patients.

As for so many aspects of American life, the years of the Second World War signified both an ending and a beginning in the story of estrogen. The war brought about a tragic ending to the life of Edgar Allen, who had combined his love of the sea with his patriotic sense of duty by joining the Coast Guard Auxiliary in Connecticut, where he had lived and worked since 1933 as professor and chairman of the anatomy department at the Yale School of Medicine. On 3 February 1943, while patrolling Long Island Sound, Allen suffered a fatal heart attack at the age of 50.[99] His partner in the discovery of estrogen, Edward Doisy, remained in Missouri and continued to work at St. Louis University School of Medicine. In the late 1930s he began a new line of research, turning his attention to the isolation and synthesis of vitamin K. For this work, Doisy won the Nobel Prize in Physiology or Medicine in 1943.[100]

As a result of the work of Allen, Doisy, and numerous other investigators, sex endocrinology became a respectable field of biomedical research. By the start of World War II it had long since ended any association with disreputable organo-therapists or rejuvenators. Looking back, one contemporary researcher described the years between 1926 and 1940 as the "heroic age" of reproductive endocrinology. In addition to the work on estrogen, scientists isolated progesterone (the second female sex hormone, extracted from the corpus luteum of the ovary) and testosterone (a male sex hormone, extracted from the testis) and discovered the gonadotrophins (pituitary hormones that control the sex organs).[101]

Many scientists, such as Doisy and Collip, established ties with drug companies, which facilitated the movement of purified and synthesized hormones from laboratory benches to pharmacy shelves. The pharmaceutical development of estrogen therapy, begun in the 1930s, really took off in the postwar years. Manufacturers achieved a level of purity and consistency in their products, thanks to the availability of sensitive biological and chemical assays. Product reliability was also required in order to obtain approval from the FDA. The World War II years marked a turning point in the commercialization of estrogen.

The explosion of estrogen products is documented in the pages of the first edition of the *Physicians' Desk Reference* (PDR) of 1947, which listed the names and makers of some five thousand medications and provided detailed information on fifteen hundred products. This detailed information came directly from the 132 manufacturers who paid for the distribution of the handbook to more than one hundred thousand doctors in the United States.[102] (The PDR is still updated and published annually. The 2007 edition is about ten times longer than the 1947 edition.) In the "Therapeutic Indications Index," the book listed medications according to the diseases and conditions for which they were indicated. Only "ethical" products—those advertised exclusively to the medical profession—were included. For "Menopausal Disorders," both artificial and natural, the 1947 PDR listed fifty-three different formulations, sold by twenty-three different companies. Several of these products were cross-listed under the categories of "Menopausal Hypertension," "Menopausal Irritability," "Menopausal Nervousness," "Menopausal Neuroses," "Menopausal Psychoses," "Menopause," and "Menopause (Mild)." The "Drug and Pharmacological Index" listed many more products—

including those made by nonparticipating manufacturers (those who chose not to supply detailed information or to subsidize the publication)—according to drug classification. There were seventy-two separate entities entered under the designation "Estrogens"; this number did not include blended products, in which estrogens were combined with barbiturate, progesterone, sedative, thyroid, or vitamin B complex. The number and variety of estrogen products on the market in 1947 are stunning compared with the mere handful available prior to the war.

Thus, by the dawn of the postwar era, collaborations among scientists, pharmaceutical manufacturers, and physicians, under the watchful eye of the FDA, had produced both natural and synthetic estrogens for replacement therapy. These drugs represent, on the one hand, the culmination of decades of experimentation by scientists in the laboratory and doctors in clinical practice, as the former sought to understand the workings of the endocrine system and the latter sought to treat transitory female complaints. On the other hand, they signal the beginning of a new age, in which medical views and cultural circumstances converged to create a climate in which female aging became the target of treatment.

# CHAPTER TWO

# From the "Neutral Gender" to "Feminine Forever"

Twenty-five years after Doisy and Allen arrived in St. Louis, another young man landed in the city to begin his scientific career. A recent medical graduate from the University of Rochester, this 28-year-old doctor moved to Missouri for an internship and residency in obstetrics and gynecology at St. Louis Maternity Hospital and Barnes Hospital of the Washington University School of Medicine. In 1947 he joined the faculty of the School of Medicine, rising from instructor to associate professor in the department of obstetrics and gynecology.[1] In the 1950s, he emerged as one of the architects of the campaign to expand the use of sex hormones from a short-term remedy to alleviate the symptoms of menopause to a long-term therapy to improve the health of older women; his ideas would have a lasting impact on the medical prescription of estrogen for the rest of the twentieth century. However, this influential body of research, reported in more than a dozen articles in medical journals from 1948 to 1958, would warrant only one of the forty-five paragraphs of his obituary in 2001, overshadowed by the work that made him a celebrity.[2] His name was William H. Masters.

Born into a well-to-do family in Cleveland in 1915, Masters attended elementary school in Kansas City, prep school in Lawrenceville, New Jersey, and college at Hamilton in Clinton, New York. Masters was a star in and out of the classroom: in addition to his scientific studies, he competed on the debate team and won varsity letters in football, basketball, and baseball.[3] In the fall of 1939, he entered medical school. It was there that he met George Corner, one of the leaders in the field of reproductive biology and the co-discoverer of progesterone, the second female sex hormone.

Under Corner's direction, Masters began work on the estrous cycle of female rabbits; this research sparked his interest in human sexuality, a neglected but potentially fascinating field of study. Although Corner moved on to the Carnegie

Institute in Washington, he continued to act as a mentor to Masters. When Masters expressed his desire to study the physiology of sex in men and women, Corner gave him a key piece of advice: wait until you are at least forty years old with an established record of academic credentials before you undertake such controversial research.[4] Masters heeded his teacher's suggestion and turned his attention to hormone replacement therapy. In the mid-1950s, having passed his fortieth birthday and having achieved a scholarly reputation, Masters joined forces with Virginia Johnson and embarked on the laboratory research that would result in the publication of the best-selling book *Human Sexual Response* in 1966, launching his second career as one of the gurus of the sexual revolution.

Masters's first career as an investigator of sex hormones and the problem of aging led him to describe ovarian failure as "the Achilles heel" of the female endocrine system.[5] Because he attributed the onset of osteoporosis, cardiovascular disease, and senility to the decline of estrogen production, he argued that long-term hormone replacement therapy could improve the physical condition and mental acuity of older women. This theory contradicted the prevailing wisdom among doctors in the 1950s, namely, that estrogen should be used only for the short-term relief of menopausal symptoms. In this regard, Masters's work on hormone replacement therapy was as revolutionary as his later investigations into the science of human sexual behavior.

From the 1930s through the 1950s, doctors incorporated estrogen products into their pharmaceutical arsenals. The dominant model, as articulated by physician-authored articles in medical journals, called for restraint in the prescription of estrogen to alleviate menopausal symptoms.[6] In reviewing the options available to doctors treating menopausal women, one Kentucky physician told his colleagues, "The sooner the estrogens can be stopped and the patient allowed to go on her own, with good health habits, food and rest, and living within her physical capacity, the better and more permanent are the results."[7] A gynecologist from North Carolina advised, "The dosage [of estrogen] should be kept at a minimum, begun early, and discontinued, even if only temporarily, as soon as possible."[8]

Medical consensus also held that only a small proportion of women needed to be treated with estrogen at menopause.[9] In this view, most women passed through the change with little or no trouble. For those who sought medical care, physicians first offered reassurance (that menopause was a natural stage in the life cycle that the patient would soon pass through) and advice on healthy living (plenty of rest, balanced diet, avoidance of stress). Sometimes husbands were called into action; one British doctor suggested to his readers, "If the husband can be made to take an interest in the problem, then he should start to take his wife for vacations without the family, and so on."[10] The next step was to try sedatives. As historian Judith Houck has noted, a prescription for sedatives did not necessarily mean that the physician thought his patient's complaints were "all in her head."[11] Rather, many doctors viewed sedatives as a more conservative approach to treating the symptoms of menopause, such as nervousness and sleeplessness, that brought women into their offices, as opposed to wholesale tinkering with the endocrine system. Hormone replacement was supposed to be prescribed only for those relatively few women who did not respond to talk or tranquilizer therapy.

Not every doctor, however, followed these recommendations of their peers, prompting some in the profession to communicate their concerns about the overzealous and injudicious application of estrogens. The first of a spate of cautionary articles in medical journals appeared in December 1939, when the *Journal of the American Medical Association* published an editorial called "Estrogen Therapy—A Warning," which concluded that "the conflicts in the reports on these substances [ethinyl estradiol and diethylstilbestrol] and the opinions of some authorities on the possible harm from estrogen therapy should warn against long continued and indiscriminate therapeutic use of estrogens."[12] Over the next several years, articles with titles such as "The Use and Misuse of Estrogens in Menopause," "Oestrogenic Therapy: Its Uses and Abuses," and "Indiscriminate Use of Estrogens in the Menopause" were published on both sides of the Atlantic.[13]

These authors sought to circumscribe the use of estrogen by enumerating the conditions for which it was indicated and the circumstances in which it was contraindicated. They stressed the importance of prescribing estrogen for a limited time and tapering off gradually, and they warned against any kind of prophylactic use of estrogen in women approaching the menopause. One doctor feared that women might become addicted to estrogen, relying on weekly shots as a panacea for whatever might ail them. This same author worried about the time and expense spent by patients undergoing long-term hormone therapy; he also castigated his colleagues who relied on estrogen treatments as time savers and money makers. "I have often heard physicians say that their office expenses were paid by injections of estrogenic hormones given by their nurses," he wrote indignantly.[14] Similarly, the authors of another article scolded doctors who engaged in wanton prescription writing with the remark, "The regrettable use of estrogen as an 'all purpose female tonic' is condemned."[15]

An unresolved issue was the relationship between estrogen and cancer. Although most writers expressed skepticism that the results of animal studies demonstrating the carcinogenic potential of estrogen could be extrapolated directly to humans, they reasoned that estrogen might cause further growth of preexisting tumors in some women, and thus they called for careful screening of patients to avoid estrogen-induced cancer. Both Edgar Allen and Edward Doisy, collaborators in the isolation of estrogen, weighed in on the cancer issue. In an article published in the *Journal of the American Medical Association* in 1940, Allen warned that overuse of the pharmaceutical applications of their discovery might lead to more cancers among older women.[16] A few months later in the same journal, Doisy refused to discount the animal data: "The increased incidence of malignant change following administration of estrogens to cancer-susceptible strains of animals serves as a definite warning."[17] However, others dismissed the estrogen-cancer link as tenuous and argued instead that "extreme caution be used before attributing carcinogenic properties to estrogens in human beings, lest an extremely valuable therapeutic agent be condemned unjustly."[18] Conflicting data and interpretations would confuse the debate over the estrogen-cancer question for the next sixty years.

What is clear from the dozens of articles written in the 1940s is the consistent view among physician-authors that estrogen was intended as a short-term palliative for menopause-specific complaints. Actual practice may have deviated from the path recommended in the medical literature, but mainstream advice endorsed the three-tiered approach to treating patients suffering from menopausal symptoms: first physician counseling, then sedatives, and finally, if all else failed, a limited course of hormone replacement. Judith Houck has characterized this medical encounter between doctors and middle-aged patients as respectful of women's experiences, because "physicians treated women as a whole and menopause as a complex process, only part of which was rooted in women's glands."[19] This situation began to change in the next decade, when William H. Masters and his co-workers in St. Louis publicized their conviction that the problem of aging lay squarely in women's declining ovaries.

William H. Masters wasted no time in getting his research program going. He made the acquaintance of William B. Kountz, who had participated in the establishment of a special geriatrics unit at the St. Louis City Infirmary in 1938 and who was named the director of clinical services of the division of gerontology at Washington University School of Medicine in 1946. Kountz attributed his interest in aging to his study of Egyptian mummies while traveling abroad as the recipient of a National Research Council Fellowship in 1930–31.[20] Masters has left no record of why he got involved in research on the aged; however, his investigation into sex hormones and aging in the 1940s and 1950s does help to explain his attention to the sexuality of middle-aged and older people in his later studies, *Human Sexual Response* and *Human Sexual Inadequacy*, published in the 1960s and 1970s.

Kountz offered Masters access to the patient population of the St. Louis City Infirmary and Infirmary Hospital, a municipal care facility for the indigent infirm and aged. The City Infirmary, occupying several adjoining buildings constructed in 1872, was known until 1910 as the City Poorhouse. It closed in 1972, when patients were transferred to a new nearby nursing home, the Truman Restorative Center, maintained for city residents who could not afford private nursing care.[21]

In 1945, Masters and his colleagues began to administer hormones to a small group of elderly female residents of this institution. The first publication of results appeared in July 1948 in the *Journal of Gerontology*. Based on thirteen post-menopausal women, aged 64–82, Masters and his co-author reported that "menstrual periods can be induced in women many years after the menopause."[22] They went on to lay out the central premise that would guide their research program for the next decade: "Administration of sex hormones to an aged woman induces a reversal of one of the inevitable changes of aging and affords an opportunity *to find out whether or not some of the manifestations of aging are reversible*" (emphasis added).[23]

Over the next six years, more women were added to the study, so that by 1951 Masters reported that over a hundred subjects had been incorporated into the pool receiving replacement therapy. Depending on which study group they were assigned to, patients submitted to periodic vaginal smears, uterine and other tissue biopsies, blood analyses, and psychological tests. When they died, their bodies were autopsied to assess the macroscopic and microscopic anatomical effects of the hormones. These women, living at the St. Louis Infirmary because they had nowhere else to go, were weighed and measured, poked and prodded, photographed and questioned for months, and sometimes years. No records of these experiments remain, beyond the accounts published in medical journals, so we cannot determine if the "volunteers" were willing or reluctant participants or even whether they understood the nature of their involvement in the research studies. However, at one conference in 1955, Masters confessed that "these people were, in essence, experimental laboratory animals."[24]

In their published papers, the St. Louis researchers paid lip service to the new "gold standard" of medical research: the controlled, double-blind, randomized clinical trial, in which subjects were randomly assigned to experimental and control groups, and neither investigators nor subjects knew whether the subject was receiving the drug or a placebo.[25] They maintained control groups, who received injections of sesame oil instead of the oil-dissolved hormone. And, in a series of experiments designed to measure the psychological effects of hormone therapy, the coordinators "blinded" both researchers and participants by keeping the identities of the subjects hidden from the examining psychologists and maintaining that "attendants, nurses, and patients were not aware that there were two distinct subgroups."[26] However, true randomization was not possible, given the small population of inmates at the City Infirmary and the even smaller number selected as subjects for experiments. No study included more than thirty women, including the controls, which precluded any sort of meaningful statistical analysis. Furthermore, as one article acknowledged, "This group was in no respect representative of subjects of similar age in the general population. Many of them had definite mental illnesses at one or another period during their lives and socioeconomic status was below average for the general population."[27] Nonetheless, Masters extrapolated from this unique population to draw conclusions about *all* older women.

The data collected from these patients included an assortment of objective measurements and subjective impressions. Try as they might, the researchers could find no physiological explanation for the clinical observations that seemed to indicate a positive correlation between hormone therapy and healthiness. In a 1951 speech, Masters admitted, "Despite six years of effort we haven't as yet the vaguest idea why the patients have made such significant gains in general well being."[28]

What evidence did Masters offer for the subjects' improvement? Of the first fifteen patients enlisted in the study, eleven had been completely bedridden at the outset. After two years of hormone therapy, "all members of the first group were ambulatory and able to take personal care of themselves. Some group members even graduated to minor daily duties such as distributing food trays in the wards at mealtimes or helping in the light floor maintenance work."[29] This image of the crippled walking again smacked of the miraculous, especially when prefaced by Masters's confession of ignorance as to the biological basis for such dramatic recovery.

However, the picture became considerably less rosy when he revealed the details of the patients' histories. Of these original fifteen women, only three remained on hormone therapy. Six had died, two "showed no demonstrable physical or mental improvement and were dropped from the series after three years of therapy," and one began to refuse medication and so she, too, was taken off the study. The last three "improved to such a degree . . . that their families, feeling that they would no longer be a major nursing problem, removed them from the institution over objection." Masters made sure to point out that all three women were once again bedridden six months after the cessation of the hormones.[30] In other words, after six years, only 40 percent (six of fifteen) showed long-term improvement in their health. In a report of a different thirteen-month study, Masters noted that two of eleven patients in the experimental group died (all of the controls survived). He ascribed one death to a previously existing intestinal cancer. Although this woman had an inoperable tumor, she continued to receive hormone injections until her death. The second patient died "without prior warning" of a heart attack. In spite of what would appear to be an 18 percent mortality rate (and 40% in the previous series), Masters concluded that "this hormone combination [1:20 ratio of estrogen to testosterone] achieves its effect by what seems to be an arrest, and possibly a partial reversal, of the aging process. This is the result of generally rejuvenated cellular and organ metabolism."[31]

Close scrutiny of the psychological studies reveals a similarly positive interpretation of ambiguous data. The investigators conceded that "from an inspection of the concrete test results, it could not be said that there was a uniform change for the better in the experimental group. However, even a casual glance at the raw data suggests that there are many positive trends."[32] For example, in the "thematic apperception test," in which subjects were asked to make up stories about a series of pictures and psychologists rated the stories in terms of mood (positive, negative, neutral), outcome (successful, unsuccessful, unresolved), and time trend (action in past, present, future, or no time trend), there was no significant difference between the experimental group and the control group. "Nevertheless," the authors asserted, "the interpretation which seems justified is that at the time of the second examination the experimental group told better stories, whether or not the mood or outcome of these stories reflected positive feelings."[33] Clearly, "better" implies a subjective judgment made by the evaluators and cannot necessarily justify the conclusion that hormone therapy produces psychological benefits.

The authors recognized that the results of mental testing were compromised by physical handicaps (visual impairment, hearing loss, poor motor control) which made it difficult for the patients to complete the assigned tasks. "Furthermore," they continued, "the entire procedure held an aura of mystery for the group, with their fantasies about the purpose of the examinations often producing a negative set toward them (e.g., the tests were being given to see if they were 'crazy,' to determine if they should be transferred, put to work, etc.). Such factors could easily cause a purposeful dissembling of true feelings and thus bias the results somewhat."[34] It is no wonder that these elderly women, living at the mercy of their caretakers at this municipal institution, feared the worst from the battery of psychological tests and hid their feelings to avoid any negative consequences. After reading about the study's methodological problems and indefinite results, the article's optimistic conclusion comes as a surprise: "If the gains here reported are substantiated in other groups and can be maintained over longer periods of time, it is believed that an avenue of treatment possessing definite psychologic benefits for older females in our culture will have been demonstrated."[35] The authors followed the consensus of the St. Louis research group, namely, that they were onto something big.

In 1951, after six years of clinical observations, Masters publicly announced the larger social implications of his research: he fully expected hormone replacement therapy to revolutionize the science of gerontology, the medical practice of geriatrics, and the social role of America's senior citizens, particularly women. He began by describing the menopause as "a period of change during which the average female retires from a reasonably active to a relatively passive existence. The concentrated activities of raising a family are gradually superceded by the less strenuous demands of spoiling the grandchildren and puttering in the garden."[36] Certainly, the luxuries of this lifestyle were reserved for middle- and upper-class women. By contrast, Masters expressed his concern that "we are progressively increasing our indigent aged population. State-supported institutions for the housing and care of this rapidly multiplying segment of our society are grossly overcrowded, and as more beds are made available to meet the ever increasing demand, our tax burden becomes progressively heavier. It goes without saying that a basically non-productive segment of society is not only a financial burden to the active population but is a tremendous waste of manpower."[37] Masters conveyed very different impressions of the problems presented by postmenopausal women depending on their economic status. His solution, however, remained the same for all, regardless of income level: long-term hormone replacement therapy for the "rejuvenation of useful mental and physical function in the aged."[38]

What about men? Masters acknowledged that endocrine deficiency might also play a role in the incapacitation of older men, but he justified the decision to limit the experiments to women according to the following rationale: "First, a more objective study of target organ results could be obtained in the female than the male. Second, females as a sex constitute more of a public health problem than the males due to their greater expected longevity. Third, they are, as a rule, generally more amenable to therapy, repeated biopsies, and examinations than are the males. And fourth, our female population is generally more static than the male."[39] His perception of women as more acquiescent research subjects and as a greater social burden contributed not only to his experimental design but also to the gendered application of his findings to medical practice. The St. Louis researchers did eventually begin to study men in 1952, but the male subjects remained a minor component of the program because, according to Masters, "it was our indirect interest."[40] The fact that Masters published several articles in obstetrics and gynecology journals also indicates his greater attention to the "problem of aging" in women.

Masters proselytized his fellow doctors at national meetings of obstetricians and gynecologists, as well as at conventions of gerontologists. He had first articulated his vision for restoring the vitality of the aged at a small gathering in South Dakota in 1951, but only physicians from that state would have read the written version of his speech in the *South Dakota Journal of Medicine and Pharmacy*. In 1954, he gave the first of several addresses, all similarly worded and subsequently published in major medical journals, on his proposal for " 'puberty to grave' sex steroid support."

He began by defining what he called "the third sex" or "the neutral gender" as "roughly all persons who have reached an average age of 60."[41] He blamed the failure of the reproductive endocrine system as "the Achilles heel" of the entire endocrine system. "We are essentially intact during our 60's," he proclaimed, "and perfectly capable of functioning . . . as efficiently as much younger persons, except that we are essentially castrates."[42] Replacing the hormones that the sex glands of older people could no longer produce would tackle "one of the greatest public health problems of the present and future . . . the rapid increase in our aging population. If we can avoid the physical and mental degenerative changes associated with senility, or at least prevent their appearance for many years beyond present expectancy, a major contribution will have been made, not only to the treated individuals but to the economy and potential manpower supply of our country."[43]

Masters included both "former males and females" in the neutral gender. Although his team had completed little research on men, he believed that his principles of hormone replacement could and should be applied to people without regard to "previous sex." He offered a rationale for unisex medication (estrogen and testosterone in a 1:20 ratio) based on saving both money and time. "The pharmaceutical houses can mass produce the material at a significant price saving for the individual treated. In addition, it is infinitely easier for the physician if the necessity for sex differentiation in steroid supportive therapy no longer applies."[44] He did, however, differentiate between the points at which the two sexes joined the third sex: "Such age conformity is of course only loosely adhered to by the new-gender members. There are obviously many women who have joined the 'neuter gender' age group before their fiftieth birthday. Equally obviously, there are a significant number of males who could not be considered candidates for the third sex even after their seventieth birthday. It is fair to generalize that the female will be a third-sex candidate roughly fifteen years ahead of the male."[45] This distinction helps to explain why hormone replacement became a therapy almost exclusively prescribed for women. If men still appeared to be manly into their seventies, then there was little need for physicians to medicate them as members of the neutral sex. Furthermore, the clinically obvious milestone of menopause provided physicians with a convenient starting point for hormone replacement in women; doctors had also been treating women with hormones to alleviate the symptoms of menopause for decades. With both the pharmaceutical products and the medical practice already in place for short-term estrogen replacement, it was not too great a leap to extend the therapy to longterm use.

Masters was careful to point out that "there is not the slightest evidence to suggest, nor has any claim been made that steroid replacement increases longevity by one single day."[46] Although one of his colleagues jokingly referred to him as "a twentieth century Ponce de Leon," Masters insisted that "steroid replacement . . . in no sense represent[s] a panacea for the problem of aging."[47] Rather, hormone replacement would enable old folks to live out their lives as "happier, better adjusted, more useful members of the 'neutral gender.' "[48] Since medical science had helped to extend life expectancy so that more and more women lived well beyond menopause, he argued that physicians ought to assume the responsibility of improving the quality of life during those later decades.

Part of a better quality of life included the avoidance or forestalling of age-related diseases, such as osteoporosis and atherosclerosis. In 1940 Fuller Albright and his colleagues at Massachusetts General Hospital identified the condition called postmenopausal osteoporosis (loss of bone density) and demonstrated the ability of estrogen to redress the balance of calcium and phosphorus (important minerals in bone formation) in older women. Since they had experimented on just three patients, the investigators were unwilling to draw conclusions about the possible clinical applications of this finding.[49] Over the next two decades, various research groups tested the relationship between estrogen and osteoporosis and confirmed both the role of estrogen in bone deposition and its ameliorative effect on existing osteoporosis.[50] In 1957 Albright's younger associates, Philip Henneman and Stanley Wallach, reviewed the records of some two hundred postmenopausal patients who had received estrogen therapy from one to twenty years.

About half of the group suffered from osteoporosis. The researchers found that women who began hormone therapy at menopause did not lose height, and those who began therapy after the onset of height loss stopped shrinking during the course of treatment.[51] They concluded that "estrogen therapy may stop the progression of osteoporosis" but cautioned that it could not restore strength to already brittle bones. Masters emphasized the former, and passed over the latter, in describing osteoporosis as "a reversible process" in building his case for long-term hormone replacement.[52]

Researchers also tried, beginning in the 1940s, to account for the observation of greater degrees of coronary atherosclerosis (thickening and hardening of the heart's arteries, also called arteriosclerosis) in men as compared to women of similar ages. Opinion was split between those who attributed the disparity to anatomical differences between males and females and those who suggested that ovarian hormones somehow protected women from developing the disease. In 1953 a group of investigators at the Mayo Clinic in Minnesota acquired hearts from autopsies of women whose ovaries had been removed and compared them with the hearts of men and women of equivalent ages. They found that the ovariotomized women showed greater atherosclerosis than the control women but less than the men. Since men, who have vastly lower levels of estrogen, exhibited the highest levels of arterial disease, and women whose ovaries were removed displayed a greater degree of arterial disease than their intact counterparts, the researchers hypothesized that estrogens must play some sort of cardioprotective role. Hormones, they reasoned, must be responsible for the delayed development of atherosclerosis in women as compared to men.[53]

This conclusion was put to use by physicians with the interests of three different groups in mind. First, it was cited as evidence to argue against the prophylactic removal of ovaries in women at risk for ovarian cancer; these obstetrician-gynecologists (with the support of cardiologists) sought to convince their peers that any possible cancer prevention was outweighed by the beneficial effects of estrogen on the cardiovascular system (and on the skeletal system, in preventing osteoporosis).[54] Second, researchers attempted to use estrogen to treat men with atherosclerosis. Although the patients demonstrated positive changes in cholesterol and lipoprotein levels, they also experienced breast development, impotence, and loss of libido, which may explain why estrogen therapy never became a popular drug for men.[55] Third, the observation that postmenopausal women have a much higher incidence of atherosclerosis than premenopausal women led to the suggestion that long-term estrogen therapy might be employed to prevent the development of atherosclerosis in older women. One study seemed to confirm this hypothesis: postmenopausal women both with and without heart disease (either angina pectoris or myocardial infarction) who took estrogen reduced their cholesterol levels near to those of healthy young women.[56] However, the sample size was small (35 women with heart disease, 58 women without) and the length of treatment was an average of only fourteen months, a fraction of the time that long-term postmenopausal hormone therapy might continue (potentially for decades after menopause) and not nearly long enough to detect slow-developing adverse effects such as cancer.

The question of the relationship of estrogen to cancer was raised as early as the 1930s. If estrogen could cause cervical cancer in mice, monkeys, and guinea pigs, why not in humans?[57] Ethical considerations precluded any direct test of whether or not estrogen would produce cancers in women, so researchers relied on observations of patients undergoing estrogen therapy. Henneman and Wallach, in their review of two hundred patients, found that, over 1,100 "patient-years" (the number of patients multiplied by the number of years of therapy for each), four women developed cancer. The investigators did not consider this number to be significant, because "these patients were all in the 'cancer age group' . . . [and] this appears to be a low incidence of genital cancer."[58] Clyde L. Randall, a gynecologist from Buffalo, New York, commented, in making his case for the importance of the ovaries even after menopause, that "the fear that estrogens . . . as used in prolonged therapy, might be carcinogenic in human beings does not often seem evidenced in clinical experience."[59] However, he offered no evidence, either from his own practice or from data collected by other researchers, to support this assumption. Masters reported that none of his experimental subjects had developed tumors during the course of treatment, but he noted, "We have picked groups of ten individuals who have had no demonstrable evidence of tumor to begin with."[60] In another article, he firmly dismissed the possibility that his prescribed regimen of hormone replacement could cause endometrial cancer (cancer of the uterus) because the addition of testosterone to the estrogen prevented abnormal growth of the uterine wall. The testosterone-estrogen combination was never widely used; two decades later, however, physicians would begin to advise the addition of progesterone in hormone replacement therapy for the same purpose: to counteract the carcinogenic effects of estrogen on the endometrium. But in the 1950s, although some doctors remained concerned about the possible estrogen-cancer association, advocates like Masters pooh-poohed the carcinogenic potential of the hormone.

The significance of the research claiming long-term protective and preventive benefits for users of hormone replacement is its influence on medical thought for the rest of the century. These small studies served as the basis for the development of the rationale for long-term hormone replacement therapy for postmenopausal women. By the time the precedent had been set for this practice, few if any thought to question the integrity of the original work. The hypotheses that estrogen would forestall osteoporosis, cardiovascular disease, and mental senility became almost axiomatic, thanks in part to the proselytizing efforts of William H. Masters.

Masters published his last article on the "neutral gender" in 1958. By then, he had turned his full attention to his true passion: the sex problems of the male gender and the female gender. He had not been alone, however, in his entreaty for long-term hormone replacement therapy; among his contemporaries were some equally vocal advocates who took up the cause of estrogen "from puberty to grave."[61]

Masters's thesis received a vote of confidence in a 1959 article in the *Journal of the American Medical Association* on the effectiveness of conjugated equine estrogens (Premarin) and ethinyl estradiol in treating menopausal symptoms, such as hot flashes, insomnia, and nervousness. Although the authors noted that long-term hormone replacement therapy was beyond the scope of their paper, they went on to comment favorably on Masters's research: "In his hands this regimen effectively influenced 75% of the persons treated, with notable physical and psychological improvement. We are entirely in accord with Masters' concepts and have confirmed his observations."[62] Another article in the same issue of the *Journal* confirmed the success of estrogen in alleviating both menopausal hot flashes and pain from postmenopausal osteoporosis. However, these investigators circumscribed their recommendation to women who presented with serious symptoms: "Prolonged, cyclic oral estrogen therapy . . . is a safe and effective therapy for postmenopausal women with disabling menopausal symptoms or postmenopausal osteoporosis."[63] They did not go as far as Masters in advocating the prescription of hormones to all women of a certain age, but they clearly saw the potential for long-term estrogen therapy in the practice of geriatric medicine.

One who did enthusiastically share Masters's views was E. Kost Shelton, clinical professor of medicine at UCLA and director emeritus of the Endocrine Clinic at Los Angeles General Hospital. The two physicians knew each other; in November 1954 they both spoke as presenters at the Graduate Symposium on Geriatric Medicine, given by the American Geriatrics Society in New York City. Although they were at very different stages of their careers (Masters was in the first decade of his professional life; Shelton, a much older man, died a few months later, in February 1955), they articulated the same vision for estrogen.

Shelton promoted long-term hormone therapy as the solution to middle-aged women's woes. Not only would estrogen replacement prevent the development of osteoporosis and coronary atherosclerosis, but it would also help maintain a youthful appearance, a positive attitude, and a happy marriage. Estrogen deprivation at menopause, Shelton warned, "is frequently accompanied by regression to a shell of the former alluring woman. The dry, atrophic, irritable, sometimes hemorrhagic vaginal tract is no longer conducive to a continuation of satisfactory intercourse. Minor sexual misunderstandings frequently arise and lead to serious family difficulties. The patient ages physically and mentally, the latter at least insofar as her ego is concerned. She becomes insecure, inadequate and ultimately careless during the most vulnerable period of her marital existence."[64] He described women of previous generations who survived into their fifties and sixties as "having fulfilled their destiny as seed-pods and . . . willing to dry up and blow away."[65] Thanks to the availability of estrogen substitution therapy, "the grandmother of today is, or should be, an entirely different person. She no longer willingly relinquishes the husband of her youth to the designing widow down the street. By reason of freedom from crippling disease, better nutrition, clever cosmetology, greater social freedom and smarter raiment she can hold her place with women even twenty years her junior—that is, provided she remains in some degree physiologically intact."[66] In Shelton's view, a woman's goal in life was "to not only capture but to hold a husband"; estrogen could and should be used to help her maintain her femininity and sexual attractiveness.[67]

Shelton scoffed at those "therapeutic nihilists" who denied estrogen to postmenopausal women: "The very person who argues that menopause is a natural phenomenon fights nature every day. He pasteurizes his milk, boils his instruments, vaccinates his stock and his children, sprays and buds his fruit trees, flies against gravity, makes new elements and splits the atom. Is the menopause any different?" The concerns about estrogen were "reminiscent of the outworn arguments against anesthesia in childbirth, against cosmetics, against everything progressive in life."[68] Like Masters, he believed that doctors had an obligation to provide their patients with the available tools of modern medicine.

Shelton also rejected the notion that estrogen caused cancer, citing the statistic that the death rate from cancer of the uterus and breast, among white women, had not changed significantly between 1930 (when few women used estrogen) and 1950 (when, according to his estimates, the number of estrogen users was in the millions). Nor was he concerned about the potential for uterine bleeding; the problem could be avoided, he claimed, by a simple adjustment of the dosage level.[69] He believed that estrogen, because it addressed the endocrinological root of aging, was a more powerful antidote than the talk therapy or sedatives prescribed by many physicians. "Phenobarbital, anticholinergic drugs, and psychotherapy, advocated by so many, may help to alleviate the hot flashes and panic reactions incident to the menopause, but *they will not postpone the aging process. Estrogen will*" (emphasis added).[70] According to Shelton, only estrogen could stave off the debilitating physical and mental deterioration that women faced after menopause.

To build his case that estrogen-starved women represented a new and pressing problem in America, Shelton distorted the meaning of the change in life expectancy from 1900 to 1950. By stating that "women now live a quarter of a century longer than in the year 1900" (life expectancy was 48.7 years for females born in 1900 and 72.4 for those born in 1950), he implied that very few women lived past the age of menopause at the turn of the century.[71] In fact, many women did survive well into their seventies and even eighties; the *average* age of life expectancy took into account females who died in infancy and during childbirth. Certainly, more women were living longer lives at midcentury, but the assumption that postmenopausal life was a modern phenomenon, requiring pharmacological intervention, was specious.

Although Shelton and Masters represented the far end of a long spectrum of opinion and practice regarding hormone usage in the 1950s, they spread their message to other physicians in their speeches at conventions and articles in medical journals. Doctors, along with the general public, could also read about "the third sex" and "hope for grandmothers" in newspapers and popular periodicals. The *New York Times,* for example, announced its account of William Kountz's speech at the first International Conference on Gerontology in Belgium in 1950 with the headline: "Hormones Rejuvenate Women Past 65, St. Louis Doctor Reports at Conference."[72] Three years later, the science journalist Waldemar Kaempffert wrote an article on Masters's research called "Aging Processes Are Arrested by the Injection of Male and Female Sex Hormones," in which he marveled over the rejuvenated inmates of the St. Louis Infirmary.[73] Estrogen appeared to be yet another miracle drug, like the recently discovered antibiotics, making its way from the laboratory bench to the pharmacy shelf.

Some of the physicians who began to tout estrogen as a wonder drug in the 1960s also depicted menopause as a disease. Allan C. Barnes, professor and chairman of obstetrics and gynecology at the Johns Hopkins University School of

Medicine and an editor of the *American Journal of Obstetrics and Gynecology*, an-nounced his view that "menopause is a pathologic rather than a physiologic phenomenon."[74] He drew an analogy between middle-aged "optic failure" and women's "ovarian failure" as universal experiences that both deserved medical treatment: reading glasses for the former and estrogen replacement for the latter. "The menopause is a disease process," Barnes asserted, "requiring active inter-vention . . . The patient should remain on this program [methyltestosterone and stilbestrol in a 10:1 ratio, 24–26 days each month] until she is 86."[75] He explicitly condemned the notion of short-term estrogen use to relieve the symptoms of menopause, because estrogen deficiency in postmenopause, like insulin defi-ciency in diabetes, would never go away.

The opinion that women should begin to take hormones to combat the disease of estrogen deficiency during menopause and continue to take them for the rest of their lives found its greatest expression in the work and writing of Dr. Robert Wilson, a New York gynecologist affiliated with the Methodist Hospital of Brook-lyn. Wilson had been in private practice for more than forty years before he published his first article in 1962, at the age of 67.[76] The following year, Wilson and his wife, a nurse, co-authored an article entitled "The Fate of the Nontreated Postmenopausal Woman: A Plea for the Maintenance of Adequate Estrogen from Puberty to Grave," in which they described, in no uncertain terms, the ravages of nonestrogenated aging. The Wilsons portrayed older women as "castrates" who "exist rather than live."[77] They blamed the loss of estrogen for causing diseases (such as hypertension and osteoporosis), psychological manifestations (such as depression, melancholia, and "a vapid cow-like feeling called a 'negative state'"), and desexualization (resulting from atrophied genitals and "loss of physical attractiveness"). They claimed that all of these conditions could be avoided if women would replace the estrogen no longer made by their aging bodies.

Wilson echoed both Shelton's analysis of the implications of increased life expectancy and Masters's assessment of the "problem" of aging from the perspec-tives of economics and public health. "In the Roman Empire," Wilson wrote, "life expectancy was about 23 years. There were very few old women . . . At the turn of this century it had risen to 48 years, but older women did not yet constitute a problem . . . 86.3 per cent of white female babies born in 1970 will live past the age of 65. This indicates the magnitude of the problem."[78] What was this problem? Since so many of these women would become "chronically incapacitated after reaching 45 years of age," they would present an economic burden on society.

Wilson broke with Masters on the issue of male aging. Whereas Masters included both "former" males and females in the "third sex," Wilson proclaimed,

"A man remains a man until the end."[79] Women, by contrast, became "only the 'part woman'" as a result of decreased estrogen production. He drew a vivid distinction between older men and women: "Our streets abound with them [older women]—walking stiffly in twos and threes, seeing little and observing less. It is not unusual to see an erect man of 75 vigorously striding along on the golf course, but never a woman of this age."[80] In case his readers didn't get the vivid imagery of "Nature's defeminization," Wilson illustrated his article with two photographs of old, unsmiling women with dowager's humps.

For Wilson, like Barnes, menopause was a pathological condition. In an article for *Connecticut Medicine* (and reprinted a month later in the *Delaware Medical Journal*), titled "The Obsolete Menopause," he wrote, "We know that menopausal women are not normal; they suffer from a deficiency disease with serious sequelae and need treatment."[81] Wilson sought to replace the qualitative subjectivity of symptom reporting by patients with the quantitative objectivity of laboratory assays. He argued that treatment should be based on cytological screening of the vaginal smear, in which the percentages of different cell types determined what he called the "Maturation Index," and later, the "Femininity Index." The dosage of estrogen (and progestin, to help slough off the endometrium) depended on how the patient's cellular profile differed from that of a healthy 20-year-old. In this way, Wilson asserted that menopause could, and should, be eliminated.[82]

Along with Masters, Shelton, and Barnes, Wilson assumed that the decrease in estrogen production not only caused difficulties during the transition period of menopause but also affected the quality of a woman's life for decades *after* the menopause. His negative portrayal of the postmenopausal female emphasized her looks and her attitude, in addition to her increased risk of disease. For Wilson, the loss of femininity was perhaps the most tragic consequence of all. His comment that "the desexed women found on our streets today . . . pass unnoticed and, in turn, notice little" both reflected and sustained the marginalization of older women in the youth-centered culture of postwar America. His proposed therapy —maintaining hormone levels from puberty to grave—was consistent with the prevailing conviction that medicine could supply the solutions to social as well as health-related problems. Estrogen replacement, Wilson implied, could end the "untold misery of alcoholism, drug addiction, divorce, and broken homes caused by these unstable, estrogen-starved women."[83]

By describing the physical, psychological, and social woes of the postmenopausal women he treated in his private practice, Wilson succeeded in capturing the attention of his colleagues who also encountered similar cases. Whereas Masters had experimented on low-income institutionalized women who were not representative of the general population, Wilson's message may have struck a chord with other gynecologists with white, middle-class clienteles. Wilson's observations and therapeutic interventions, building on the work Masters did in the closed setting of the St. Louis Infirmary, expanded long-term hormone replacement from an experimental model into a clinical practice.

In 1966, Wilson reached an even larger audience with his best-selling book *Feminine Forever*, which sold more than 140,000 copies in its first year, and through the tireless promotional efforts of his Wilson Research Foundation, which relied on financial support from the pharmaceutical industry to carry out its mission of educating women about the ravages of menopause and the miracle therapy of estrogen.[84] The media picked up Wilson's message, prompting articles in magazines from *Vogue* and *Good Housekeeping* to *Newsweek* and *Science Digest*.[85] How women responded to this information is considered more fully in chapter 4, but sales figures reveal a significant rise in the use of estrogen. From 1966 to 1975, the annual number of estrogen prescriptions almost doubled,[86] and the market value of noncontraceptive estrogen almost quadrupled.[87] One survey of women in the Seattle area found that half of the menopausal and postmenopausal women interviewed had taken estrogen for an average duration of ten years.[88] Although this rate of estrogen use was certainly among the highest in the country, it does demonstrate the trend of doctors prescribing, and women agreeing to take, estrogen for long periods of time.

By the early 1970s, the range of medical opinion about the prescription of estrogen for midlife and older women had expanded to include not only the short-term treatment of severe menopausal symptoms but also the long-term prevention of the debility and disease associated with aging. This is not to say that the entire medical community wholeheartedly embraced these new indications; there were still many physicians who questioned the wisdom of years of sex hormone replacement. However, those who advocated estrogen "from puberty to grave" received the attention of their peers and gained followers from among the ranks. The International Health Foundation sponsored workshop conferences in Geneva, Switzerland, on "Ageing and Estrogens" in 1972 and "Estrogens in the Postmenopause" the following year. In 1976, the first International Congress on the Menopause was held at La Grande Motte, France. The consensus of opinion among American and Western European physicians recorded in these proceedings indicated strong approval of the more comprehensive application of estrogen in older women's health care.[89]

This sea change in medical views on estrogen, from caution to enthusiasm, came about as the result of several interacting forces. The zeal of certain promoters was infectious. Masters, Shelton, Barnes, and Wilson offered both case studies and statistical data that seemed to be compelling evidence for the benefits of long-term hormone replacement. Case studies, although anecdotal, had been a long-standing method of reporting observations in medical journals, and therefore readers may have overlooked the idiosyncrasies of individual patient's circumstances. When assembled into larger groups (Wilson's data base of 304 women, for example), the findings seemed even more impressive.[90] The research studies were the most persuasive of all. With the patina of scientific impartiality, these investigations offered convincing testimony of the effectiveness of estrogen not only in treating menopausal symptoms but also in reversing or forestalling diseases such as osteoporosis, cardiovascular disease, and mental senility. Although these studies were deficient by modern standards in terms of randomization, control of variables, and statistical analysis, they were accepted at the time as adequate and objective. They received no criticism in the contemporary medical literature; on the contrary, the results and conclusions of this research provided the early rationale for long-term estrogen use, as marketed by pharmaceutical companies and prescribed by physicians in the 1960s.

The increased popularity of long-term hormone therapy for older women followed on the heels of that other long-term hormone therapy for younger women: oral contraception. The anovulant pill was approved by the FDA in 1960, and within five years it became the most popular method of birth control in the United States, used by 26 percent of married women under forty-five and prescribed by 95 percent of gynecologists.[91] Although some physicians expressed concerns about the long-term use of these potent drugs in healthy women, this minority opinion was overshadowed by the overwhelming acceptance of the pill by the medical profession and the public alike. By the mid-1960s it was not uncommon for a woman to plan to take the pill for twenty or thirty years, interrupting contraception only when she wanted to become pregnant. Both doctors and women grew conditioned to the idea of ingesting hormones daily to prevent pregnancy, and long-term preventive drug therapy became viewed as normal. To continue taking hormones through menopause and for decades after menopause fit this model of long-term preventive therapy, for estrogen allegedly prevented, or at least delayed, the development of several age-related diseases. While some people objected to interfering with a healthy woman's fertility during her reproductive years, few disputed the rationale of drug therapy to keep a woman healthy during her postreproductive years.[92]

Thanks to the myriad of newly available wonder drugs—not just hormone therapies but antibiotics, antihistamines, and tranquilizers as well—patients came to expect their doctors to write prescriptions to treat what ailed them, and doctors willingly obliged these patients.[93] A prescription for medication signaled the conclusion to the medical encounter, especially for those who could afford visits to the doctor's office. Increases in health insurance coverage also expanded the potential clientele for estrogen therapy. By 1970, 57 percent of Americans had health insurance that covered outpatient doctor visits, up from just 35 percent in 1963. Similarly, the proportion with outpatient drug coverage rose from 26 percent to 46 percent in the same seven-year period.[94]

Physicians' inclination to prescribe estrogen was further encouraged by the aggressive marketing techniques of pharmaceutical manufacturers. Since these companies were not permitted to advertise prescription drugs directly to the public, they undertook a variety of efforts to make their estrogen products known to doctors. Starting in the 1930s, they supplied the drugs free of charge to researchers, who then acknowledged the companies' generosity in the resulting articles published in medical journals. Manufacturers also advertised in the pages of medical journals, sent product information by direct mail to physicians, and employed special salesmen—called detailmen—to visit physicians and deliver the information in person. In the late 1960s Ayerst Laboratories was spending a million dollars a year to advertise Premarin.[95] These efforts helped Premarin, the most popular brand of estrogen, to become one of the top five prescription medications in the United States by 1975.[96] The role of pharmaceutical manufacturers in the rise of hormone replacement therapy from the 1940s to the 1970s deserves a close look.

# CHAPTER THREE

# Selling Estrogen to Doctors

Photograph number one: An attractive woman is seated on a lawn chair in a suburban backyard on a sunny afternoon. She has nicely coiffed gray hair; her face is unlined. Her husband, also gray-haired and good-looking, is sitting next to her. They are both smiling, perhaps in response to something a little girl in the foreground is saying or doing. Another little girl runs toward the couple. A younger woman, most likely the mother of the two girls and the daughter of the older couple, approaches with a tray of lemonade-filled glasses.[1]

Photograph number two: The same older woman is standing under a chandelier in the foyer of a theater, perhaps at intermission. She is wearing a stylish suit and white gloves and holding a playbill. She is chatting with two attractive well-dressed men, and smiles winningly at one of them.[2]

Both of these images appeared in full-page advertisements for Ayerst's Premarin in several issues of the *Journal of the American Medical Association* and *Obstetrics and Gynecology* from 1966 to 1968 as part of the company's advertising campaign called "The Time of Her Life." The caption for each read, simply, "Help keep her this way." At the bottom of the ad, Premarin was described as "specifically designed for estrogen replacement in the menopause . . . and later years."

Purchasing advertising space in medical journals was just one of several means used by pharmaceutical manufacturers to market prescription drugs. Drug companies sent out sales representatives, known as "detailmen," to meet face-to-face with doctors to provide information about the latest developments and to answer questions about the product line. At the start of the 1950s, Ayerst employed 184 detailmen in the United States; by the end of the 1960s, its American sales force numbered 550.[3] All told, there were some fifteen thousand detailmen working in the United States in the late 1960s.[4] Several studies in the 1950s and 1960s reported that the majority of physicians found the information provided by these company representatives to be useful.[5] Surely they also appreciated the complimentary samples, gifts, and free lunches bestowed by the drug sales-men. In a survey of its members in 1973, the American Medical Association found that more than half (52.2%) listed detailmen as one of the sources influenc-ing their prescription habits.[6]

Companies also sent information about their products by direct mail to physi-cians. These lavishly illustrated promotional materials were designed not merely to inform but also to sell; products were presented in the most positive light. In a 1948 article called "The Use and Abuse of Estrogen," two doctors cast a critical eye on these direct mail offerings: "If one depends on the beautifully embossed brochures which exhort the practitioner with every mail, one falls, unhappily, into the security of the illusion that there are neither contraindications nor side effects in the use of estrogens. Such pamphlets, while providing succinct clinical reviews and emphasizing the multiple commercial forms of estrogen, are often full of omissions."[7] Although physicians may have complained about the content and volume of commercial literature they received, most of them opened their mail and gave it at least a cursory review.[8] More often than not, however, these pam-phlets probably ended up in the wastepaper basket, because few have survived to become part of the historical record. Since the substance of these mailings was often replicated in the published advertisements, in condensed form, the lost ephemera of pharmaceutical promotions can be gleaned from the related ads, which have been saved in the bound volumes of medical journals.

The messages imparted by these ads reached their target audience. A 1971 study of 5,347 internists found that 75 percent of those surveyed read the adver-tisements in their medical journals; of these readers, 80 percent thought the ads were effective in communicating information about drug products.[9] In the 1940s, the New York marketing firm of Murray Breese Associates sponsored studies to evaluate the relative effectiveness of different kinds of pharmaceutical advertis-ing. The results indicated that direct mail was slightly more successful than journal advertisements in capturing physicians' attention; however, the cost of producing and distributing direct mail was three to five times greater than the cost of publishing a comparable advertisement. The author of the survey cau-tioned that the findings were based on interviews with doctors (measuring their recall of products encountered in both types of media); sales data, which were not considered, would be a more valuable assessment of advertising success.[10] An-other study sponsored by Modern Medical Publications at the University of Il-linois a decade later used similar methods to determine that doctors reported journals to be their primary source of information about new drugs and their second most significant source of knowledge for existing drugs (after detailmen). This study also confirmed the results of the 1940s series: journal ads delivered more bang for the buck, in terms of relative cost effectiveness.[11]

That pharmaceuticals had become big business is evidenced both by the number of studies of pharmaceutical advertising undertaken in the postwar period (one review counted nine investigations between 1952 and 1958) and by the enormous increase in drug sales in America (from $300 million in 1939 to $2.3 billion in 1959).[12] Manufacturers wanted to be sure that their advertising dollars were being spent wisely.

In the 1940s and 1950s, at least twenty-five drug companies advertised estrogen products in the pages of major medical journals, such as the *Journal of the American Medical Association* and the *American Journal of Obstetrics and Gynecology*.[13] Clearly, it was worth competing for the market in menopause. In the mid-to-late 1960s, thanks to the publicity created by Robert Wilson's *Feminine Forever*, the annual number of estrogen prescriptions soared, rising from 15.5 million in 1966 to 28 million in 1975.[14] By 1975, annual estrogen sales reached $85 million. Savvy and persistent advertising enabled Ayerst Laboratories to capture the majority of the market for its Premarin brand. Most sources agree that Premarin accounted for around 70 percent of estrogen prescriptions for menopausal and postmenopausal use by the 1970s.[15]

Ayerst achieved this success with Premarin in part by shifting its advertising in the 1960s to include indications for long-term use. By promoting Premarin both to treat the short-term symptoms of menopause and to prevent the long-term ravages of aging, Ayerst tapped into the potentially enormous market of older women in America. However, Ayerst did not initiate this new campaign until the idea had become almost commonplace in the medical literature. Although Masters published several articles advocating long-term sex hormone therapy in the 1950s, the pharmaceutical manufacturer waited almost a full decade before incorporating these indications into its ads. Only after Wilson began to publish in the early 1960s did Ayerst include the suggestion that Premarin could be used in the "later years." The industry took a backseat to the clinicians and researchers who drove the crusade to keep women on estrogen after the menopause. Once drug manufacturers became convinced that doctors would prescribe—and women would take—estrogen for long periods of time, they eagerly took up the cause.

Well before postmenopausal hormone therapy became popular, when drug companies marketed estrogen products mainly for menopause, they illustrated their advertisements with drawings and photographs of potential patients. The character and composition of these representations changed over time, in accordance with prevailing medical and cultural conceptions of the role and status of older women in American society, and the images both borrowed from and contributed to physicians' perceptions of their female patients. Women in these ads were depicted sometimes as youthful and sometimes as elderly, sometimes as homemakers and sometimes as workers, sometimes as distraught (unmedicated) and sometimes as euphoric (medicated). Thus estrogen advertisements not only reflect the transition from short-term palliative treatment to long-term preventive therapy but also reveal shifts in contemporary opinions of what sorts of women might be in need of estrogen replacement. Furthermore, these ads divulge the tactics used by pharmaceutical manufacturers to convince doctors to choose their products. Advertisements in medical journals, in conjunction with clinical studies and research reports published in those same journals, helped first to introduce and then to reinforce the medical practice of prescribing estrogen first to midlife and then to older women.

The first advertisements for estrogen focused on the product, not on the patient. Manufacturers emphasized the efficacy, potency, stability, and chemical purity of their wares. If any pictures accompanied the text, they featured images of the hormone product, in boxes, bottles, pills, and ampoules. These early ads appeared during the Great Depression, and several touted price reductions as another reason to choose one formulation over the others on the market.

The other consistent feature of estrogen ads in the 1930s and early 1940s was an appeal to science; manufacturers boasted of the scientific standards by which their hormones were produced and the clinical evidence that supported the drugs' effectiveness. A 1934 Schering ad proclaimed, "Clinical proof of the efficacy of Progynon has been given by leading physicians who have subjected it to searching tests." The copy continued, "It is standardized by the Allen-Doisy method."[16] In 1937 Ayerst, McKenna and Harrison assured readers of the *American Journal of Obstetrics and Gynecology* that "the value of Emmenin in the treatment of menopausal symptoms has been ably demonstrated by many clinicians."[17] That same year, Schering backed up its claims with footnoted references to clinical studies, naming the physician-researchers who "proved" the efficacy of Progynon and citing the articles published in medical journals.[18] Other manufacturers quickly followed suit. One of the first ads for Premarin consisted of a collage of twenty excerpts cut out from medical journal articles. Superimposed over these snippets was a prominent caption in a black box with white lettering that read, "The rapidly-growing bibliography of 'Premarin' provides ample evidence of its high therapeutic effectiveness as a medium for oral estrogenic therapy. Reprints of several recently-published papers are available on request."[19] The ad implied that there were too many articles to list all of the publication information, but if the physician wanted to read the literature for himself, Ayerst would be happy to supply it. This and other similar notices favored published reports as evidence of a drug's merit; if a peer-reviewed journal agreed to publish an article that demonstrated the clinical success of a particular brand of estrogen, then the findings presented must have been true.

Illustrations of women in ads for estrogen began to appear in the early 1940s, indicating an emphasis on the patient (and the physician's ability to help her). Although most of these pictures were simple sketches (often of nudes), an ad for Estrogenic Substance manufactured by George A. Breon and Company included two photographs, which combined the representation of a patient in need of medical care (one photo showed a pensive woman with downcast eyes and her hands clasped in front of her mouth) with the authority of science (the other depicted a mouse strapped to a board undergoing surgery at the hands of two white-coated scientists). Titled "In the Difficult Days of the Menopause," this ad assured physicians that they could take advantage of Breon's scientific expertise (and the sacrifice of the poor mouse) to soothe their menopausal patients' woes.[20]

Breon used the same female model, in a similar pose, to depict the multiple problems faced by menopausal women during World War II. In this ad, the model is brooding over a book of ration coupons, with the "problem" identified as "ADD ration coupons, green stamps and brown, high prices, no help TO hot flushes, headache, insomnia common at the menopause." The text read, "No mathematician is equal to that one. The woman of 45 to 50, responsible for family welfare, cannot run a well-ordered house, in these disjointed times, if weighted with menopausal nervous instability. But the clinician—he knows that the harassments, so big to such women, can be smoothed out."[21] The message encoded here reassured physicians of their ability, with the aid of Breon's estrogen, to control the vagaries of their patients' physical and mental condition.

The physician was the focus of another World War II era advertisement run by Ciba Pharmaceutical Products for Di-Ovocyclin. Illustrated with drawings of a clock, a list of "calls to make," and an ampoule of the drug, the ad addressed itself to the busy homefront doctor: "TIME IS VALUABLE . . . Save it! *Conserve time and use it wisely. More and more patients will come to you as their physicians are called into the armed forces."*[22] Since Di-Ovocyclin injections lasted longer, the doctor could reduce the number and frequency of patient visits. This ad did not try to convince physicians to prescribe estrogen to their menopausal patients. It assumed that they had already incorporated the use of hormone replacement therapy into medical practice; all Ciba had to do was persuade them to use this more "time-conserving, economical" formulation.

Some advertisers acknowledged the increasing number of women in the work force during the war and targeted their advertisements accordingly. In 1940, 24 percent of women 45–64 and 18 percent of women 55–64 worked for wages; in 1945, those figures had leapt to 37 percent and 27 percent, respectively.[23] E. R. Squibb and Sons incorporated Rosie the Riveter into a multi-ad campaign for its estrogen products during the war years. One pictured a young-looking woman dressed in a uniform, carrying a lantern and walking along a train platform. The caption announced, "She's doing a Man's job . . . *but she's still a Woman.*" The ad recommended both Amniotin and DES, with the following rationale: "Like so many women in their forties, she's joined other adults of the family in the task of winning the war. She's thinking more of passengers than of tea parties . . . more of switches than of sewing. But her thoughts are also on another problem—of how the menopause will affect her." It is interesting that the patient targeted here was premenopausal; the implication was that the physician would not hesitate to prescribe estrogen at the first sign of menopausal discomfort. Squibb's products were also portrayed as modern marvels: "Two decades ago women were inclined to fear their forties. Today they may face them with confidence, secure in the knowledge that adequate estrogen therapy will often relieve the distressing vasomotor symptoms of the menopause." The wartime economy made the physician's role that much more important in keeping women "physically and emotionally fit, and 'on the job.' "[24]

A second ad showed a photograph of the woman war worker (in street clothes) in conversation with a physician at his desk. Titled "The Battle of Nerves," the ad told the story of this woman's plight. "To the woman of middle age who is trying to do her share in war work, the battle of nerves, she finds, is often more debilitating than the details of her job." Again, physicians were ensured a central role in solving patients' problems. Moreover, Squibb offered them a choice of products, to be selected according to their professional inclination and expertise: DES or Amniotin, "for physicians who prefer *natural* estrogenic substance."[25]

The illustration in another ad in the series featured a drawing of a woman (also young-looking), this time lying in a hospital bed. Titled "Accidents and Absenteeism," the ad copy warned, "In women—particularly the conscientious middle-aged who try to stay on the job—the nervous symptoms associated with the menopause may directly affect their efficiency, and contribute to accidents of one kind or another."[26] Without proper medication, it implied, menopausal workers endangered themselves and hindered the war effort.

Although Squibb's ads were progressive in the sense that they recognized that women made up a significant proportion of the homefront work force, their message did not disrupt the traditional doctor-patient relationship. As in other ads both before and after World War II, the physician was cast as the leading actor, with estrogen in an important supporting role. Estrogen enabled the doctor to skillfully and successfully manage the patient's menopause. Without expert medical care and the prescription of hormone therapy, middle-aged women would remain distressed; in the unusual circumstances of World War II, they might not fully contribute to the war effort, and might even prove to be a liability, if their menopausal symptoms were left untreated. Squibb's admen recognized the cultural currency of Rosie the Riveter and capitalized on wartime sentiment to promote their company's estrogen products.

As the war came to a close, advertisers abandoned the tactic of portraying women as important participants in the national economy. Even before the atomic bombs were dropped on Hiroshima and Nagasaki to end the war with Japan, Ayerst ran an ad titled "Remote Control." It pictured a woman sitting in a chaise longue on the beach, sporting sunglasses and reading a magazine. "Menopausal patients who join the "gone-for-the-summer army" present a problem . . . for without proper treatment many may experience a recurrence of symptoms." The solution? An adequate supply of Premarin tablets. "Your menopausal patient may be "gone for the summer" . . . but she remains under your care."[27] Although Rosie the Riveter may have traded in her welding tools for a beach umbrella, she could still benefit from the supervision of her physician in the "management" of her menopause.

In the late 1940s and 1950s, the full-page ad campaigns for estrogen products that ran repeatedly in the pages of medical journals often depicted menopausal women as mercurial and capricious. These representations fell into two categories: some images showed photographs of unhappy-looking women or graphic illustrations intended to symbolize menopausal angst (such as a frayed rope) before medication, while others showed calm-looking women or representations of their serenity (such as a butterfly) after medication. Some offered hopeful captions ("The Glory of Autumn," "The Calm of Eventide"); others were more ominous ("Forever Anger," "The Breaking Point"). The message in both kinds of ads was the same: estrogen could help ease not only the physical discomforts but also, and perhaps more importantly, the mental turbulence brought about by menopause.

The suggestion that estrogen could promote "a sense of well-being" became central in the promotion of several companies' products. In 1947 Ayerst began a concerted campaign to highlight the positive emotional effects of Premarin. Underneath a photograph of a smiling, elegantly dressed couple whirling around on a dance floor, one ad noted, "An increasing number of investigators are commenting on the general "sense of well-being" which is usually experienced by menopausal patients following 'Premarin' administration."[28] Another ad commented, "The woman in the climacterium [menopause] may be disturbed by disquieting thoughts and foolish fears. Such mental anguish is oftentimes allayed when the physical symptoms associated with declining ovarian function have been relieved." Moreover, "there is a plus in Premarin . . . the gratifying sense of well-being usually experienced by the patient following administration of this naturally occurring, orally active estrogen."[29] The distinction here is important to point out: the company contended not only that its product would alleviate physical symptoms and the mental distress they caused but also that it would bring about tranquility and contentment, a vague claim at best. Squibb also touted "the far-reaching effects of Amniotin therapy."[30] Its advertisement announced that "Amniotin provides menopausal therapy *beyond the mere relief of vasomotor symptoms.* Amniotin does more than relieve climacteric flushes and sweating. The patient experiences a heightened feeling of well-being, improved strength and vigor, and 'a greater sense of general relief, exclusive of the amelioration of hot flashes.'" This last phrase, in quotation marks, was followed by a footnote to a reference in the *Journal of Clinical Endocrinology,* in an effort to validate the claim as scientifically based. Ayerst, too, began to back up its claims for well-being and "tonic effects" by citing specific articles from medical journals in its ads for Premarin.[31]

A woman's self-esteem was considered to be vulnerable during menopause. Abbott ran an ad for its Estrone that showed a sketch of an older woman weeping over a sketch of herself at a younger age, "the girl *she* left behind." Thanks to estrogen shots, "Arrival of the menopause need not cause a woman to feel that she must face a future devoid of femininity and charm."[32] In a similar vein, Schering asked, "At this time in her life must she endure . . . feelings of insecurity . . .?" Its product, Estinyl, would allow the menopausal woman to "enjoy a more normal life."[33] Even the Cereal Institute capitalized on women's fears of losing their good looks. Its ad in the *American Journal of Obstetrics and Gynecology* suggested that physicians advise their patients to eat cereal "to prevent or correct undue weight gain during the menopause."[34]

Many of the ads in this time period depicted menopausal patients as relatively young women, with blonde or brunette hair and trim figures. The models employed by Ayerst—photographed variously as gazing cheerfully out of an open window, or pulling on a pair of flirty white gloves, or carrying a spring bouquet in one arm and waving to someone outside the frame, or pausing at the edge of a swimming pool to flash a giant smile—could easily have been in their thirties.[35] Breon proclaimed "a woman's right to enjoy her forties."[36] These ads implied that menopause began before a woman reached the age of 50; however, they also portrayed menopause as a temporary transition. Following the contemporary conservative medical consensus, pharmaceutical manufacturers marketed estrogen as a short-term palliative to be used for a few months or, at most, a few years. The old ladies that William Masters rejuvenated did not make their way into hormone replacement advertisements in the 1950s.

Several of the ads suggested that a woman's menopausal miseries might affect other lives besides her own. Abbott promoted its injectable Estrone with a drawing of " 'Mrs. Jekyll' and 'Mrs. Hyde.' " The text read: "Many 'Mrs. Hydes' undergo such radical personality changes during the menopause that they seem like entirely different women to their families and associates."[37] Merrell captioned its tableau of a living room confrontation between a women and her family, "tempestas in utero." The woman stands with her hands on her hips, clearly chastising her seated husband. Their teenaged daughter sits on the arm of his chair, a protective hand resting on his shoulder. Ignoring almost fifty years of biomedical research that located hormone production in the ovaries, the ad attributed menopause to uterine dysfunction. "When the family 'ship of state' is rocked by the inner turmoil of menopausal unrest," it advised, "the use of estrogenic therapy to calm the 'uterine tempest' is clearly indicated."[38]

Postwar ads continued to assert the centrality of the physician in managing menopause. Doctors were told that, with the aid of Abbott's Sulestrex tablets, they would be "still solving their [patients'] problems right through the menopause."[39] "Do 'human relations' lie within the physician's sphere?" asked an advertisement for Wyeth's Conestron. "From earliest times the doctor, like the spiritual adviser, has been the repository of the family's deepest problems. Today, many of these can actually be treated with new scientific accuracy and improved medical therapy. For example, the frequent marital and family dislocations brought about by the approaching menopause."[40] This statement echoed messages imparted in several other contemporary ads: estrogen therapy was scientific, menopause affected a woman's marriage and family, and the physician held the key to facilitate the patient's adjustment.

Ayerst employed another approach to convince potential clients—all physicians, of course, since estrogen was sold by prescription only—to order Premarin for menopausal women. Instead of referring to the success of the therapy from the perspective of the patient, one ad announced that "Doctors, too, like 'Premarin.' "[41] It featured a photograph of three physicians having a collegial discussion, with one seated at a desk, one perched on a file cabinet, and a third lounging with his legs outstretched. "The doctor's room in the hospital is used for a variety of reasons," the ad explained. "Most any morning, you will find the internist talking with the surgeon, the resident discussing a case with the gynecologist, or the pediatrician in for a cigarette [!]. It's sort of a club, the room, and it's a good place to get the low-down on 'Premarin' therapy." The recommendation of one's colleagues was as significant as the gratitude of one's patients.

Finally, pharmaceutical companies continued to tout the properties of the estrogen products themselves. Some recommended oral therapy over injections, for greater convenience and less discomfort. Others claimed that natural estrogens produced fewer side effects than synthetic versions. Lakeside illustrated the "natural source" of its estrogens with a photograph of horses' backsides, complete with attached urine collection vessels.[42] (In an article written for *Clinical Obstetrics and Gynecology*, Dr. Allan Barnes scoffed at the claim that one type of estrogen was more "natural" than another: "The blind worship of estrogens from a natural source, however, when such a source is usually equine, implies that one's practice is largely composed of horses.")[43] Both Ayerst and Schering advertised combined estrogen-testosterone products (perhaps on the advice of William Masters); Ayerst promoted this combined formulation not only for menopause but also to speed the healing of fractures in the elderly and to treat the male climacteric.[44]

Although estrogen manufacturers vied for their share of the menopause market, the real competition in the 1950s and 1960s came not from within but from without. Early in the decade, manufacturers of sedatives began to advertise their drugs for the relief of menopausal symptoms. The ads took direct aim at estrogens, using similar tactics (photographs of distressed-looking women, references to articles in medical journals) to assert the superiority of sedatives in menopausal therapy. In a two-page spread for Butisol Sodium, McNeil illustrated one page with a large photograph of a woman sitting at a desk, pencil in one hand, her head resting on the other, looking distractedly and disconsolately into the distance; on the other page, it let a quotation from a medical report do the talking: " 'Mild hot flashes, nervousness, irritability, headache and insomnia . . . can often be adequately overcome by mild sedation . . . Where this can be accomplished . . . by these means without the aid of added hormones, it is indeed worthwhile.' "[45] Smith, Kline and French reminded readers of its ad for Dexamyl that " 'the great majority of menopausal women require no endocrine treatment at all.'" This quote (also from a medical journal) was followed by the assurance that, for those women who did encounter troubles (as represented by the bent figure of a woman silhouetted behind smoked glass), Dexamyl would "relieve depression," "relieve anxiety," and "restore self-esteem."[46]

These ads borrowed from the contemporary medical practice of a three-tiered approach to treating menopause. Prior to the 1960s, doctors first tried counseling their menopausal patients; if that didn't work, they prescribed sedatives. Only if these methods failed did they resort to hormone therapy. As discussed in chapter 2, sedatives were intended as palliatives to alleviate the overt symptoms of menopause, as opposed to the more radical intervention of estrogen therapy, which produced profound systemic alterations.

Barbiturates (such as Butisol) had been available since 1903, but their effects were powerful: they induced both sedation and muscle relaxation. Moreover, they were addictive and potentially lethal if taken in too large a dose. The appearance of the first of the so-called minor tranquilizers (generically known as meprobamate, sold as Miltown by Wallace and as Equanil by Wyeth) in 1955 gave physicians a drug to treat anxiety with minimal side effects; sales of these popular pills were soon overtaken by the second group of minor tranquilizers, Librium and its cousin Valium, in the early 1960s.[47] Although these also turned out to be habit-forming, they were initially hailed as miracle drugs for the stresses of modern life.[48] The minor tranquilizers appeared to be another link in the chain of pharmaceutical successes, joining contemporary wonder drugs such as antibiotics, antihistamines, and the recently approved polio vaccine.

Since many of women's menopausal complaints—anxiety, tension, depression, insomnia—resembled those addressed by the minor tranquilizers, manufacturers included menopausal women as potential consumers of their products. Wallace not only promoted Miltown to "relieve tensions of the menopause," it also marketed a combination product of Miltown and conjugated equine estrogens called Milprem. Described as "an important advance in menopausal therapy because it replaces *half* control with *full* control, because it treats the *whole* menopausal syndrome, because *one* prescription manages *both* the psychic and somatic symptoms," Milprem covered all of the therapeutic bases.[49]

Like the estrogen ads, tranquilizer ads showed both cheerful and distraught women, signifying before medication and after medication, respectively. However, the women pictured in the tranquilizer ads were older than their counterparts in contemporary estrogen ads. They often had gray hair, wrinkles, and jowls. These unflattering portraits contributed to negative images of midlife and older women in American society: not only were these menopausal women perceived to be tense, anxious, and irritable, but they also were depicted as physically unattractive. The significance of this imagery, in the context of the budding cult of youth in the 1960s, will be considered more fully in chapter 4.

One manufacturer fought back directly against the incursion of the tranquilizers into menopause territory. In the early 1960s Ayerst reminded readers of *Obstetrics and Gynecology* that "when you treat the menopause . . . consider that current medical opinion favors estrogens."[50] The ad copy conceded that sedatives and tranquilizers might be "helpful in some ways," but nothing could match the efficacy of estrogen. A decade later, Ayerst explicitly defined menopause as an estrogen deficiency disease. The company ran a two-page spread that said, "Premarin helps relieve emotional symptoms of menopausal estrogen deficiency because it gives her back something she's lost . . . no tranquilizer can do that!"[51] This ad implied that sedative therapy only treated the outward symptoms of menopause without addressing their root cause; estrogen therapy, by contrast, replaced the hormones purported to be essential to a woman's well-being and, in so doing, enabled the patient to achieve emotional and physical balance.

More importantly, the classification of menopause as a deficiency disease meant that all postmenopausal women were, by definition, estrogen deficient and therefore in need of replacement therapy. This, of course, was the model promoted by Masters and Shelton and later by Barnes and Wilson. By the late 1960s, with increasing support for the "estrogen from puberty to grave" recommendation in both the medical literature and the popular press, Ayerst led the pharmaceutical industry in incorporating postmenopausal hormone replacement therapy into advertisements for estrogen products. This strategy had the immediate effect of reducing the competition from tranquilizers; its contribution to long-term consequences, namely, the medicalization and commercialization of postmenopause, would develop over the next several decades.

Ayerst's first advertisement to explicitly promote Premarin for long-term hormone replacement therapy appeared in 1964. The ad featured the outline of a woman on a black background, with vivid anatomical renditions of only her spine and pelvis (in white) and her heart and reproductive organs (in red). The large-print caption read, "long preferred in the menopause . . . and throughout the later years," and the small-print copy referenced Wilson in claiming, "A fuller recognition of estrogen as the 'metabolic strength' of women—of its beneficial effect on practically every system, organ, and tissue of the body—provides a scientific basis for the wider acceptance of the concept that estrogen administration should be continued beyond the limits of the actual menopause for its protective influence on vital processes, notably cardiovascular, bone and protein, and cellular metabolism."[52] The manufacturer followed the clinical advocates in touting the multiple benefits of estrogen supplements.

As mentioned at the beginning of this chapter, Ayerst's "The Time of Her Life" ad campaign also pitched estrogen replacement for postmenopausal women, using an elegant model to represent the older Premarin patient. The women in another late 1960s Premarin ad, however, were considerably less glamorous. This one included three photographs: a gray-haired bespectacled woman in sensible oxford shoes reading a book in a rocking chair, three stout matrons (two seated, one standing and serving) at a table with coffee and cake, and a close-up of another gray-haired granny peeling potatoes into a basin on her lap. "Estrogen replacement may help her . . . even years after the menopause," read the ad, which urged doctors to prescribe estrogen to "allow her to age physiologically rather than pathologically."[53]

The rest of this recommendation revealed an extraordinary message: even though estrogen deficiency may be asymptomatic, most women would eventually suffer "metabolic consequences" in postmenopause and would therefore benefit from taking Premarin. Several paragraphs of text affirmed that estrogens would help prevent loss of skin elasticity, lowered resistance to genitourinary infection, urinary dysfunction, osteoporosis, and depression. By picturing much older women and by referencing several common age-related conditions, the ad suggested that physicians expand the patient base for estrogen to include women in their fifties, sixties, and seventies.

Ayerst made it clear that women could be introduced to the estrogen regimen at any time. A 1970 ad captioned a photo of an elderly woman with the following words, "Born too soon to have benefited from estrogen replacement therapy during her menopausal years—but, even now, it is not too late to institute treatment for her postmenopausal syndrome."[54] For the "elderly, debilitated patient," Ayerst recommended Premarin with Methyltestosterone. The comment that this combination product "may often mean the difference between relative incapacitation and days of small joyful activities" could have been copied from one of Masters's articles fifteen years earlier.[55]

As Ayerst began to promote both short-term and long-term indications for Premarin for midlife and older women, it expanded both the size and scope of its advertisements, so that some journal ads approached the length of journal articles. One four-page spread offered eleven references, plus a chart of the results of four clinical studies, to make the case for understanding menopause as "an estrogen deficiency state."[56] In 1975 an ad that opened with the words, "When women outlive their ovaries . . . ," ran to eight pages, with six women (some who looked younger than 50, others who were most likely past 70) telling their personal menopause and postmenopause horror stories.

The last page of this ad included the FDA-approved indications for Premarin, based on a 1972 review of the drug undertaken by the National Academy of Sciences–National Research Council (as part of an effort to evaluate all drugs that had been approved by the FDA before the 1962 Kefauver-Harris Amendments to the 1938 Federal Food, Drug, and Cosmetic Act, which thereafter required drug manufacturers to demonstrate not only the safety but also the effectiveness of products prior to marketing them). These indications listed the drug as "effective as replacement therapy for naturally occurring or surgically induced estrogen deficiency states associated with the climacteric, including the menopausal syndrome and postmenopause" and "probably effective for estrogen deficiency-induced osteoporosis" when used in conjunction with other therapies, such as calcium.[57] Although the notice cautioned that "final classification of this indication requires further investigation," Ayerst had no compunctions about incorporating this "probable effectiveness" against osteoporosis into its advertising.

To explain to physicians how Premarin could prevent bone loss, Ayerst ran an ad that detailed the pathology of postmenopausal osteoporosis with X-rays and medical illustrations of diseased bones. The company relied on scientific evidence to convince doctors to "start her on Premarin . . . keep her on Premarin." It included an appeal to aesthetics in noting that Premarin would slow the development of the dreaded "dowager's hump."[58] Ayerst also sold Formatrix, a combination of Premarin, testosterone, and vitamin C, as a prophylactic against osteoporosis. Advertised with the threatening dual image of an upright middle-aged woman and her bent-over future counterpart, the ad warned of "the shape of things to come, unless her postmenopausal osteoporosis is treated early."[59]

Ayerst's crusade to expand the market for Premarin was carefully calculated. Evidence of this coordinated campaign comes from articles in *Scoreboard*, Ayerst's in-house publication for its sales force. The magazine offers a behind-the-scenes glimpse at the efforts of this particular pharmaceutical company to maximize sales of its signature product. In the spring of 1970 the company asked its detailmen to complete profiles of the physicians in their districts, recording such details as how often and for what indications they prescribed estrogens and whether they had any reservations in using estrogens. Based on these compiled profiles, the company generated three customized brochures, each designed to address one of the three most common concerns. "In short," an article in the magazine reported, "instead of the Salesman's using the 'shotgun' approach in dealing with the doctor's thinking, the profile cards and brochures enable him to zero in on the target and hit the mark with just one rifle 'bullet.' "[60] The hunting metaphor described one facet of the relationship between the salesman and the physician; other stories sought to portray it as more akin to that of teacher and student.

A November 1971 article touted the "progressive dialogue between the Ayerst Sales Representative and the physician" in providing ongoing education about the benefits of Premarin in menopause and postmenopause.[61] This report described the evolution of a new Premarin advertising campaign: "To have some idea of what went into the development of each ad, one only needs to examine the parts to realize that for years Ayerst has been building up its ammunition to be able to say about PREMARIN what it is saying today."[62] Designed to replace "The Time of Her Life" theme, the new campaign moved away from both the woman and her subjective symptoms toward what Ayerst believed was "objective medical evidence." "In 1970 the estrogen deficient woman was the focus of the ad. Now she's still in several of the ads, but is subordinate to the objective symptoms."[63] This new tack was intended to shore up the recommendations for Premarin with clinical proof of its therapeutic merit in treating an expanding list of indications.

*Scoreboard* also acted as a forum in which the sales force could share their experiences. Individual detailmen wrote in to the magazine with anecdotes from the field that served as testimonials to the success of Premarin. A salesman from Tacoma, Washington, reported correspondence from a doctor's wife thanking him for her supply of Premarin. " 'Thank you so much for my 'lil ole lady' pills,' " the woman wrote, on a card adorned with "a kitten saying 'Purrr-r-r for PRE-MARIN.' "[64] A representative from Washington, D.C., sent in a "fan letter" from a grateful doctor's secretary, who wrote: "I honestly don't know if I ever would have made it the past few years without PREMARIN and, when a gynecologist I know suggested a 'cheaper brand,' I told him, thanks, but no thanks, I wasn't switching when I had such a good thing going with PREMARIN. I don't know what that little yellow pill contains. All I know is it keeps me feeling content."[65] And the following endorsement of Premarin came from Houston: "The pharmacist, startled at receiving a script for 1 x 1,000 PREMARIN 1.25, asked the woman if she knew the size of the prescription she had given him. She explained that her doctor had sold her on the idea that Estrogen Replacement Therapy was life-time therapy and, since she didn't want to run out of anything so important, couldn't he please write for 1,000 at a time!"[66] These accounts served as inspirational tales to motivate the salesmen to get more and more physicians and pharmacists onto the Premarin bandwagon. In the words of one detailman, the goal of his pitch was to get doctors "talking more like a PREMARIN prescriber should."[67]

Although Ayerst was the most prominent advertiser of estrogen for meno-pausal and postmenopausal use, a few other manufacturers also promoted their products in the pages of medical journals. These companies sought to emphasize the differences between their estrogens and Premarin. Thus Abbott promoted its Ogen as natural, because it was "made from pure estrone, chemically and biolog-ically indistinguishable from that produced by the human body" (as opposed to Ayerst's Premarin, which came from horse urine).[68] Smith Kline and French plugged its product's low cost: "She may need estrogen replacement therapy for a long time. Perhaps years. So drug cost is important to her. A prescription for 'SK-Estrogens' can save her money. And she can count on each refill costing less, too."[69] Despite these appeals, by 1975 the pharmaceutical industry had conceded the majority of the estrogen business to Ayerst and its conjugated equine estrogens.

Between the 1940s and the 1970s, estrogen advocates in the medical profes-sion and their corporate cousins, the drug companies, brought about a seismic shift in the medical conceptualization of menopause and postmenopause. In the 1940s and 1950s, menopause was understood to be a potentially rocky transition between two relatively tranquil periods: the reproductive years and the postmeno-pausal years. Manufacturers responded to this view by marketing estrogen prod-ucts "to calm the storms" or "to quiet the troubled waters" during the "change of life." Although estrogen ads went along with the medical consensus that meno-pause was a natural and normal stage for women, they encouraged pharmaceuti-cal treatment of its symptoms; in so doing, they contributed to the medicalization of menopause.

The years of menopause and postmenopause were not only medicalized but also pathologized in the 1960s by Robert Wilson and his followers who character-ized menopause as an estrogen deficiency disease. The drug companies, espe-cially Ayerst, were quick to incorporate this model into the promotional materials for estrogen, because it greatly expanded their potential market. If menopause marked the beginning of the estrogen shortage, then postmenopause could be defined as a chronic state of deficiency, in need of pharmaceutical replacement (and, of course, medical management). Because postmenopause lasted for the rest of a woman's life, the medical, cultural, and economic implications of this premise were enormous.

The ads for estrogen in medical journals presented midlife women as troubled and troubling, incapable of dealing with their jobs, their families, or even the simplest aspects of daily life. Whereas earlier ads depicted menopausal women as youthful in appearance, later versions expanded their imagery to include post-menopausal grandmother types. The graphic representations of women in estrogen ads in the 1960s and 1970s reinforced Wilson's characterization of older women as defeminized and superannuated. These portrayals both reflected and shaped physicians' own experiences with and attitudes toward their older female patients. They also drew from a cultural milieu that was increasingly dismissive of old people as it celebrated American youth.

From these ads, physicians could imbibe the message that the woes of female aging were attributable to declining estrogen production and treatable by pharmaceutical estrogen replacement. There were big profits to be made in selling estrogen to older women, for both the companies that manufactured the drugs and the physicians who prescribed them. By capitalizing on the new medical indications for estrogen as a long-term replacement therapy, the pharmaceutical industry helped to commercialize menopause and female aging. Whereas previous generations of women had managed to grow old on their own, menopausal and postmenopausal women in the 1960s were supposed to purchase medical services and pharmaceutical products. As the ultimate consumers of estrogen, women had to be convinced that they needed these hormonal treatments. Thus the next step for estrogen advocates was to go public and popularize the problem of estrogen deficiency and the solution of replacement therapy.

# Selling Estrogen to Women

The man who brought hormone replacement therapy to the attention of the American public traced his professional interest in menopause to two key events in his boyhood in Ramsbottom, an industrial town in Lancashire, England, in the early twentieth century. First, he remembered his mother's "tragic decline" during menopause from "that vital, wonderful woman who had been the dynamic focal point of our family into a pain-racked, petulant individual."[1] Second, he recalled witnessing the recovery of the body of a woman who drowned herself in a reservoir in his town. Neighbors attributed her suicide to "the change of life." Haunted by the memory of the bloated corpse and stung by his mother's "senseless rages," he carried these memories with him when he immigrated to the United States and enrolled as a medical student at the Downstate campus of the State University of New York.

After graduating in 1919, this young doctor set up practice as an obstetrician-gynecologist in Brooklyn, New York. The first four decades of his career were unremarkable. Then, in 1966, he vaulted onto the national scene with the publication of his book *Feminine Forever*. Overnight, the septuagenarian Robert A. Wilson (about the same age as Brown-Séquard when he announced his rejuvenation with testicular extracts three-quarters of a century earlier) became the latest prophet of the miracles of sex hormone supplements to combat the ravages of aging.

Wilson devoted one of the eleven chapters of his book to autobiographical remarks, "not for the sake of self-aggrandizement, but rather to give my readers some feeling of personal contact with the doctor whom they have consulted through the pages of this book."[2] Thus his audience learned about his mean menopausal mother, his radical autodidactic father, and his first love, eight-year-old Madge Morris, who kissed him in the schoolyard and began his "lifelong involvement with women—or rather the idea of feminity [sic]."[3] Wilson reassured his readers that he was faithful to his wife, indulging his wanderlust with a classic

Mercedes 540 K cabriolet. He confessed that he dealt with the frustration of years of professional indifference to his self-proclaimed medical breakthrough by comparing himself to other unappreciated pioneers, such as Edward Jenner, discoverer of vaccination, Louis Pasteur, creator of the germ theory, and George Papanicolaou, inventor of the Pap smear. Wilson recounted that the more time he spent on his research and writing, the less time he had for his private practice and the income it generated. When he had to sell his beloved Mercedes, he felt that he had "indeed joined the company of the martyrs of science."[4]

Wilson addressed his financial concerns by establishing the Wilson Research Foundation in 1963. He secured funding from pharmaceutical companies to support the foundation's goal of educating both the medical community and the general public about hormone replacement therapy. In addition to relentlessly pursuing press coverage of the hormone replacement solution to the problem of estrogen deficiency at and after menopause, the foundation sponsored an annual conference on HRT, sent representatives to give talks at meetings of medical societies and women's clubs, and published its own pamphlets to be distributed to women in doctors' waiting rooms.[5] These pamphlets, like the book *Feminine Forever*, reiterated in laymen's terms the information conveyed in the half dozen articles Wilson had published in medical journals in the early 1960s. Wilson's popular publications, combined with the foundation's successful efforts to interest journalists and magazine writers in reporting on HRT, brought about a dramatic transformation in the messages presented to women about the management of their menopausal and postmenopausal years.

Popular information about menopause and hormone replacement therapy in magazine articles tended to follow the vicissitudes in medical thinking on those topics from the 1940s through the early 1970s. Leaflets available in doctors' offices, put out by pharmaceutical companies as well as the Wilson Research Foundation, conveyed a consistent message of the necessity of estrogen. What women read about estrogen echoed what they heard from their doctors. In the years prior to the efflorescence of the women's health movement, promotional pamphlets, magazine articles, and a handful of books constituted the main source of educational materials on the symptoms, treatment, and aftermath of menopause. The provision of health information fell to physicians, pharmaceutical manufacturers, and journalists who were content to take their cues from medical experts.

The consistency between the prevailing medical consensus and articles intended for the general public is evident in the era's premier lay publications on health and medical issues, *Hygeia* (1923–1950) and its successor *Today's Health* (1950–1976). Published by the American Medical Association, and read by patients in doctors' waiting rooms and by subscribers in their homes, these magazines represented the mainstream medical consensus on a wide variety of topics. A look at the pieces on menopause from 1940 to 1970 illustrates how doctors wanted women to understand the role of hormones as therapeutic agents.

Two articles in 1941, published two months apart, reveal how the initial excitement over DES shaped the portrayal of the menopausal transition. The first, published in September just prior to FDA approval of DES, reassured readers that the large majority of women experienced little or no discomfort at menopause; for the unlucky 15 percent who did have debilitating symptoms, it referred somewhat cryptically to "new palliative remedies." These ovarian hormones, "available in the chemically pure state," could, under a doctor's supervision, relieve hot flashes.[6] The author was probably referring to costly injectable products such as Amniotin, Progynon, and Theelin. The second article, published in November, described the evolution of estrogen therapy, which had culminated recently in the successful synthesis and clinical testing of DES. This article presented a much more dismal picture of menopause, reporting that 40 percent of the eight million women going through menopause each year suffered "physical and mental torment." However, with the availability of the "sensational drug" DES, "millions who never received help before will be helped now, safely and at relatively slight expense."[7]

Whereas earlier articles focused on the psychological aspects of menopause, physical signs took priority in the late 1940s. "Once the physical symptoms are taken care of, the mental and emotional upsets take care of themselves," thanks to hormone treatment, described as "medical science's newest aid to women."[8] One author enthused over women's "renewed youth and vitality" on hormones, claiming that "often there is a seemingly magical transformation, with a thrilling new self confidence in addition to the greater energy and sense of well-being."[9] She scoffed at women who refused to take estrogen, asking rhetorically, "Who wants to go back to the good old days of kerosene lamps, unpaved streets, wood-burning ranges, cotton stockings—and the 'fearsome' forties?"[10] By 1947, estrogen (in the form of Premarin or DES), perhaps with a side of Phenobarbital ("in cases of extreme nervousness"), had become the modern solution to the age-old problem of menopause.

In the next decade, *Today's Health* reflected the turn toward a more conservative approach to estrogen that was espoused in the professional literature: hormone replacement therapy was warranted for a small proportion of women, for a limited duration, and only under a doctor's supervision. Articles echoed the concern expressed in the pages of medical journals about the misuse and overuse of estrogens at menopause. However, whereas physician-authors writing in medical journals accused their peers of overprescribing estrogen, *Today's Health* laid the blame for estrogen abuse solely on the shoulders of the women who took it.

These articles for the general public constantly reiterated the importance of the physician during the menopausal transition. Women were advised to seek medical counsel, even if they had no symptoms. For those who did experience problems, "medical science, through the doctor, can alleviate menopausal symptoms and produce a gratifying sense of well-being."[11] In other words, estrogen was a valuable tool in the medical arsenal, so long as it was dispensed judiciously by a wise physician. The medical community did not intend to alienate the pharmaceutical industry; that the two worked in close cooperation is represented by the striking similarity between any number of contemporary advertisements for estrogen in medical journals and the illustration in a 1956 *Today's Health* article titled "Easing Those Difficult Years"—a large photograph of a well-dressed middle-aged woman seated across a desk from her physician while he took notes (or perhaps wrote out a prescription).[12]

Since *Today's Health* functioned as the mouthpiece of the American Medical Association, it did not publicize the ideas of the radical fringe of the medical profession. Thus the "estrogen from puberty to grave" recommendations of Masters and Shelton never graced the pages of the magazine in the 1950s. However, the negative depiction of menopause advanced by these estrogen advocates, and their colleagues Barnes and Wilson, became incorporated into mainstream portrayals by the mid-1960s. For example, an article titled "The Change and What Husbands Should Know About It" (adapted from a book by a physician called *How Not to Kill Your Wife*), insisted that menopause was not a disease, then went on to describe in detail its symptoms, one or more of which were said to be experienced by 90 percent of all women during the course of their climacteric. Besides hot flashes (affecting 60–70%) and headaches (50%), women might suffer from rheumatism, flatulence, dyspepsia (described as "faulty digestion, which may cause pain in the stomach, distension, nausea, bringing up mouthfuls of acid or sour gas, and even vomiting"), colic, coarsening or darkening of the skin, and hemorrhagic menstrual flow. The prospect of worry-free sex as a positive outcome of menopause must have seemed small comfort in the face of this long list of potential ailments. The author did note, in passing, the availability of "mild drugs or hormone pills," but the crux of his message was that maintenance of a positive attitude (on the part of both husband and wife) was the key to survival during menopause.[13]

By 1970, perhaps in an effort to counteract the expansion in negative imagery, *Today's Health* ran an article called "Menopause—Something to Look Forward To?" This article also presented a long list of symptoms (including weight gain, memory loss, thinning hair, and formication, described as "the sensation of bugs crawling on the skin"), but assured readers that "most of these symptoms, whether or not they are truly menopause-related, usually can be alleviated by one medical treatment or another."[14] These drugs included not only estrogen but also progesterone, androgen, testosterone, thyroid hormone, tranquilizers, sedatives, stimulants, vitamins, and aspirin. "Physical comfort and mental ease," the author attested, were essential components of a smooth menopause, and a visit to the doctor was a sure way to achieve the former.[15]

The message in these two periodicals was consistent over the thirty years between 1940 and 1970: menopause, while a natural stage of life, was best navigated with the guidance of a physician. The American Medical Association advocated medical attention, but not necessarily medication, for the menopausal years. In 1976 *Today's Health* claimed an audience of 3.2 million (no statistics are available for earlier years), a significant number but far below those of other mass circulation magazines, which claimed five to ten times as many readers.[16] Furthermore, it disseminated only one viewpoint—that of middle-of-the-road medicine.

More people encountered information about health and medicine from magazines such as *Reader's Digest, Time, Newsweek, Ladies' Home Journal, McCall's*, and *Good Housekeeping*. Articles on hormone replacement therapy in these popular magazines served as a significant source of information for women. In 1978 a national study of American families sponsored by General Mills confirmed the educational role of popular media when it found that 25 percent of respondents named popular magazines and newspapers as their main suppliers of information on health-related topics.[17] Although some women went to their doctors for advice and relief during menopause, many others turned to the newsstand for information. Representing a wider range of editorial stances, these publications might have offered a greater variety of articles on menopause and hormone replacement therapy. However, popular magazines did not pay much attention to menopause; the few articles that were published on the subject tended to follow the latest medical ideas. Journalists reported what they heard at medical conferences or what they read in medical journals. Oftentimes the authors were physicians themselves, reporting on their own personal experiences with menopausal patients. The most vocal doctors, such as Robert Wilson, earned the most media exposure. Whether an article presented Wilson's radical model or the more mainstream approach to menopause, it usually promoted the medicalization of this life course transition, perhaps with medication, and certainly with medical supervision.

One of the earliest articles to address the use of estrogen appeared in 1939 in *Independent Woman,* the magazine of the National Federation of Business and Professional Women's Clubs. The author divided menopausal women into two groups: the 60 percent who experience "little actual suffering" and the 40 percent with "acute physiological and emotional crises." The former were told to keep busy and focus their energy and attention outward; the worst things they could do would be to reflect on their lives or to vocalize their concerns about aging. The latter were offered an easier way out: sex hormone therapy, to replace their missing estrogen. When the article was reprinted, in condensed form in *Reader's Digest,* many more Americans learned about the science of female endocrinology and the medical management of menopause.[18]

*Reader's Digest* also published an abridged version of the November 1941 *Hygeia* article on the new miracle drug, DES, bringing news of the inexpensive synthetic hormone to millions of readers.[19] Seven years later, readers of the *Digest* learned from none other than the famous science writer and champion of scientific medicine Paul de Kruif that "medical authorities now endorse the use of the extraordinary female hormones," available in synthetic (DES) and natural (Premarin) forms. Writing specially for this mass-market publication, de Kruif recounted the efforts of "hormone hunters" (adapting the phrase from his bestselling book *The Microbe Hunters*) to relieve the "melancholy sickness that blights the happiness of some women at their change of life."[20] Praising not only the scientists who isolated estrogen but also the doctors who administered the hormone, de Kruif made it clear that medical supervision was essential to safe and successful treatment.

Other magazine writers credited modern medicine with taking the trepidation and misery out of menopause by generating knowledge about its physiology and therapy for its symptoms. A *Good Housekeeping* writer used the metaphor of spring cleaning to speak to her largely female audience: "Modern medical science has swept away the cobwebs of superstition, opened wide the windows of the mind to the light of facts, and developed proven therapeutic methods for treating symptoms."[21] Dr. James Scott told readers of *Ladies' Home Journal,* "You need not fear the menopause."[22] Although he estimated that 75 percent of women experi-

enced some symptoms, only 10 percent would have them severe enough to warrant treatment, and they could relieve their discomfort with daily pills "for not
more than [the cost of] a pack of cigarettes." There was no need to "grin and bear
it"; just as anesthetics eased childbirth, so too would estrogen ease menopause.
He assured women that the prescription-only requirement for estrogen was not a
ruse to increase doctors' incomes; rather, it was necessary to provide the most
effective and expedient therapy.[23]

As in the medical literature, the tone in popular articles on hormone replacement therapy turned from enthusiastic to cautious in the 1950s. *Reader's Digest*
quoted a gynecologist as saying, "Hormones are excellent, but reassurance is
even better" for the small percentage of women who had acute menopausal
symptoms.[24] A *Better Homes and Gardens* writer criticized the indiscriminate use
of estrogen: "Some years ago, estrogen hormones were made available for use,
when necessary, in women severely disturbed by menopause. Always, the idea of
some simple panacea—some cure-all in pill or shot form—has a strange fascination. And estrogen got an immediate and eager acceptance. Too eager and too
misinformed!"[25] He blamed not only "the minority of medical men involved" but
also women and their husbands who pestered their doctors for the treatment.

Significantly, because of the striking contrast with the medical consensus that
arose in later decades, the author insisted that "menopause does not produce . . .
chronic degenerative diseases such as cancer, heart disorders, and arthritis."[26]
Menopause itself was not a disease, nor did it cause the diseases associated with
aging. Like other contemporary popular writers, he strived to make the point that
menopause was merely a natural transition from reproductive to nonreproduc-
tive capability. While estrogen might be necessary for a short while to ease the
changeover, it was no magic elixir. However, at the same time magazines were
publishing these guarded recommendations, physicians in St. Louis and elsewhere around the country were conducting research that would change medical,
and then popular, conceptions of menopause, aging, and hormones.

William H. Masters and his colleagues did a good job of publicizing their
results at scientific and medical meetings; newspapers and periodicals that sent
journalists to those conferences often published articles on the latest findings
described within the professional community. The *New York Times* followed the
progress of the St. Louis group, reporting in 1950 that "sex hormones are making
old persons feel younger and healthier,"[27] and in 1953 that "Drs. Masters and
Grody infer that this combination of hormones [estrogen and testosterone]
achieves its effect by apparently arresting and partially reversing the aging process."[28] Specialized popular science periodicals, such as *Science News Letter* and

*Science Digest,* also reported on sex hormone research in the 1950s with attention-grabbing titles such as "Sex Hormones Fight Age," "Female Hormone Sparks Minds of Older Women," and "Sex Hormones Make Old People Feel Better."[29]

Only two mass circulation magazines picked up the anti-aging story. In 1954 *Newsweek* ran a brief article on E. Kost Shelton and his insistent campaign to keep women on hormones for decades after the menopause. Titled "Hope for Grandmothers," the article quoted from Shelton's speech at the annual AMA convention, in which he referred to women in their fifties as dried-up seed pods and offered them the prospect of hanging on to their husbands, with the help of estrogen replacement.[30] The following year *Coronet,* a monthly periodical similar to *Reader's Digest,* published a five-page article on the miracles taking place at the St. Louis City Infirmary. The author could barely contain her enthusiasm, commenting that "Dr. Masters' treatment offers nothing less than promise of the prolongation of man's prime" and noting that there may be "millions of potential beneficiaries of hormone replacement."[31] After describing the dramatic results of Masters's experiments, in which bedridden, disheveled female patients became ambulatory and took pains to improve their appearance, she concluded that the age-old dream of rejuvenation was close at hand. Although these articles sang the praises of the anti-aging potential of hormone therapy, they were exceptions to an otherwise much more cautious literature.

Not until the 1960s, when the Wilson Research Foundation began its aggressive public relations campaign, did popular magazines warm to the possibility of long-term hormone replacement. Some journalists parroted Wilson's notion that menopause was an estrogen deficiency disease; others, although unwilling to let go of the idea that menopause was a natural process, presented menopause in a much more negative light than in previous years. However, there was a solution at hand, and many of these writers presented estrogen as something close to a miracle drug. One doctor, writing in the *Ladies' Home Journal* in 1965, asserted, "Despite baseless fears, estrogen is usually safe and effective in relieving the distressing symptoms of menopause. It is perfectly natural for women to wish to slow up the aging process and to remain more attractive. They don't hesitate to use contact lenses for failing eyesight, color rinses for drab-looking hair or caps for their teeth. Then why should they put up with the discomforts that afflict about half of them in middle age, when the menopause begins?"[32] Readers of this article were assured of the safety of this drug and encouraged to avail themselves of its beneficial effects. Another article in *McCall's* a few months later, called "Pills to Keep Women Young," relied on "eight distinguished doctors who have done important work with estrogen" to "tell of its exciting results."[33] These articles relied heavily on doctors as experts, promoting estrogen as a panacea for the woes of menopause and aging.[34] But nobody made the case for estrogen more persuasively than Robert A. Wilson.

His 1966 book *Feminine Forever* was an expanded version of the claims he had set forth three years earlier in his article "The Fate of the Nontreated Postmenopausal Woman: A Plea for the Maintenance of Adequate Estrogen from Puberty to Grave," in the *Journal of the American Geriatrics Society.* Wilson decided to cut out the medical middle man and appeal directly to women. Although he described menopause and its aftermath in the most pessimistic terms, he presented himself as an advocate for women. Wilson acknowledged the validity of older women's physical and psychological symptoms, and he criticized his fellow physicians who ignored their patients' complaints and concerns at midlife. He promised that his plan of long-term hormone replacement therapy would guarantee continued physical attractiveness and sex appeal, and he unabashedly defended "a woman's right to be feminine."[35] While some scholars have condemned Wilson as misogynistic, historian Judith Houck offers a more nuanced interpretation. Without denying the cruelty of his imagery in portraying women as castrates, she contends that Wilson sympathized with the plight of older women in America's youth-centered culture and offered them the opportunity to feel like valued and valuable participants in contemporary society.[36]

Wilson drew stories from his private practice to illustrate the physical, emotional, sexual, and professional problems of menopausal and postmenopausal women. He attributed all of these trials and tribulations to women's decreased production of estrogen, which robbed them of their femininity, their health, and oftentimes their happiness. He emphasized that the psychosocial troubles were not the fault of the women but the unfortunate by-product of the deficiency disease called menopause.

After more than a hundred pages of sad tales of older women (and their husbands), simplistic descriptions of female physiology and endocrinology, heroic recounting of Doisy and Allen's isolation of estrogen, and the efforts of physicians to apply this scientific discovery to medical therapeutics (including Wilson's own forty-year involvement in clinical research and practice), the author finally agreed to "divulge precise and specific details of the treatment that holds such abundant promise."[37] First, however, he made sure to emphasize the danger of self-medication and the absolute necessity of a doctor's supervision.

Wilson's treatment plan was clear-cut: replace the missing estrogen. As part of the physician's complete physical examination of the patient, Wilson advised, he should perform a diagnostic test based on an adaptation of the Pap smear. This test, which Wilson dubbed the "femininity index," measured the relative proportions of different types of cells in the vaginal lining. The goal of hormone replacement therapy was to restore the femininity index back to the "normal" ratio found in women during their reproductive years.[38]

Wilson arbitrarily set the vaginal cellular profile of women in their twenties as the standard for women of all ages, rather than comparing women of a given age to their peers. The norm for a 55-year-old woman might have been based on the average pattern displayed by a group of 55-year-olds, not on that of women thirty years her junior. Because older women's femininity indexes fell outside the norms established by Wilson, he concluded that they needed replacement hormones to achieve the levels they had maintained during their reproductive years. By defining young, premenopausal women as normal, Wilson implied that older, postmenopausal women were abnormal.

Wilson reinforced this classification scheme by basing it on a quantifiable measurement; the numerical femininity index wore the patina of scientific objectivity. He recommended that "every woman over thirty . . . have her Femininity Index checked once a year. This quick, painless test may prove a turning point in her life, an assurance of continued health and happiness."[39] One of Wilson's critics objected to the implications of this cavalier advice, describing patients who came to him, "asking for the 'youth pill' and they say 'Check my estrogen level.' From what they're read, they think it's as simple as driving into a gasoline station and having their oil checked."[40] Clearly, a number of women bought the message that their femininity could be revealed by the cells in their vaginas and also that they could be rejuvenated by hormone replacement therapy. They understood that postmenopausal meant abnormal, and they looked to medical science to bring them back to normal.

Estrogen, according to Wilson, could turn back time. "The skin becomes supple again, the muscles regain their tone and strength, the breasts are restored to almost their former fullness and contours, the genitals again become supple and distensible, skin cracks and genital inflammations heal. Bones that have become more brittle regain most of their former strength."[41] Magazine writers picked up this message and brought it to an even wider audience; whereas *Feminine Forever* was read by hundreds of thousands, articles in popular periodicals reached millions.

*Vogue* allowed Wilson to speak for himself in its June 1966 issue.[42] Introduced as "a famous and distinguished gynecologist," Wilson accompanied his article—a very abridged version of his book—with five "key hormone charts" for women in five age groups. For each cohort, he estimated the percentage with estrogen deficiency: 5 percent of those aged 17–29, 40 percent of those aged 30–39, 85 percent of those aged 40–51, 90 percent of those aged 52–69, and 100 percent of those aged 70–85. Each chart listed the reported symptoms of estrogen deficiency, additional findings based on physical examinations, typical values for the femininity index of hormone-deficient women, and possible treatment (varying regimens of varying doses of estrogen with or without progesterone). While estrogen deficiency in women in the youngest age group was probably pathological in origin, Wilson blamed menopause for the symptoms in the other four groups. Those 30–39 could avoid the longer and even more depressing lists of symptoms in the charts of the three older groups by taking birth control pills and then switching over to the menopausal estrogen treatment. In Wilson's scheme, any woman at any age could benefit from hormone replacement.

Readers of this fashion magazine had already encountered news of estrogen in an article published the previous summer, a transcript of an interview with "the distinguished director of one of New York's most respected hospitals," whose name was never revealed. The introduction described estrogen in sensationalist hyperbolic terms: "There is a pill that appears to play a vital role in the prevention of aging in women . . . women can live happier and longer, and, during this longer, happier life, they can retard, dazzlingly, many of the physiological signs of aging . . . the only people who stand to be affected by this revolution in hormone therapy may be: all the women on earth."[43] By the early 1970s, hormone replacement therapy had become incorporated into fashion and beauty magazines' recommendations for the over-40 crowd. *Harper's Bazaar* asserted in its "Over-40 Guide on Health, Looks, Sex" that "prevalent medical opinion is that the safety and benefits of estrogen therapy have been convincingly demonstrated."[44]

However, not all articles in popular periodicals were so unreservedly positive. Other writers adopted a more equivocal stance, presenting the range of medical opinion on long-term hormone therapy. *Good Housekeeping* divided physicians into four groups according to their views on estrogen: conservatives, moderates, liberals, and radicals. The conservatives rejected the notion that menopause was an estrogen deficiency disease and preferred reassurance and sedatives to treat symptoms; however, some would prescribe small doses of estrogen for short periods of time to treat hot flashes. The moderates accepted that some women benefited from supplemental estrogen to treat their menopausal symptoms; many in this group had no problem with continuing hormone therapy for years or decades, if necessary. The liberals believed that all women needed estrogen during and after menopause; Allan C. Barnes, for example, said "the only excuse for calling the menopause normal . . . is that every woman goes through it."[45] Robert Wilson and a few others were considered the radicals because of their use of hormones to continue or induce menstruation in midlife and older women, as part of their program to eliminate menopause entirely. "What should women do?" asked the author. Not surprisingly, she advised her readers to consult with their doctors and to "seek a physician who is alert to changing viewpoints." While she did not advocate a participatory role for women in the doctor-patient relationship, she did acknowledge that a woman "has a right to know that some doctors will and some doctors won't [prescribe estrogen]."[46]

*Reader's Digest* also cautioned its audience to exercise restraint in the medical management of menopause and postmenopause, telling readers that "no pill can make one young again. Nor can a pill make one feminine . . . nor is it a cure-all for the stresses and strains of a woman's life."[47] The author (in an article condensed from *U.S. Lady*) downplayed the number of women who truly needed estrogen at menopause (about 15%) and who were hormone deficient after the menopause (about 25%). She abhorred the recent rush to estrogenate, blaming "enthusiastic articles and sensational advertisements" for inveigling women into believing in this mythical fountain of youth.

This author quoted the *Medical Letter* (an independent nonprofit publication that began reviewing medications and therapies in 1959) as a respected and objective arbiter of medical claims. *Time* also cited this publication in a review of *Feminine Forever* and its impact on gynecological practice. The *Medical Letter* recognized that estrogen could relieve immediate symptoms of menopause (such as hot flashes), and it conceded that the hormone played some role in treating osteoporosis (but not necessarily in its prevention); however, it rejected as premature the assertions that estrogen "promotes either the appearance or the feeling of youthfulness, or prevents or lessens the complications of arteriosclerosis," and it recommended against the prescription of hormone replacement for all women.[48] The mass circulation magazines appealed to the authority of the specialist journal in an attempt to temper what they perceived to be a potentially dangerous fad.

By the mid-1970s, it appeared that the fad was not going to fade away. To assess the prevalence of hormone replacement therapy in medical practice, *McCall's* sent questionnaires to gynecologists, two hundred of whom responded. These participants were not a representative sample of American gynecologists: they were all women over the age of 40, "with a double interest in hormone therapy."[49] Both the results of the survey and the manner in which they were reported in this popular women's monthly expose a snapshot of the standing of estrogen in the eyes of doctors and magazine writers in 1975.

Among this group of female physicians, estrogen was, in a word, popular. Ninety-eight percent of the gynecologists had prescribed HRT for patients, and 58 percent of those who were menopausal used it themselves. Eighty-four percent believed estrogen alleviated psychological problems, such as depression and anxiety, and 51 percent recommended HRT as a preventive against osteoporosis and atherosclerosis. The author of the article raised the cancer question and gave what had come to be the typical answer: the majority of gynecologists surveyed refused to prescribe estrogen to women with a prior history of cancer, but "there is *no* evidence that estrogen causes cancer in humans . . . there is even some evidence that estrogen may *protect* women against breast cancer." The final analysis reflected the ambivalence within the medical profession. Was menopause an estrogen deficiency disease? Some said yes, others said no, and ultimately "the decision whether to use replacement hormones still remains one for an individual woman and her doctor to make."[50]

Thus, in leafing through their favorite women's monthly or news weekly, women learned about the existence of estrogen and the claims made for its anti-aging effects, but they received confusing and sometimes conflicting recommendations about the wisdom of long-term therapy. Those women who took magazine writers' advice and went to see their doctors encountered another source of information in the waiting room: pharmaceutical brochures for estrogen products.

Drug companies promoted the medical management of menopause through "informational" pamphlets, which were distributed to doctors' offices to be handed out to patients. These brochures touted the benefits of hormonal treatments and urged women to consult their doctors at menopause "whether you are having symptoms or not."[51] Women received a very one-sided message from these booklets: menopause, although a natural transition, caused unpleasant effects. According to one manufacturer's guide to menopause, 90 percent of women experienced symptoms such as "physical and emotional turmoil, sagging and flabbiness of the breasts, tendency to put on ugly weight, . . . embarrassingly copious perspiration, excruciating headaches, causeless crying spells, . . . senseless, addle-headed anxiety."[52] In case these vivid descriptions didn't come across clearly, they were accompanied by cartoon illustrations of a shapeless woman demonstrating the various conditions (sad, depressed, confused, angry, fat, cold).

The pamphlets went on to offer a solution to these miseries: hormone replacement therapy, which might potentially be taken for the rest of one's life. One cartoon-illustrated guide explained that the decrease in estrogen production caused these symptoms, represented by a picture of a factory door ("Ovary Mfg. Co.") with the shade pulled down, a lock on the handle gathering cobwebs, and a sign that said, "Closed Forever."[53] The next page showed a giant bottle labeled "Hormones" with a faucet, and reassured readers that replacement hormones would prevent menopausal symptoms.[54] Several pages later, under the heading "Does estrogen cause cancer?" the text said, "No. In fact, there is now some evidence suggesting a LOWER incidence of cancer in women taking estrogen." It went on to comment that only mice given massive doses of estrogen developed cancer; illustrating this point was a cheerful-looking cartoon woman, her head thrown back and one foot kicking up behind her, clutching a bottle labeled estrogen in one hand, popping a pill into her mouth with the other, saying, "Only in mice."[55] The simplistic text and the caricature sketches left no room for doubt or concern about hormone replacement.

This message was presented even more persuasively in the pamphlets published by the Wilson Research Foundation. "Feminine . . . FOR LIFE" used scientific language and illustrations of vaginal cells to explain the femininity index and its use in determining when (not whether) estrogen replacement therapy should begin.[56] "The Complete Woman" listed detailed descriptions of the "consequences of the menopause," including tissue weakness, excess of cholesterol in the blood, abnormally porous bones, joint diseases, psychic manifestations, and menopausal negativism. This publication made it clear that the medical discovery and prescription of "lifelong" hormone replacement easily trumped the inadequacies of Mother Nature.[57] "Mistrust without Logic" addressed the cancer issue, rejecting the evidence that suggested estrogen might be carcinogenic and postulating instead that estrogen might have a preventive effect.[58] Sitting in her doctor's waiting room, a menopausal woman could easily become convinced of the benefits of hormone replacement therapy from the reading material supplied by this allegedly impartial "non-profit medical research foundation" and by the pharmaceutical companies, which took pains to downplay their roles (identifying their sponsorship in tiny type, for example) in the publication of their educational booklets.

How did women respond to the messages they heard from their doctors or read in magazine articles and promotional pamphlets about the use of estrogen during and after menopause? This is a difficult question to answer. Evidence of women's attitudes and opinions on hormone replacement therapy can be gleaned from surveys of midlife women in the 1950s, 1960s, and 1970s. Unfortunately, the small sample size of these groups—never more than a few hundred people— precludes making any sort of generalization from the collected responses. Some of the questionnaires, such as the one developed by Bernice L. Neugarten of the Committee on Human Development at the University of Chicago in the early 1960s, ignored estrogen entirely, focusing only on attitudes toward menopause.[59] In the more qualitative studies, which allowed space for additional comments, some individuals described their thoughts about and experiences with hormone replacement therapy.

These individual responses reveal a gamut of interactions, both positive and negative, with doctors and hormones. Some women swore by estrogen injections or tablets; others were considerably less enthusiastic. Some women reported a good rapport with their physicians, although many others resented them for not taking menopausal complaints seriously.[60] Some seemed to parrot the messages they garnered from popular magazines; others felt baffled by the conflicting recommendations they received from different sources. Neugarten found that most middle-aged women, whether they feared menopause or dismissed it as inconsequential, wanted to talk about it and expressed a desire for more information and greater communication.[61] Presumably, this interest in menopause education involved learning more about possible therapies, including hormone replacement.

Another way to gauge women's responses to changing medical and popular advice on estrogen is to look at changes in the number of prescriptions filled each year. New prescriptions for Premarin, which had remained level at about 1.6 million per year from 1958 to 1965, doubled in 1966, the year in which Robert Wilson published *Feminine Forever*.[62] That same year, according to the National Prescription Audit, retail pharmacies dispensed a total of 16 million prescriptions (both new and refills) for non-contraceptive estrogen products, or enough to treat 2.6 million women (assuming each prescription covered a two-month supply of pills, the standard quantity dispensed). This figure is probably an underestimate, because the National Prescription Audit did not count prescriptions from hospitals, supermarket pharmacies, mail-order firms, or discount distributors. Furthermore, it included only oral and vaginal preparations, since doctors usually did not write prescriptions for injections administered in their offices (in 1983, the only year for which such data exist, injectables constituted 23 percent of all estrogen doses).[63]

The number of estrogen prescriptions rose dramatically from 1966 to 1975. Retail pharmacies sold 19.5 million in 1968 and 25 million in 1972.[64] New prescriptions of Premarin reached five million in 1974, for some 800,000 first-time users.[65] In its 1972 promotional pamphlet called "Let's Discuss the Menopause and Your Need for Estrogens," Syntex Laboratories estimated that seven million women were taking estrogens "on a more or less regular basis."[66] That number must have represented worldwide users of menopausal estrogens; if not, it was a big exaggeration. However, by 1975, the commissioner of the Food and Drug Administration, Alexander M. Schmidt, testified before Congress that enough estrogen (specifically, Premarin) was sold in the United States to treat five million women per year.[67] The National Prescription Audit reported a total of twenty-eight million prescriptions for non-contraceptive estrogens dispensed in 1975.[68]

Public health surveys revealed vast discrepancies in estrogen use according to geographic region. In 1967 the U.S. Department of Commerce estimated that 13 percent of women between the ages of 45 and 64 used hormone replacement therapy. In Massachusetts, however, only 8 percent of women reported estrogen use in 1972. A survey done the following year in Washington found that 51 percent of women in two counties in that state used estrogen during and/or after menopause; average duration of treatment exceeded ten years.[69] Thanks to the willingness of doctors to prescribe and women to use estrogen, Premarin sales in the Pacific Northwest increased at an annual rate of 25 percent in the late 1960s and early 1970s.[70] These inexplicable variations in regional prescription rates notwithstanding, the total amount of estrogen consumed by American women was certainly on the rise.

What accounts for the increased popularity of hormone replacement among menopausal and postmenopausal women in the late 1960s and early 1970s? While women may not have bought into the notion of menopause as a deficiency disease, they did appreciate estrogen's ability to relieve their menopausal discomforts.[71] Thanks to the promotional work of the Wilson Research Foundation and media coverage of the expanded possibilities of hormone replacement therapy (not only to treat immediate symptoms of menopause, such as hot flashes and painful intercourse, but also to prevent the later development of osteoporosis and heart disease and straying husbands), more and more women encountered information about estrogen and were encouraged to take advantage of this modern medicine.

Estrogen's profile was further raised by its use in another product widely advertised to aging women: skin creams. From the 1940s through the 1970s, cosmetics manufacturers marketed creams, lotions, and oils containing estrogen; these products were advertised directly to the public in the pages of newspapers such as the *New York Times* and the *New York Herald Tribune* (as topical cosmetics, they could be sold without a doctor's prescription).[72] The ads imparted a clear message:

use of estrogen-containing creams would make a woman's skin look younger. And young-looking skin, it almost went without saying, was beautiful skin.

Indeed, by the postwar years, youth had become synonymous with beauty. The cosmetics industry had helped to bring about this association in the 1920s, transforming the nineteenth-century distinction between the appearances of mothers and daughters.[73] A 1923 advertisement for Pompeian beauty products depicted a mother seated before a mirror with her daughter looking on and commenting admiringly, "Mother, you're looking younger every day!"[74] Only the caption revealed that the speaker was in fact the younger of the two females; otherwise, they looked to be about the same age. In the 1920s cosmetic companies bought advertising space in popular magazines, such as *Vogue, Good Housekeeping, Time*, the *Saturday Evening Post*, and the *New Yorker*, in which they touted the age-defying benefits of various facial preparations. "A woman is only as old as her complexion," announced an ad for Boncilla products.[75] Edna Wallace Hopper urged readers of the *Ladies' Home Journal* to "stay a girl, cultivate your beauty and retain your youth," by purchasing her line of creams and powders.[76] Marie Barlow made clear the reason for applying facial treatments: "for men demand youth in women's faces!"[77]

Similar ads continued through the 1930s. In spite of economic hard times during the Great Depression, women were encouraged to purchase products to keep up their appearances. "Age comes fast after 30," Dorothy Gray warned readers of *Harper's Bazaar*.[78] In another ad for its emollients, this company instructed women to "watch for signs of age in these 3 places," outlining the alarming signs of age in the skin on one's nose, on one's throat, and around one's eyes.[79] Even soap companies got into the act, abandoning earlier claims of superior cleanliness in favor of promises of "youth's charm" for Palmolive and "youth and beauty" for Ivory.[80]

Thus by the 1940s women had become accustomed to turning to cleansers, creams, and cosmetics to fight the outward signs of aging. The addition of estrogen, according to the persuasive phrasing of advertisements, further enhanced the ability of beauty products to help women maintain a youthful countenance. Companies relied on the midcentury fascination with science to sell their estrogen creams; at the same time, they underscored the naturalness of the hormones used in these preparations. The makers of Endocreme boasted that the lotion contained Activol, "its own laboratory-developed counterpart of an essential, skin-vitalizing substance all women possess."[81] An ad for Helena Rubenstein's "famous" estrogenic hormone cream described it as "scientifically prepared with natural estrogen."[82] Just as many injectable and oral estrogens were promoted to physicians in the 1940s based on the high scientific standards of their laboratory preparation as well as the natural sources from which they were derived, so too were estrogen skin creams marketed to women on the grounds of both science and nature.

Estrogen creams initially received the endorsement of science in medical reports in the 1940s and early 1950s. In one instance, endocrinologist Max A. Goldzieher published a favorable article in the *Journal of Gerontology* in 1946 claiming that topically applied estradiol and DES resulted in the regeneration of skin cells. Although he experimented on just five patients, he justified the validity of his findings based on the use of controls (for each woman, the estrogen cream was rubbed onto one thigh and forearm and the same ointment base, minus the estrogen, was rubbed onto the other side).[83] Other investigators announced similar results, based on biopsies of skin treated with estrogen cream.

By the mid-1950s, some in the medical profession began to doubt the extrapolation of these clinical findings to advertising promises for noticeably younger-looking skin. In experiments reported in the *Journal of the American Medical Association,* a skeptical New York physician took before-and-after photographs of women, aged 35–65, who used commercially available estrogen creams on one side of their faces and control creams (obtained from the manufacturers) on the other. Although some of the women commented that both sides of their faces felt less dry, a dermatologist could discern no cosmetic or clinical differences in the sides treated with and without estrogen.[84]

The American Medical Association weighed in on the debate in 1961. Its Committee on Cosmetics issued a statement opposing the use of drugs, such as estrogen and antibiotics, in cosmetics.[85] The committee criticized estrogen creams for being at once dangerous and ineffective. On the one hand, the seemingly low concentration of estrogen in skin cream could still cause harm, as evidenced by a 1952 medical report of damage resulting from overabsorption of estrogen. On the other hand, low concentrations of estrogens probably had no biological effect, thus discrediting the claims made by manufacturers.[86] The Food and Drug Administration also expressed its skepticism of " 'mysterious miracle ingredients' that made 'far-fetched promises that appeal to natural desires to be young and attractive.' "[87]

These declarations from the nation's drug regulatory agency and its largest medical organization had little effect on the marketing of estrogen skin creams. In fact, three weeks after the *New York Times* reported on the AMA and FDA statements, the Helena Rubenstein Salon ran a large ad in that same newspaper promoting its Ultra Feminine cream, which contained estrogen and progesterone, "the only face cream that reverses two ways in which your skin grows old!"[88] In advertising estrogen creams, manufacturers went beyond the alleged qualities of their products to tap American women's fears of aging. These tactics built on the well-established foundation of youth as the standard for beauty established decades earlier. They also preyed on women's economic and emotional depen- dence on men. An ad for Dorothy Gray's Cellogen Cream pictured a woman looking worriedly at a man, presumably her husband, flirting with another woman. The caption read, "Does Your Husband Look Younger Than You Do?" Cellogen Cream, with 10,000 international units of natural estrogenic hormones, could help women to "side-step the tragedy that overtakes so many wives."[89] This potential scenario—of a husband fleeing his wife's wrinkled skin and finding comfort in the arms of a smoother-skinned woman—may have frightened women whose lives revolved around their husbands.

Helena Rubenstein also manipulated women's fears of aging in dozens of advertisements that ran from the 1950s well into the 1970s. Many of her ads featured a large title and several paragraphs of text explaining the virtues of estrogenic hormone creams. Often the ads were accompanied by a large photo- graph, usually of Helena Rubenstein herself, showing off her flawless skin or working beside a chemist in her company's laboratory. Intended to look like newspaper articles, these pseudo-journalistic pieces ran headlines such as "How long can a woman look young?," How to Take Years Off Your Age," "Why Some Women Look Young at 50 While Others Look Old at 35," and "New Wonder Drug Cosmetic Restores Young Look to Aging Skin."[90]

The "wonder drug" campaign drew explicit connections between the world of cosmetics and the world of medical science, using female sex hormones as the link. Helena Rubenstein's description of estrogen and progesterone as "nature's way of making a woman feminine" paralleled the rhetoric of Robert Wilson. Another campaign resembled a clinical report, complete with statistical evidence of Ultra Feminine's success: "Just Released! Responses from 5,157 Users of Helena Rubenstein's Ultra Feminine! . . . Overwhelming endorsement given by 94% of these users of Ultra Feminine, the only face cream that works within the skin to replenish loss of vital estrogen and progesterone, *natural youth* sub- stances!"[91] Advertisements for estrogen skin creams reinforced the image of estrogen as an anti-aging agent and made the twin recommendations—for pre- scription oral hormone replacement therapy and for over-the-counter topical hormone-supplemented skin treatment—seem complementary. The aggressive promotional tactics of Helena Rubenstein and other cosmetics manufacturers worked not only to increase sales of their estrogen skin care products but also to increase public perceptions of estrogen as the latest weapon in an individual woman's battle against the appearance of growing old.

Women's anxieties about looking old must be understood in the context of the celebration of youth that began in the years after the Second World War. Baby boom culture extolled not only children but also their parents and the lifestyle of the nuclear family. In the 1950s the marriage rate was higher than ever, the age at which people married was lower than ever, and the rising birth rate had reversed a 150-year trend of decline. Often beginning in their teens, many women defined their lives in terms of their roles as wives and mothers. Older women who had already raised their children felt superannuated in a society that equated woman-hood with motherhood. There was a sizeable literature on the emotional trauma of the "empty nest," in which psychologists and psychiatrists offered varying interpretations of the types of personalities most likely to succumb to depression at middle age and the options for coping with this particular stage of the life course. Many of the articles on menopause also addressed these midlife issues, echoing the professional rhetoric and encouraging women to take up hobbies, to volunteer in their communities, or even to work part-time after their children left home. This prescriptive literature conveyed the message that, after so many years of caring for their families, women should channel their energies into serving others and perhaps even attending to themselves.[92]

Magazine writers tried to convince women that life continued after the age of 40 or 45.[93] These articles may have been persuasive in the 1950s, but when the youth movement gained momentum in the 1960s, with its slogan "Don't trust anyone over thirty," it became hard for middle-aged people not to feel left behind. By 1971, *McCall's* declared that " 'middle-aged' just may be the dirtiest ten-letter word in the English language . . . In the verbal arsenal of America's more militant youth, 'middle-aged' is the ultimate put-down, a patronizing epithet designed to make an entire generation wince."[94] The author sympathized with the plight of this generation, "caught in a culture that literally worships youth and conspires to hide the elderly in old-folks' homes."[95] His recommendations reiterated those of previous decades—women should find something to do outside the home—but with the added sensibility begat by the women's liberation movement. Thus he gave as examples women in their forties who began careers in law and politics, fields closed to earlier generations of women (and to these very same women when they were younger). He explained that greater independence might lead to changes in marriages, and he noted that some "liberated" middle-aged women felt free to engage in extramarital sex. More likely, however, and of greater concern to middle-aged women, was the possibility that their husbands would stray from the marital bed.

Although magazines ran articles about marital sex (enough for *Newsweek* to take notice in an article called "Sex in the Magazines" in 1959), the media coverage of the so-called sexual revolution of the 1960s tended to focus on the sexual attitudes and behaviors of young people.[96] As the divorce rate rose in the 1960s (by 62% for couples married less than five years, by 37% for those married more than five years, and by 50% for those married twenty to thirty years, the group most relevant to this discussion), middle-aged and older women felt keenly the problem of remaining sexually viable in a society that valued youth as the symbol not only of beauty but also of sexuality.[97]

Susan Sontag articulated the predicament faced by American women in a 1972 article titled "The Double Standard of Aging." She wrote: "This society offers even fewer rewards for aging to women. Being physically attractive counts much more in a woman's life than in a man's, but beauty, identified, as it is for women, with youthfulness, does not stand up well to age . . . Women have a more intimate relationship to aging than men do, simply because one of the accepted 'women's' occupations is taking pains to keep one's face and body from showing the signs of growing older."[98] Sontag was not the only one to recognize the plight of the older woman. In 1970 Pauline Bart, a sociologist at the University of Illinois, wrote about depression in middle-aged women in a cleverly titled article, "Portnoy's Mother's Complaint," which was reprinted the following year in the feminist collection *Women in Sexist Society*.[99] In defending her choice of topic, she argued that "a nation's humanity may be measured by how it treats its women and its aged as well as by how it treats its racial and religious minorities. This is not a good society in which to grow old or to be a woman."[100] She went on to explain the circumstances that led to clinical depression; in her words, "there is no bar mitzvah for menopause."[101] The absence of a ceremony to mark the transition from children-at-home to children-grown-up, or even a road map to guide the way through the uncertainty of this period, left many women feeling lost. Bart looked to the women's liberation movement to provide the guidance and support for women to realize their full potential, beyond the narrowly defined gender roles prescribed by society.[102]

However, the women's movement was in large part a movement of young women, for whom the problems of aging seemed remote. *Sisterhood Is Powerful*, the seminal 1970 anthology edited by Robin Morgan, included just one piece— out of a total of seventy-five—on aging. The author of this essay bemoaned her invisible status in American society, at the age of just 43. Older women realized the disparity between their interests and those of their younger sisters, and they formed groups to address their concerns. One group of women in New York, calling themselves the Over-45 Committee, began a newsletter in 1971, *Prime Time*, "for the liberation of women in the prime of life." Its purpose was to address issues such as job discrimination, advertising bias, problems faced by displaced homemakers (divorced women with no experience beyond being wives and mothers), self-image (and the social pressure to look youthful), and the role of older women in the larger women's liberation movement. In 1974 the editors published several readers' accounts of their menopause experiences in a "Menopause Speak-Out."[103] Articles and letters in subsequent issues continued the conversation.[104] While these women objected to the medicalization of menopause and the social discrimination against postmenopausal women, these concerns were just a few of the multitude of "ageist" injustices they felt they faced.

Another subset of feminists interested specifically in health issues spun off from the larger women's movement in the late 1960s and early 1970s; however, this women's health movement paid scant attention to menopause and postmenopause in its first years. In the initial edition of *Our Bodies, Ourselves* in 1971, the members of the Boston Women's Health Course Collective did not include anything on menopause or aging. As young women in their twenties and thirties, they were more concerned with issues directly affecting their current lives: birth control, abortion, pregnancy, childbirth, and sexually transmitted diseases.[105] In the second printing of the first edition, the authors included a note with the table of contents acknowledging this omission and promising to address menopause and aging (along with several other omitted topics) in the next edition. Sure enough, the 1973 edition, published by Simon and Schuster, included chapters on relationships, lesbians, nutrition, exercise, rape and self-defense, women's diseases, and menopause.

Unlike the rest of this ground-breaking book, which questioned the intervention of medicine in life processes such as childbirth and looked beyond the medical profession to lay sources of information, the chapter on menopause tended to echo the advice espoused by doctors. The bibliography of further reading for the chapter listed just nine sources, including five books by medical doctors, a clinical guide published by Ayerst Laboratories (the makers of the Premarin brand of estrogen), and an unpublished paper by an undergraduate at the University of Southern California. The authors commented on the "pitifully little research [that] has been done into symptoms and cures for symptoms of a physical experience more universally shared by women even than childbirth."[106] Their discussion of the advantages and disadvantages of estrogen therapy described (as did publications authored by health professionals) the symptoms estrogen could alleviate and the side effects the drug might induce. In drawing an analogy between estrogen and birth control pills, the authors allowed a measure of skepticism to seep into their analysis, noting that "you're damned if you do, damned if you don't take estrogen—the trick is to find the right amount, if any, for you, to get regular check-ups twice a year if you are taking estrogen, and to find a doctor who is aware of both the positive uses and the potential risks of estrogen, who will be very careful about what s/he prescribes."[107]

The Boston Women's Health Book Collective recognized that their information on menopause was woefully inadequate, and they set out to rectify the situation. In 1974 they turned to one of their most trusted sources—women themselves—and sent out a questionnaire to five hundred women, with sixty questions about symptoms, treatments, support, sex life, hysterectomy, pregnancy, and general personal information.[108] In an attempt to understand how women learned about and prepared for menopause, the survey's final question asked, "In what ways do you feel that your own education about menopause has been good or bad?"

The replies to this question varied enormously. While a few women considered themselves to be well-informed, many more commented on the paucity of available information. Respondents described their education as "very bad," "too sparse," "negligible," "inadequate," "non-existent and difficult to get."[109] One woman replied angrily, "What 'education'? We don't get any on the subject, and doctors are the *last* to admit you're going through menopause or are starting it!"[110] Another echoed the sentiment that physicians withheld information: "It has been scanty because doctors don't always tell and there isn't enough material published."[111] Several described themselves as self-educated, reading up on the subject in whatever resources they could find: "I read in the newspapers and magazines, and once sent away for a booklet."[112] These responses reflected the limited availability of educational materials on menopause in the early 1970s. Women were not privy to a wide range of balanced information about potential therapies; they had to make choices based on what their doctors chose to tell them and what they gleaned from scattered reports in the popular press.

Furthermore, they had to make these health care decisions in the midst of a barrage of dominant cultural representations of beauty and persuasive cultural messages about the lowly status of aging women in American society.[113] Within the context of a culture that valued youth as the standard for beauty, particularly for females, women sought ways to thwart the aging process.[114] Hormone replacement therapy, then, was simply one more weapon in an arsenal that included vitamins, hair dyes, and cosmetics. Many American women in the first half of the 1970s, particularly those older women who did not yet fully participate in the women's liberation movement, still felt bound by traditional social roles and cultural expectations for females. While women both young and old had begun in the previous decade to speak out against women's oppression in both public and private life, there were many who remained ignorant of or opposed to the possibility of major changes in the social, cultural, and political situation of women in America. Surely, estrogen helped many women to cope with the unpleasant physical symptoms of menopause, such as hot flashes, night sweats, and painful intercourse. And those who claimed that estrogen helped them to "feel" and "look" young (or younger than their non-estrogenated selves) sincerely believed in its restorative powers. The fact that clinical evidence did not support the contention that estrogen decreased wrinkles or improved mood mattered little to the estrogen users who thought that they looked and felt better.

Moreover, most of the available information sources supported the decision to take estrogen during menopause and to continue taking it for years afterward. The magazine article that introduced a woman to estrogen, the pharmaceutical brochure she read in her doctor's waiting room, and the advice of her trusted physician all presented a generally positive image of hormone replacement therapy. With few risks and a whole host of immediate and potential benefits, estrogen seemed like modern medicine's solution to the modern woman's problem of aging.

Although the nascent women's health movement objected to the medicalization of any stage of women's lives, it lacked the resources to devote much attention to addressing the issue of the rising use of hormone replacement therapy by menopausal and postmenopausal women. Feminists, like many journalists, advised women to exercise caution and restraint when considering whether or not to take estrogen. Overall, however, both health activists and magazine writers tended to defer to the expertise of physicians on this matter . . . until shocking news in December 1975 forced everyone to reconsider their views on estrogen.

# CHAPTER FIVE

# From Hero to Villain

## Estrogen and Endometrial Cancer

Estrogen turned from hero to villain upon the publication of two studies that presented conclusive evidence that estrogen users were more likely to get endometrial cancer (cancer of the uterine lining) as compared to nonusers. The appearance of these two reports in the prestigious *New England Journal of Medicine* ensured that they would be taken seriously by the medical profession. The editors of the *Journal* underscored the gravity of the findings by commissioning two editorials on the subject of estrogen and cancer, published along with the research articles in the same issue on 4 December 1975.

The first study compared two groups of older women in the state of Washington, 317 diagnosed with endometrial cancer and 317 diagnosed with other gynecological cancers (such as ovarian or cervical cancer).[1] The investigators found that 152 of those with endometrial cancer, as opposed to 54 of the controls, had used estrogen, which translated to a 4.5 times greater risk for women exposed to prescription estrogen. Upon further refinement of the data, they found that the risk grew with increased duration of estrogen use. The study did not take into account dosages, specific estrogens, or treatment schedules, but it did look at the potentially confounding factors of hypertension and obesity. Surprisingly, women *without* the predisposing risk factors of hypertension or obesity who took estrogen had the highest likelihood of developing endometrial cancer. The authors stopped short of announcing that estrogen caused the disease; in the circumspect style that had come to characterize scientific prose, they noted a pattern of increasing numbers of endometrial cancers in women without typical physiological predisposition to the disease "coincident with" a pattern of increasing use of prescription estrogen, and they proposed that "an apparently significant relative risk" of developing this type of cancer was "associated with estrogen administration."[2]

The other study matched endometrial cancer patients with control subjects (without cancer) and found that a greater proportion of the cancer group had used conjugated estrogens (e.g., Premarin).[3] Furthermore, it appeared that the longer a woman took conjugated estrogens, the greater her risk of endometrial cancer (up to fourteen times for women who took the hormones for more than seven years). One of the authors, Harry K. Ziel, an obstetrician-gynecologist at the Kaiser Permanente Medical Center in Los Angeles, initiated the study because he had noticed that a significant number of his patients with endometrial cancer took this form of estrogen.[4]

Investigators matched the record of each cancer patient with the records of two controls, based on age, area of residence, and duration of membership in the Kaiser Foundation Health Plan. They also controlled for age at menopause, excessive weight, and parity (number of children). When they looked at three other drugs (Valium, thyroid hormone, and reserpine—used to treat high blood pressure) taken by patients and controls, the risk ratios of developing endometrial cancer associated with these medications were each less than 1.0, compared to the 7.6 risk ratio for estrogen. By working with preexisting medical records and a stringent set of requirements, the researchers claimed to have avoided both information bias (in ascertaining use of estrogen) and selection bias (in choosing patients and controls). They concluded that "the evidence for a connection between the use of conjugated estrogens and the development of endometrial cancer seems rather persuasive," but they cautioned that further study was needed to confirm their results in other groups of women and for other types of estrogen before any policy decisions could be made.[5]

The accompanying editorials attempted to locate the estrogen-cancer studies in a broader frame of reference. The first one acknowledged the legitimacy of the epidemiological findings, especially in light of anecdotal evidence in humans and animal studies that demonstrated estrogen-induced cancer. However, the author played down the five-to-fourteen-fold increase in the risk of endometrial cancer for estrogen users by comparing it to the three-to-nine-fold increase in risk of endometrial cancer for obese women and the seventeen-fold increase in risk of dying from lung cancer for pack-a-day cigarette smokers. "Although such comparisons may seem specious and falsely comforting," he wrote, "they are testimony to the hypothesis that one might take estrogens for valid medical indications with a potential risk of cancer comparable to the self-abuse of overeating or smoking."[6] This physician was unwilling to deny estrogen to women who needed it for hot flashes and genital atrophy; he also mentioned the possible, but as of yet unsubstantiated, role of estrogen in preventing osteoporosis. He dismissed the alleged protective benefit of estrogen against cardiovascular disease as discredited, but, for the valid indications, he concluded that "there is every reason to believe that an effective and safe type of hormone replacement therapy can be made available."[7]

The second commentator was more circumspect in his evaluation of the future of estrogen therapy. He raised several unresolved issues: Was there a difference between the different formulations of estrogen in predisposition to endometrial cancer? Did dosage matter? Would the addition of progesterone to the regimen help the uterine lining to shed and thus prevent the development of cancer? What other treatments might be offered to the menopausal woman suffering from hot flashes? He lamented the uncertain position of physicians who had to weigh the probable adverse effects against the possible beneficial effects of estrogen in deciding whether to prescribe the hormone for their menopausal and postmenopausal patients. Given the murkiness of the situation, the author concluded glumly, "There is little choice but to remain in the dark for a few years more."[8]

Both of the research studies were retrospective, case-control studies that matched a group with a disease (here, endometrial cancer) with a control group that did not have it and compared the proportions of each group who used a drug, in this case estrogen. Retrospective studies were not considered to provide as strong evidence as prospective studies (which followed two groups—one taking the drug, the other not—for several years to see what proportion of each developed the disease in question). Although the investigators had attempted to control for confounding factors (such as obesity and hypertension), many physicians took issue with the reported methodologies. In April the *New England Journal of Medicine* printed some letters criticizing the reliability of the research in the estrogen-endometrial cancer studies. One correspondent postulated that the clinical indications for which the patients received estrogen might have presented some sort of predisposition to the development of cancer, and he faulted the investigators for ignoring this potentially significant factor.[9] Another pointed out that the long latency period between exposure to a carcinogen and the development of a cancerous tumor made it unlikely that relatively recent estrogen use could have been responsible for the endometrial cancers in the study group.[10] A third asked why the pathology specimens on which the cancer diagnoses were based had not been independently reviewed and dismissed the reports as "in the minds of many, poorly controlled and of faulty design."[11]

These letters offered an early glimpse of the division within the medical community into those who believed the findings of the case-control studies and those who did not. Critics picked on several variables that could have skewed the results of the research. Some went further and doubted the ability of retrospective studies in general to identify causal relationships; at best, they argued, a well-designed retrospective study could only show an association between estrogen use and the development of endometrial cancer. These clinicians were reluctant to cast aside what they perceived to be a valuable therapeutic aid just because of a tenuous linkage based on unconvincing data. Their rejection of the statistical data in favor of their own firsthand experience grew into a key issue in the larger debate over hormone replacement therapy, namely, the role of epidemiology in the practice of medicine. How were doctors supposed to incorporate population-based findings into their treatment of individual patients?

The results of the studies published in the *New England Journal of Medicine* in December 1975 were corroborated by two more research reports published in the same journal the following June. One, a case-control study of the female residents of a predominantly white, affluent retirement community near Los Angeles, used medical records, pharmacy records, and interviews to test the hypothesis that estrogen use caused endometrial cancer.[12] The investigators found the risk of developing endometrial cancer to be eight times higher for women who used any type of estrogen and 5.6 times higher for those who used conjugated estrogens. The other set of damning, albeit circumstantial, evidence came from an epidemiological study of eight areas of the United States that showed a sharp increase in the incidence rates of endometrial cancer in the 1970s, especially among middle-aged women (presumably the population most exposed to prescription estrogens).[13]

The concern that estrogen use contributed to rising rates of endometrial cancer applied, obviously, only to women with intact uteruses. The risk of endometrial cancer did not pertain to women who had undergone hysterectomy, the most commonly performed surgery in the United States, regardless of sex, in the 1970s.[14] The annual number of hysterectomies increased from 427,000 in 1965 to a peak of 725,000 in 1975, then decreased slightly to level off at around 650,000 in the 1980s. By 1985, 19.1 percent of all American females over the age of 15 had had a hysterectomy. The large majority (about 65%) of these surgeries were performed on women in their thirties and forties, with another 23 percent performed on women over 50, so that more than a third (37.4%) of women had had their uteruses removed by the age of 60.[15]

The increasing rate of hysterectomy among American women in the 1960s and early 1970s meant that the overall incidence rate of the development of endometrial cancer had in fact increased even more than national surveys indicated, because those surveys were based on the total female population rather than on the much smaller population of females who still had their uteruses. Statistically, ignoring hysterectomized women led to an underassessment of the relative risk of endometrial cancer for nonhysterectomized women. When the at-risk population was corrected for hysterectomy, the occurrence of this disease increased by 20–45 percent, which lent support to the hypothesis that use of prescription estrogen by midlife and older women had contributed to a rise in the number of endometrial cancers.[16] The exclusion of women without uteruses in epidemiological studies of estrogen use also raised questions about the applicability of research findings to this cohort. What did the results mean for women without uteruses? Women who had had hysterectomies, especially those whose ovaries were also removed, were often prescribed estrogen to combat hot flashes and genital atrophy; the younger the woman at the time of the operation, the more likely she was to take the hormone. Could these women continue to take estrogen to relieve their menopausal symptoms? Should they take estrogen to prevent osteoporosis? For the moment, medical attention focused exclusively on the carcinogenic threat to women with uteruses, disregarding this other group of estrogen users.

In the summer of 1976 the *New England Journal of Medicine* published two articles on the effects of estrogen therapy that applied to all estrogen users, with or without uteruses or ovaries. The first addressed the relationship between estrogen and the risk of myocardial infarction (heart attack). Comparing the hospital records of 336 postmenopausal women who had heart attacks with those of 6,730 women admitted to hospital for other reasons, and controlling for confounding factors such as age, diabetes, and cigarette smoking, the researchers found no statistically significant association between estrogen and this particular manifestation of heart disease.[17] In other words, estrogen therapy did not increase a woman's risk of heart attack; equally importantly, the data indicated that estrogen use did not protect women from heart attack. This finding challenged the long-standing hypothesis, based on studies done in the 1950s, that estrogen played some sort of cardio-protective role. Proponents of estrogen "from puberty to grave" had included cardiovascular disease as one of the menopause-related conditions that lifelong hormone replacement could forestall. Now, the wisdom of this recommendation was called into question.

Another of the estrogen advocates' assertions—that hormone replacement did not increase the risk of breast cancer and might even protect against it—was the subject of the second study, published as the lead article in the *Journal*'s 19 August issue.[18] A team of epidemiologists reviewed the medical records of 1,891 female patients of one private practice in Louisville, Kentucky, who had taken conjugated estrogens for at least six months during the period from 1939 through 1972. Although not a true prospective study (which would have followed these women forward from their initial estrogen prescription to see who developed the disease), this investigation did look for the number of estrogen users who developed breast cancer and compared it to the number expected based on rates of breast cancer in the general population. Estrogen users developed 30 percent more cases of breast cancer, for a relative risk of 1.3, which was considered to be barely statistically significant. This seemingly good news was tempered by the more ominous finding that, after ten years of follow-up observation, women seemed to lose the protective effects conferred by oophorectomy (removal of the ovaries) and nulliparity (having no children). Furthermore, estrogen use doubled the risk of breast cancer development in women who already had benign breast disease; women who developed benign breast disease after they began estrogen therapy had a seven times higher risk of developing cancer. Thus the results of this investigation mirrored those of the myocardial infarction study: while estrogen did not seem to be a direct cause of breast cancer, it certainly did not protect against development of the disease.

Noticeably absent from the controversy about the health effects of estrogen was the voice of its staunchest supporter, Robert Wilson. Indeed, after the publication of *Feminine Forever* in 1966, there is little trace of Wilson in either the medical literature or the popular press. No records exist to document the remainder of Wilson's life after the peak of his influence in the mid-1960s. Wilson would have been 80 years old in 1975; it may be that he had retired from his professional career and public life. It may be that he chose, or was pressured, to retire from the debate because of evidence collected about his research foundation by the American Medical Association's Department of Investigation.[19] Although no formal charges were ever filed, Wilson's proselytizing campaign caused some concern in the nation's largest medical society. It may also be that a new generation of clinical scientists demanded more rigorous standards in medical research. Those trained in the decades after World War II learned a healthy respect for statistics and came increasingly to rely on the participation of biostatisticians in the design and evaluation of clinical studies.[20] Less tolerance for shoddy experimental methods meant less acceptance of conclusions and recommendations based on data collected from those experiments. Physicians like Wilson continued to see patients in medical practice, but their amateur investigations were no longer recognized as valid contributions to modern medical research.

The reports on estrogen and its associations with endometrial cancer, breast cancer, and myocardial infarction accentuated the distance between epidemiological research and everyday medical practice. What was the vast majority of the medical profession—especially those general practitioners, internists, and obstetrician-gynecologists who saw menopausal and postmenopausal patients in clinics and private practices—to make of this research? Several professional journals hurried to give advice to their readers, publishing editorials that evaluated the significance of the latest findings and offered recommendations for clinicians. Most of these articles counseled prudence in balancing the very real benefits of estrogen against its potential risks. A new journal called *The Female Patient* led its March 1976 issue with an editorial that reassured physicians that they had both the tools and the knowledge to make informed decisions about their patients' care. In "Estrogen Controversy: A Rational Approach," the editorial director, Hugh R. K. Barber, wrote, "It is for practitioners to determine which of their patients will really be benefited by estrogen, and which the most endangered."[21] However, he conceded that "years may pass before large enough studies of large enough groups of women can be carried out and fully evaluated."[22] Abraham Lilienfeld, professor of epidemiology at Johns Hopkins University, came to a similar conclusion in his editorial in *Postgraduate Medicine: The Journal of Applied Medicine for Physicians Providing Primary Care:* "Clearly, further studies are necessary in order to determine properly the overall balance between benefits and risks of estrogen use. In the meantime, the physician must take a prudent course of action."[23]

Robert W. Kistner, an obstetrician-gynecologist at Harvard Medical School, was much less circumspect in his appraisal of the estrogen-endometrial cancer link. In an editorial for *Obstetrics and Gynecology*, he announced, "From studying the effects of estrogen, progesterone, and other progestins on the endometrium in animals and women for 20 years, I am convinced that estrogens *per se* are not a cause of endometrial cancer in the human female."[24] Although he began with the caveat that "this statement is a firm personal judgment," he went on to state, "I believe that I express the opinions of many members of the American College of Obstetricians and Gynecologists. We refuse, pending further pathological and statistical study, to accept the conclusions of two recent papers."[25] Kistner rejected the conclusions of the epidemiological research, arguing instead that the increased incidence of endometrial cancer in estrogen users could be attributed to preexisting carcinomas or to misdiagnosis (the misreading of hyperplasia as carcinoma) in women who consulted their doctors about uterine bleeding due to continuous estrogen administration. In a rather snide dismissal of women's concerns and physicians' diagnoses, he concluded, "Endometrial cancer is a disease of those women who have easy access to physicians."[26]

The American College of Obstetricians and Gynecologists (ACOG) did not go as far as Kistner did in condemning the recent reports. Its technical bulletin on estrogen replacement therapy, issued in October 1976, sat squarely on the fence: while "no firm conclusions are warranted regarding the carcinogenicity of estrogens . . . the exercise of caution . . . is advocated."[27] ACOG leaders recognized the diversity of opinions and the general state of confusion among its members on the practical applications of the research findings; its annual clinical meeting the following spring featured a session titled "The Great Debate—Estrogen Replacement Therapy for Postmenopausal Women," with Nathan Kase advocating continued use of hormones and Roy Hertz arguing for a more conservative approach.[28] Unfortunately, no transcript exists, so both the substance and the outcome of the debate remain unknown. However, the prominent position given to this session in the meeting program suggests a high level of interest in the topic among practicing obstetricians and gynecologists. These doctors faced the modern task of serving the needs of their patients while keeping up with the cutting edge of their specialty.

Two more case-control studies published in the spring of 1977, one in the *American Journal of Obstetrics and Gynecology* and the other in *Obstetrics and Gynecology*, lent support to the indictment of estrogen in the elevation of the risk of endometrial cancer. The first study reviewed all the cases of endometrial cancer in Olmsted County, Minnesota, from 1945 to 1974 and found that women who had used estrogen for at least six months had a 4.9 times higher risk of developing cancer than the nonuser controls. For women who took estrogen for more than three years, the relative risk rose to 7.9.[29] The second used the records of a single private practice in Louisville, Kentucky, matching 205 patients with endometrial cancer with 205 controls who had hysterectomies for benign uterine disease. Estrogen users were 3.1 times more likely than nonusers to have endometrial cancer. This investigation also found that the risk increased with increasing duration of use; women who had taken estrogen for more than ten years had an 11.5 times greater risk.[30] Both studies reported that higher doses of estrogen led to higher relative risks of endometrial cancer.

Although the evidence appeared to be damning, investigators hesitated to express wholesale disapproval of estrogen. The Louisville team tried to weigh the danger of endometrial cancer against the improved quality of life afforded to many women by estrogen therapy. They estimated that the proportion of hysterectomized women would approach 50 percent in the near future, removing half the population from any risk of the disease. They also noted the relatively high cure rate for endometrial cancer (70.9 percent survived five years and 55.8 percent survived ten years) and the potential, with early diagnosis, for the cure rate to reach 95 percent.[31] Their implication that hysterectomy (either before starting hormone therapy or after the detection of a tumor) provided a viable solution to the dilemma of whether or not to administer estrogen to an older female patient reveals a curious disregard for the inherent value of a postmenopausal woman's uterus. It also indicates the extent to which menopause had become medicalized: these doctors did not question the widespread practice of major abdominal surgery among older women, but instead condoned the procedure as part of the risk-benefit calculus for estrogen replacement.

In the fall of 1977, the *Journal of the American Medical Association* (*JAMA*) published a special review article on the benefits and risks of estrogen therapy for postmenopausal women. The American Medical Association (AMA) draws its members from every medical specialty; in the late 1970s, the organization represented 45.5 percent of the doctors in the United States (approximately 212,000 of 467,000 total).[32] Physicians who belong to the AMA receive a free subscription to *JAMA* as a membership benefit; many nonmembers read copies borrowed from colleagues or in libraries. An article in the most widely disseminated medical journal in the country was ensured a broad audience.

The authors evaluated the putative benefits and risks of low-dose estrogen replacement therapy for both short-term and long-term indications.[33] They confirmed that estrogen was highly effective in relieving two major symptoms of menopause: hot flashes and genital atrophy. They dismissed claims that estrogen could improve emotional health as "unpredictable and difficult to evaluate objectively."[34] After consideration of both clinical and laboratory investigations into the possibility that estrogen might prevent atherosclerosis, they rejected both the alleged problem ("the decreased estrogen secretion of postmenopausal women does not constitute an additional risk factor for CHD [coronary heart disease]") and the proposed solution ("estrogen treatment of postmenopausal women does not reduce the incidence rate of CHD").[35] They also credited estrogen with preventing bone loss in young women with surgically induced menopause (removal of both ovaries) and with reducing—with short-term treatment only—bone deterioration in older women with osteoporosis; however, they presented a mixed bag of evidence both for and against the prophylactic use of estrogen to forestall the development of osteoporosis in postmenopausal women.

On the risk side of the equation, they reviewed studies on breast cancer, endometrial cancer, myocardial infarction and stroke, gallbladder disease, thromboembolic (blood clotting) disease, and hypertension. Only endometrial cancer and gallbladder disease were more likely to develop in estrogen users as compared to nonusers; the rest of the data was too ambiguous to make a definite determination. The authors also commented on the range of unpleasant side effects experienced by estrogen users, such as uterine bleeding, weight gain, breast tenderness, edema, nausea, and heartburn.

In the final analysis, the authors recommended low-dose, cyclic estrogen therapy for menopausal women suffering from hot flashes, provided the patient was able eventually to wean herself off the medication. They also condoned short-term (one to three years) estrogen therapy to slow down the rate of bone loss in selected patients. The only group for whom they recommended longer-term estrogen therapy was young women in premature or surgical menopause. Significantly, they advised against the use of estrogen in asymptomatic postmenopausal women. In other words, they rejected the prophylactic prescription of estrogen to prevent or forestall the development of age-related diseases.

The drug companies paid heed to this change in medical opinion and revised their promotional materials accordingly. Just a few months after the *New England Journal of Medicine* articles of December 1975, Ayerst moved quickly to backtrack from its earlier far-reaching promises for the therapeutic advantages of estrogen. In 1976 Premarin's ad campaign was reduced to three pages from the eight-page spread that ran the previous year. Gone were the testimonials from patients who complained, "I was always so irritable . . . snapping at the children . . . almost anything could annoy me. I knew I was doing it, but I couldn't seem to stop myself."[36] The new ad listed only hot flashes and night sweats as indications for Premarin.[37] Instead of Premarin "for the emotional symptoms of the menopause related to estrogen deficiency," the tag line now read "for the classic symptoms of the menopause."[38]

Another four-page spread for Premarin devoted the entire first page to the following quotation from Barber's editorial in *The Female Patient:* "The height of professional integrity is to prescribe estrogen when needed and indicated, and to withhold it if there is no indication or when a contraindication to its use exists."[39] The subtext of this strategy could be read as an effort to ally the manufacturer with the physician, against the sensationalist media and hysterical women: the advice to withhold estrogen implied that wise and well-informed doctors could and should ignore their patients' misinformed entreaties for the so-called "youth pill." A campaign that ran in 1979 reinforced the central role of the physician in treating menopausal patients: "The one ingredient we can't put in Premarin . . . is your counsel." The text went on to underscore the enduring scientific legitimacy, and clinical necessity, of estrogen: "Because the effectiveness of Premarin in controlling moderate to severe vasomotor symptoms due to estrogen deficiency remains unchallenged, it's more important than ever that patients obtain this benefit with the fewest possible problems."[40]

Readers flipping through the pages of the *JAMA* issue with the special review article on estrogen would also have come across a multipage advertisement for Premarin, which exhibited the manufacturer's modified claims for its product.[41] This ad began with a full-page photograph of an older, gray-haired woman clad in a cardigan sweater looking dejectedly out the window of her large, well-appointed home. The caption asked, "*THE MENOPAUSE*: Does she just have to live with it?" The next page explained that Premarin was "effective treatment" for "distressing menopausal symptoms," namely, flushes and sweats, atrophic vaginitis, and postmenopausal osteoporosis. The third page, illustrated with a male, white-coated physician consulting with a female patient in his office, bolstered the confidence of physicians by telling them, "You'll know in a matter of one or two weeks if Premarin helps . . . You know the principles of good management . . . You'll know when it's time to stop." The next page featured a beautiful, smiling woman; thanks to the physician's judicious prescription of Premarin, "She doesn't just have to live with it." This advertisement made it clear that there was still an important place for Premarin in medical practice, for the treatment of uncomfortable menopausal symptoms.

Ayerst also made sure to remind clinicians of the therapeutic value of Premarin for young women after removal of their ovaries. One photograph of an abdominal incision was followed by another of a young mother with her young son; the caption read, "Surgical menopause: The sign . . . without the symptoms." After several paragraphs of text explaining why and how to use Premarin to treat oophorectomized women, the ad concluded: "PREMARIN. When the menopause is premature . . . and when it isn't."[42] This drug company was not about to take the bad news about estrogen lying down; it responded with a carefully coordinated advertising campaign that acknowledged the circumscribed indications for post-menopausal estrogen use and, at the same time, reminded its customers—the physicians who wrote the prescriptions—of the millions of women who still fit the profile of a Premarin patient.

Other manufacturers seized the opportunity to tout the benefits of their estrogen products over Premarin, the clear market leader. Organon proclaimed the "end of the age of 'natural' estrogens" in introducing Genisis, its new synthetic brand of conjugated estrogens. A horse grazing in a field represented the old-fashioned way of making hormones; a winged Pegasus indicated "the dawn of a new age" of conjugated estrogens "free of extraneous natural residues." This ad tried to dispel the myth of the "natural" that surrounded urine-derived estrogens such as Premarin; however, it incorporated similar rhetoric in its description of the product's "snow-white pure *uncoated* tablets."[43] Mead Johnson also emphasized the superiority of the source of its new Estrace, "the major *human* estrogen micronized for oral effectiveness." Produced in the laboratory, this product left "no 'barnyard' aftertaste," as some conjugated estrogens apparently did.[44] These ads hearkened back to those estrogen ads of the 1940s that focused on the product, not on the patient.

Ayerst probably feared the competition from other manufacturers less than the negative publicity from the medical reports linking estrogen to endometrial cancer. In 1976 alone, dollar sales for menopausal estrogens decreased 11 percent and the number of prescriptions dropped 19 percent.[45] Oral contraceptives (combined estrogen-progestin products) also sustained a significant drop in sales, owing to widespread press coverage of the pill's adverse health effects. However, other steroid hormones were thriving; sales of corticosteroids, for example, rose considerably during the same period.[46] Estrogen, it appears, was suffering from a damaged reputation.

The assault on estrogen did not sit well with some members of the medical community, who took it upon themselves to advocate on its behalf. Robert B. Greenblatt, a well-known endocrinologist and professor emeritus at the Medical College of Georgia in Augusta, penned two articles in defense of estrogen therapy for postmenopausal women, one in *Geriatrics* in November 1977 and another in the *Journal of the American Geriatrics Society* two months later.[47] By publishing in these journals, he hoped to appeal to physicians whose patients were candidates for long-term hormone replacement therapy.

Greenblatt criticized the data supporting a causal association between estrogen and endometrial cancer. First, he raised the issue of ambiguity in diagnosing the early stages of endometrial cancer. Second, he questioned the design of the experiments and selection of controls, writing off the validity of the statistical studies as "garbage in, garbage out."[48] Third, he pointed out that incidence rates of endometrial cancer rose not only in the United States but also in Norway and Czechoslovakia, where few women took estrogen. He acknowledged that estrogen could stimulate the growth of tumors, but he claimed that the addition of progestin (synthetic progesterone) for five to seven days to the monthly regimen would prevent endometrial carcinoma by causing the uterus to shed its lining (this proposal is discussed at length in chapter 8). The benefits of combined hormone therapy, he insisted, far outweighed any "putative" risks. Most notably, osteoporosis carried a much higher risk of death than endometrial cancer, and the potential for collapsed vertebrae and fractured hips posed a grave threat to the health of older women. Greenblatt castigated the authors of the *New England Journal* studies for "unduly alarm[ing] physicians and women the world over," and he beseeched his fellow physicians not to abandon hormone therapy for patients "in need of it."[49]

Two Yale biostatisticians entered the debate with a critical review of case-control methods as applied to the study of estrogen and endometrial cancer. Their report was published as the lead article in the 16 November 1978 issue of the *New England Journal;* in the interest of full disclosure, the authors listed Ayerst Laboratories as one of their financial supporters.[50] The authors, Ralph Horwitz and Alvan R. Feinstein, performed two studies at Yale–New Haven Medical Center, one using conventional methodology to select cases and controls and the other using an alternative approach. They found that the former resulted in a relative risk of almost 12 for estrogen users as compared to nonusers, while the latter generated a relative risk of just 1.7. The discrepancy, they contended, could be attributed to a selection bias that produced a falsely elevated risk for estrogen users. This report successfully passed the peer-review process, but the *Journal*'s editors felt it necessary to accompany the article with an editorial that questioned the validity of Horwitz and Feinstein's assumptions and conclusions. The authors of the editorial, a physician and a scientist from Harvard's School of Public Health, pointed out that the Yale investigators' method of choosing cases and controls produced a different selection bias, one that resulted in a falsely *low* relative risk.[51]

The letters published in March 1979 in response to the article and the editorial represented a mixed bag: some supported the findings of the article, some backed the logic of the editorial, some were unconvinced by either side.[52] One author called the controversy "disturbing" and postulated, perhaps only partly in jest, that the dispute was really a Yale-Harvard contest in disguise.[53] This particular debate notwithstanding, the pendulum had swung decisively in favor of a causal relationship between prescription estrogen and endometrial cancer.

In January and February, the *New England Journal* published two more large studies that offered convincing corroboration of the estrogen-cancer link; these articles were followed by a third in *JAMA* in July. The first, a large retrospective study conducted in the Baltimore area, directly tackled the charges of biases and other design flaws in case-control methodology. The investigators considered and ultimately set aside the criticisms, claiming instead that their study confirmed previous findings. They reported a sixfold increase in risk of developing endometrial cancer among estrogen users as compared with nonusers, and a fifteen-fold increased risk for those who used estrogen for more than five years.[54] The second study reported that a decline in the incidence of endometrial cancer among women in a large group health plan in the Seattle area (from July 1975 to July 1977) paralleled a decrease in the number of prescriptions written for estrogen.[55] The good news was that a woman's risk of developing endometrial cancer began to drop upon cessation of replacement therapy; this result was upheld by the third study, which also found elevated risks for both conjugated and other forms of estrogen and for both cyclic and continuous use among women in King County, Washington. These investigators recommended only short-term (less than one year) estrogen therapy to relieve menopausal symptoms. While long-term, albeit low-dose, estrogen therapy had been promoted as the best way to stave off the effects of osteoporosis, they warned that use of estrogen for longer than three years greatly increased the risk of cancer. However, they refrained from total condemnation of long-term preventive therapy, calling instead for more studies to sort out the risk-benefit calculus of estrogen use and these two diseases.[56] The authors commented that there were "few who doubt[ed] that women who use noncontraceptive estrogens are at risk for endometrial cancer."[57] Also, by mid-1979, there were few who would advocate without reservation the long-term prescription of estrogen for postmenopausal women.

After almost four years of heated debate, the Medical Intelligence section of the *New England Journal of Medicine* reviewed the current thinking on postmenopausal estrogen in an article titled "Estrogen-Replacement Therapy—Help or Hazard?" The two authors, from the Division of Reproductive Endocrinology and the Department of Obstetrics and Gynecology at Duke University in North Carolina, began by acknowledging that "few controversies in the practice of medicine have generated as much discussion as has the role of estrogen-replacement therapy in the climacteric woman."[58] They went on to rehash the evidence for and against the risks and benefits of estrogen use; anyone who had kept up with the literature in recent years would have found nothing new in this article. On the other hand, a medical Rip Van Winkle who fell asleep a decade earlier would have been stunned to read the advice on estrogen as of September 1979. After enjoying several years of "feminine forever" promises and "puberty to grave" recommendations, estrogen returned to the place it had occupied in the medicine cabinet of the 1940s and 1950s, as a short-term palliative prescribed only for the severe menopausal symptoms of hot flashes and genital atrophy. Along with the acceptance of a causal relationship between estrogen use and endometrial cancer came a refutation of earlier claims that estrogen relieved psychological symptoms or prevented atherosclerosis and an ambivalence about the merit of estrogen in treating or forestalling osteoporosis. Although many others had said the same thing many times before, the authors felt compelled to remind their readers, "No drug will prevent the natural course of aging; estrogen should not be given in a futile attempt to do so."[59]

The *New England Journal* review appeared in the same week that the National Institutes of Health (NIH) held a conference to develop a consensus report on estrogen use and postmenopausal women. This conference was part of a program begun in 1977, in which researchers, physicians, and consumers were brought together to discuss and to reach an agreement on the safety and effectiveness of various drugs, devices, surgical procedures, and medical practices. The seventeen previous conferences had addressed issues such as breast cancer screening, dental implants, and supportive therapy in burn care.[60]

The goal of this Consensus Development Conference was to assess the relative risks and benefits of estrogen therapy and to provide some guidance for the practicing physician. Over two days in September, the fifteen-member panel heard reports on the existing therapeutic, epidemiological, and sociological evidence; they engaged in discussion among themselves and with members of an audience that included feminist activists, consumer lobbyists, pharmaceutical company representatives, and government regulators; and they then met privately to develop the consensus statement. According to one account, the deliberations were so involved that the group did not complete its task until five o'clock in the morning.[61]

The resulting document, printed as a seven-page booklet and distributed under the auspices of the Department of Health, Education, and Welfare (the cabinet department that oversaw NIH), came to the same conclusions as did the *New England Journal* review, with a few notable differences. First, the government-sponsored panel hesitated to recommend the addition of progestin to the hormone replacement regimen, because the risks of combination therapy had not yet been studied. Second, and here the influence of the lay participants was clearly evident, the group agreed that "the patient should be given as much information as possible about the evidence for the effectiveness of estrogens in treating specific menopausal conditions and the risks that their use may entail. Patients must be kept continually informed of new findings as they arise."[62] This statement represented a marked departure from the medical literature on estrogen therapy, which rarely considered the patient as a participant in health care decision-making.

Contrary to the wishes of many physicians, the debate over the risks and benefits of estrogen use in menopause and postmenopause did not remain within the medical community. In the wake of the discovery of the link between estrogen use and endometrial cancer, health feminists worked to expand women's access to balanced information on hormone replacement therapy. If women were to be more involved in medical decision-making, they would need more information about the pros and cons of different treatment options. As in previous years, the media functioned as the main disseminators of the latest medical findings and opinions to the public. An understanding of how and what women learned about estrogen after 1975 must begin with a look at the coverage of the estrogen debate in newspapers and magazines and on television.

# CHAPTER SIX

# Enter the Feminists

## Informing Women about Estrogen

Viewers who tuned in to the NBC Nightly News on Thursday, 4 December 1975, learned about the latest witnesses to be subpoenaed in the ongoing case of the kidnapping and murder of Jimmy Hoffa, reported missing four months earlier. Stories followed on a tax bill in the House of Representatives, a Senate Intelligence Committee report on CIA involvement in Chile, the retirement of longtime Republican Senator Hugh Scott, and November's wholesale prices. The main news story that evening was President Ford's visit to the People's Republic of China. The president, along with his wife, his daughter Susan, Secretary of State Henry Kissinger, and two hundred news reporters, had arrived in Peking on the first of December. John Chancellor and Tom Brokaw reported on the final day of meetings between the Chinese and American leaders. After two commercials for microwave ovens and stories on Indonesian terrorists in the Netherlands and Israeli air raids on Lebanon, John Hart (sitting in for the regular anchorman, John Chancellor) reported from the New York studio that women who took estrogen ran five to fourteen times the chance of getting uterine cancer as women who did not. The segment included an interview in Los Angeles with Harry K. Ziel, co-author of one of the *New England Journal of Medicine* articles. Ziel incriminated the Premarin brand of conjugated estrogens and then called for more research. Back in New York, Hart noted that the FDA would decide later in the month whether to increase restrictions on the use of estrogen.[1] NBC was the only network to pay significant attention to the estrogen-endometrial cancer discovery. ABC mentioned it just briefly, sandwiched between reports on over-the-counter sleep aids and an artificial sweetener.[2] Those who preferred the avuncular style of Walter Cronkite on CBS heard nothing about estrogen at all.[3]

Readers of the *New York Times* that day saw, on the bottom right-hand corner of the front page, next to a triptych of photographs of First Lady Betty Ford dancing at a Peking school, the headline, "Estrogen Is Linked to Uterine Cancer."[4] Continued on page 55, the article described the findings of the Seattle and Los Angeles studies, quoting from both Harry K. Ziel and Donald C. Smith, director of the Seattle study. It also gave a brief history of the rise in popularity of estrogen in the 1960s as a long-term replacement therapy, noting that "millions of women—particularly those in the upper socioeconomic brackets—have been using estrogens, not just during the months of menopausal discomfort, but for years afterward."[5] Since many American newspapers subscribed to the New York Times News Service, this story probably appeared widely across the country, although not necessarily on the front page.

The next day, the *Times* health reporter Jane Brody wrote about her spot check of a dozen gynecologists around the country in the wake of the estrogen-endometrial cancer association. All claimed that the news would have little effect on their prescribing habits. Those who prescribed long-term estrogen replacement still believed the benefits outweighed the risks; those who limited estrogen use to short-term relief of menopausal symptoms continued to regard the medication as an important and necessary treatment for many women. Buried in the middle of the article, however, was an interesting revelation: most of the doctors said that "the patient should participate in the evaluation of benefit versus risk and that such participation meant that the women must be told of the possible hazards of estrogen therapy."[6] Although this opinion may not have reflected those held by other American physicians, it did reflect a change in some doctors' attitudes toward their patients.

Over the next four years, nationally televised news broadcasts reported on selected scientific findings as they were released, particularly those that associated estrogen use with cancer. Local stations also got in on the act. For example, Channel 2 News, New York's CBS affiliate, ran a special "Survival Report" on 3 February 1976 called "Estrogen, Birth Control, and Menopause: Are You Taking the Bad with the Good?" In the first part of this two-night report, the news program sent its Washington correspondent to talk to women, doctors, and FDA officials about estrogen therapy.[7] To attract viewers, the station ran a large advertisement that covered almost a quarter of a page in the *New York Times*.[8] The second part of the report looked at oral contraceptives; since the pill contained estrogen, as well as progestin, it also came under scrutiny in the mid-1970s.

The *New York Times* followed the estrogen story fairly closely as it developed over the course of the latter half of the 1970s. In addition to reports on medical studies, it covered the various actions taken by the FDA on the matter of estrogen products.[9] In October 1977 the *Times* reported that the singer Kitty Kallen, who had recorded several hit songs in the 1950s, won $300,000 in her lawsuit against Ayerst Laboratories and the dermatologist who prescribed estrogen for her in 1966. Kallen, hoping to revive her career, had consulted Dr. Norman Orenreich, who gave her Premarin pills to treat the wrinkles on her face. Kallen had a prior history of blood clots, dating from the birth of her son, and she developed more clots after taking Premarin.[10]

A couple of weeks after the Kallen story, Jane Brody wrote a long article, "Why Has Estrogen Fallen on Such Difficult Times?" The subheading read, "Just a Few Years Ago, It Was Hailed as a Miracle for Women." The lesson Brody drew from recent research findings was that "too much of a good thing can sometimes be bad."[11] She blamed both women and their doctors for overly enthusiastic acceptance of estrogen as a postmenopausal cure-all. Two years later, after the NIH Consensus Development Conference on estrogen use and postmenopausal women in September 1979, Brody reviewed the status of "the 'feminine' drug" with respect to cancer, heart disease, osteoporosis, menopausal symptoms (hot flashes and vaginal dryness), emotional problems, and other disorders (gallbladder disease and uterine bleeding). She stressed the NIH panel's call for women themselves to make the decision about whether to take estrogen; "an informed choice by the woman is preferable to a dictate from a physician who, after all, does not have to live the woman's life."[12]

Magazine writers also became more cautious in their approach to estrogen, as they charted its decline in status from miracle drug to potential carcinogen. Readers of women's magazines in the winter of 1976 encountered articles with titles such as "Are Hormones Giving You Cancer?," "Can Estrogen Therapy Cause Cancer?," and "What's behind the Scary Rise in Uterine Cancer."[13] *Consumer Reports* called its exposé "Estrogen Therapy: The Dangerous Road to Shangri-La."[14] By the end of the decade, several articles in mass media publications had called into question the safety of estrogens.[15]

People who read magazines for information about estrogen found more than just the latest medical findings; many of the articles also gave advice on what menopausal and postmenopausal women should do, and these recommendations varied significantly from those given prior to 1975. Whereas the readers of *Harper's Bazaar's* 1973 "Over-40 Guide on Health, Looks, and Sex" were told, "It is up to your doctor to decide . . . whether you should have estrogen and if so how much, how often, for how long," articles a few years later encouraged women to become active participants in the decision whether or not to treat menopause with hormones. A 1977 article in *McCall's*, titled "Estrogen: The Rewards and Risks," asked, "On what basis can a woman make her decision about the use of estrogen therapy if she cannot rely upon her physician's knowledge?" It went on to reply, "Fortunately, for a woman to educate herself about menopause and the use of estrogen replacement therapy is not all that difficult."[16] In 1979, *Harper's Bazaar* posed a similar question: "If a woman cannot rely on the knowledge and judgment of her physician, how is she to decide what is the best treatment to follow?"[17] The rest of the article explained the role of hormones in the female reproductive cycle and the benefits and risks of hormone replacement therapy. It concluded by advocating that doctors and patients work together as partners in determining the best course of action: "Doctors prescribing estrogen replacements in spite of remaining questions should certainly inform their patients of the possible hazards they face. If you decide to take estrogen replacements, insist that you have the smallest dosage during the shortest span of time." In 1973, the balance of power in the doctor-patient relationship lay with the physician; by the end of the decade, it appeared that female patients, encouraged by journalists to arm themselves with information, had begun to tip the scales, certainly not entirely in their favor but at least toward a more equitable distribution of influence.

This shift in journalists' treatment of estrogen and other menopause-related topics reflects not only the change in medical consensus but also the emerging influence of the women's health movement, an important outgrowth of the broader women's movement of the 1960s. As a result of feminist initiatives (with contributions from the consumers' movement and the patients' rights movement as well), the equation that knowledge equals power was applied to the doctor-patient relationship. All patients, not just women, were urged to become better informed, so that they could be active participants in, not passive recipients of, health care and medical treatment. The women's health movement of the 1970s contributed to a change in the tone and message as well as an expansion in the quantity and variety of information sources on menopause, postmenopause, and hormone replacement therapy.

The women's health movement was part of the women's movement that took shape in the United States in the 1960s. This second wave of feminism (so called to acknowledge the first wave of feminist activism in the early twentieth century, whose best-known accomplishment was the passage in 1920 of the Nineteenth Amendment, which gave women the right to vote) challenged gender-based discrimination and injustice in politics, economics, business, law, medicine, society, and culture. Women rose up to demand their equal rights in the Constitution (endorsing the Equal Rights Amendment), in the workplace (arguing for equal pay), in universities (advocating for the abolition of quotas in admission to schools of law and medicine), and in the home (insisting that their husbands participate in housekeeping and childcare). The movement drew strength from the spirit of the 1960s, a decade in which so many social institutions and cultural conventions were subject to intense questioning.

The origins of second wave feminism can be traced to the dissatisfactions of two distinct groups of women.[18] The more conservative wing of the women's movement resulted from the repeated setbacks faced by professional working women. It was spearheaded by the writer Betty Friedan, whose 1963 best seller *The Feminine Mystique* identified housewives' malaise as "the problem that has no name" and criticized the media for its prescriptive imagery of women's role in society. Friedan founded the National Organization for Women (NOW) in 1966, which worked to achieve equal rights and equal opportunities through legislative and judicial channels. The women of NOW and other mainstream groups, such as the Women's Equity Action League and the National Women's Political Caucus, sought to reform the existing system, to make a place for women in all aspects of public life.

The more radical branch was born out of the frustrations experienced by younger women active in the civil rights movement and the student movement.[19] Repeatedly forced into subservient positions in male-dominated organizations such as the civil rights group Student Non-violent Coordinating Committee (SNCC) and the student group Students for a Democratic Society (SDS), these activist women realized that they were experiencing their own version of oppression when their male counterparts asked them to make the coffee or type up the meeting notes or provide sex on demand. Angered by future Black Power leader Stokely Carmichael's pronouncement that "the position of women in SNCC is prone," which summed up the gender inequity among civil rights workers, these female activists began to call for women's liberation. This branch of the women's movement instigated a more trenchant evaluation of male-female relations, challenging the traditional assumptions of patriarchal society in private life as well as in public life and arguing that "the personal is political."

It is important to recognize that the divisions between the reform and radical wings were often blurred. As Susan Brownmiller has observed, the mere coming together of small-town women to talk about their experiences as women was considered an act of radicalism, while several women who became prominent in the radical movement in large urban centers got their starts as members of NOW.[20] A striking characteristic of the women's movement was the sheer number and variety of activities and efforts that were spawned in its name in the late 1960s and 1970s: rallies and marches, strikes and sit-ins, childcare and housework cooperatives, self-defense classes, rape and domestic violence crisis centers and hotlines, rock bands and art exhibits. Perhaps its most significant creation was the consciousness-raising group, in which women gathered to share their personal experiences. These personal experiences, in turn, served as the basis for women's claims for social and cultural change.

In many of these consciousness-raising groups, which sprang up all over the country in the late 1960s, the discussion turned to women's bodies. Some became particularly interested in issues of women's health and decided to focus exclusively on that subject. One group of women who had met at a workshop on women and their bodies at a women's liberation conference in Boston in 1969 decided to do their own research on topics such as women's anatomy and physiology, sexuality and relationships, pregnancy and childbirth, birth control and abortion. After several months of research and dozens of meetings, the group, initially called the Boston Women's Health Course Collective, began to offer their course to other women. The course, and the research papers on which it was based, became so popular that the Collective engaged the nonprofit New England Free Press to print the papers as a book. In December 1970, *Our Bodies, Ourselves* was born.[21]

Three thousand miles away, in Los Angeles, a housewife and mother of six named Carol Downer joined her local chapter of NOW and began to work to legalize abortion in California. In the course of learning about abortion procedures, she became frustrated by the monopoly held by physicians over knowledge of women's bodies. Using a mirror, a flashlight, and a plastic speculum, Downer explored her own body and found her own cervix. In April 1971, she took off her pants, lay on a table, and demonstrated cervical self-examination to a group of thirty women who had gathered at a bookstore to talk about reproductive control and reproductive rights. A few months later, Downer and Lorraine Rothman, who had developed a menstrual extraction kit so women could control their periods, went on the road, traveling to twenty-three cities across the country to demonstrate their techniques and share their knowledge. The movement for self-help soon led to woman-controlled health care when, the following year, Downer and Rothman opened the Los Angeles Feminist Women's Health Center.[22]

In Washington, D.C., another group of women had been discussing the problem of illegal abortion in particular and the relationship of women to the health care system more generally. In January 1970 several members of this collective, called D.C. Women's Liberation, heard that a Senate subcommittee was holding a hearing about the safety of the birth control pill, so they went up to Capitol Hill to see what was going on. When they got there, they were appalled to find that all of the senators were men and all of the people testifying were men. With no female scientists and no female pill users on the program, the women of D.C. Women's Liberation sitting in the audience took it upon themselves to raise their hands and ask questions. After the leader of the hearing, Senator Gaylord Nelson, kicked them out of the room, they met Barbara Seaman, the New York journalist who had written the book that had prompted the hearings in the first place.

Seaman's publisher, Peter Wyden, had brought her book, *The Doctors' Case against the Pill*, to the attention of Senator Nelson, chairman of the Senate Subcommittee on Monopoly of the Select Committee on Small Business, which had been conducting investigations of the pharmaceutical industry since 1967. Many of the doctors Seaman had interviewed for her book were called as expert witnesses to testify before the Senate subcommittee, although neither Seaman nor any of the women she had interviewed were invited to participate. This gender imbalance infuriated the members of D.C. Women's Liberation. After their initial spontaneous outburst, the women quickly organized themselves and held militant demonstrations at every subsequent day of the hearings on the pill. When their questions remained unanswered, they decided to hold their own hearings. Their quarrel extended beyond the immediate problem of the safety of the birth control pill to encompass the entire existing system of traditional medicine. The members of D.C. Women's Liberation rejected reformist feminism because it did not address what they perceived as the underlying ills of a patriarchal system. These feminists wanted women to participate as equal partners in the provision and consumption of health care services.[23]

A group of women in Chicago took a giant step in this direction, moving beyond meetings and demonstrations to direct action in 1969 when they started "Jane," an organization that helped women get safe, albeit illegal abortions. For these feminists, the marches and lawsuits of reformist organizations such as the National Association for Repeal of Abortion Laws (NARAL) offered little to women with unwanted pregnancies who needed immediate solutions. Initially, members of the Jane collective acted as go-betweens to coordinate the process of finding reputable underground abortion providers. They referred several women to one practitioner who was willing to work with Jane. It turned out that this man was not a physician, as he professed to be. Realizing that the technical steps of abortion did not require a medical degree, some members of Jane decided to learn to do the procedure themselves. The Jane collective eventually took over the entire process of abortion provision: scheduling services, counseling women, transporting them to and from the abortion site, doing the abortions, monitoring postprocedure care, and, above all, supplying their clients with complete information at every step. For four years, until the service disbanded after the Supreme Court's *Roe v. Wade* decision legalized abortion in the United States in 1973, Jane helped some eleven thousand women get the abortions they wanted.[24]

All of these events and accomplishments, taken together, make up what historian Sandra Morgen has called some of the key foundational stories of the women's health movement.[25] Although the women's health movement paralleled the broader women's movement in terms of the diversity of interests and activities of its constituents, all of its efforts shared a commitment to informed medical consumerism. Both the woman-run health centers that arose from the self-help movement and the multiple editions of *Our Bodies, Ourselves* emphasized the sharing of knowledge. The members of the Jane collective worked hard to ensure that their clients were kept informed at every stage of the abortion process. Barbara Seaman dedicated her writing career to learning about health care and communicating that information to other women. In the early 1970s she met Belita Cowan, who had been researching the effects of diethylstilbestrol (DES) in Ann Arbor, Michigan. Seaman and Cowan, along with Alice Wolfson of D.C. Women's Liberation, realized that the interests of women in health care would be best served by a Washington-based organization that could monitor federal health agencies and lobby on Capitol Hill. In 1975 these three feminists joined forces with Phyllis Chesler, a psychologist, and Mary Howell, a physician, to found the National Women's Health Network. The Network soon expanded its mission to become an information clearinghouse, dedicated to educating both Congress and the general public about women's health issues.[26]

Feminist health activists blamed estrogen for causing too much unnecessary death and disease. In 1968 researchers had linked the estrogen in birth control pills with potentially fatal blood clotting. In 1971, DES, which had been given to women to prevent miscarriage since the 1940s, was found to affect some of the daughters with vaginal cancer and reproductive tract abnormalities. Four years later, when the news broke about the link between menopausal estrogen therapy and endometrial cancer, adding yet another adverse health effect to estrogen use, feminists vented their anger at the medical profession for overprescribing these harmful hormones, the federal government for underregulating them, and the pharmaceutical industry for misrepresenting them in medical advertising. The first public action taken by the National Women's Health Network was a demonstration, which they called a memorial service, held on the steps of the Food and Drug Administration building in Rockville, Maryland, to commemorate the women who had died estrogen-related deaths from taking the pill, DES, or HRT. The visibility of the women's health movement and the dissemination of its message (by both feminist activists and journalists influenced by feminism) that "knowledge is power" encouraged many women to look beyond their doctors for information about health care issues.

One woman who sought information about menopause ended up writing a book to help educate others. Rosetta Reitz came to research menopause out of frustration, because her doctor was "too busy" to answer her questions and the few books she was able to find on menopause portrayed it negatively and simplistically, as resulting from either physiological hormone deficiency or the social stresses of being middle-aged.[27] An active member of New York Radical Feminists, Reitz gathered midlife women at her home for weekly consciousness-raising sessions; those intimate discussions—about menopause, aging, love, sex, health, work, children—provided Reitz with the information she wanted to share with a broader audience.[28] Sustained by the energy of the women's movement and her own personal interest, Reitz published *Menopause: A Positive Approach* in 1977.

Reitz chose to write a book about what she had learned because writing and books had always played a central role in her life. Born in 1924 in upstate New York, she was the sixth and youngest child of Polish Jews.[29] In spite of the hardship of losing both her parents by the age of 13 and the disruption of World War II, Reitz entered the University of Wisconsin at Madison in the early 1940s, where she majored in sociology. After graduating, Reitz moved to New York City and got a job at the Gotham Book Mart on West 47th Street, a celebrated gathering place for American literati. After two years, she opened her own bookstore in Greenwich Village. Called the Four Seasons Bookshop, the store attracted well-known authors such as Anaïs Nin, Ralph Ellison, Saul Bellow, and e. e. cummings.

In the early 1950s, Reitz did what so many other American women did in the postwar years: she got married and had children (three) in quick succession. However, unlike most of her middle-class contemporaries, she continued to work. After a dozen years of marriage, when her daughters were eight, nine, and ten years old, Reitz left her husband. Although she loved the bookstore, she needed to earn a better living, so in 1965, at the age of 41, she went to Wall Street to learn the stock market trade. Rejected by every training program because she was "too old" (and too female), she worked as a clerk, studied on her own time, and earned her broker's license. Shut out of the old-boy's network of the investment world, Reitz found it difficult to make a steady income as a stockbroker. She began writing for the *Village Voice*, an alternative newsweekly. Initially, she contributed pieces on jazz and food; her experience in writing the Dining In/Dining Out column helped her to publish a book called *Mushroom Cookery* in 1965. Reitz later moved over to run the classified ads department; it was this position that enabled her to begin her quest for information about menopause.

Reitz had learned the techniques of consciousness-raising by attending meet-ings of New York Radical Women, starting in 1969. Younger members of the group were not particularly interested in menopause, so Reitz sought out older women. The classified ads in the *Voice* ran on a page called the "Bulletin Board." When the Bulletin Board had an empty space, Reitz entered her own advertise-ment: "If you're a woman interested in talking about menopause, call . . . " She listed her direct telephone line at work, and soon the calls came in. Reitz held a series of "menopause workshops" in her apartment; as the group grew in num-ber, other women volunteered their homes for meetings. Women came from all over the area, from New Jersey and Brooklyn and the Bronx, as well as Manhattan. Some women were well-to-do, others were not. Some were well-educated, others were not. However, the one thing all of the women had in common was that they were all white; not a single African American chose to participate. The absence of women of color in these self-selected menopause discussion groups mirrored the racial divisions that plagued the larger women's movement as well.

Reitz drew her information about menopause from the combined experiences of hundreds of women and her inspiration from the ideology of the women's health movement. The first three sentences laid out her premise: "This is a revolutionary book. It looks on menopause as a positive experience. It is revolu-tionary because its emphasis is exactly opposite to what we have been taught, consciously and unconsciously, about our bodies."[30] Later in the book, she ex-plained what feminist health activists were trying to accomplish: "The self-help movement is not seeking to replace the medical establishment, just change it. The first step is for women to reclaim their own bodies; next, to remove fear through information. An informed medical consumer is a more dignified one."[31] Reitz railed against the scary, negative language used by male physicians and scientists to describe menopause as "linguistic conveniences derived from a narrow scien-tific conception imposed on our experience by people who are observing it preju-dicially, from a distance."[32]

In order to help midlife women take back their bodies and the language used to describe them, Reitz sought to demystify menopause and its myriad symptoms. She disputed the notion that menopause was a disease, counseling her readers instead to accept physiological and psychological changes as a natural part of the life cycle. She also rejected the idea that women became estrogen deficient at menopause, devoting an entire chapter to the dangers of hormone replacement therapy. Reitz's alternative prescription was simple: good nutrition, physical exercise, regular sex, and healthy relationships. And to counter the "cultural invisibility and cultural undesirability" heaped upon older women, she counseled self-love and self-affirmation. Reitz hoped that women would embrace their menopause and grow with it, rather than dreading the transition and undertaking desperate, medicalized measures to stave off its effects.

At first, Reitz had trouble getting her book published; several editors tried, unsuccessfully, to convince her to change the title to something more euphemistic. Reitz insisted on *Menopause: A Positive Approach*, reasoning that the word *menopause* had to be "brought out of the closet" in order for people to stop fearing it. Once it was on the market (published by Chilton Book Company), the book was selected by the Book-of-the-Month Club, which helped boost sales; later, Penguin Books brought it out in paperback, keeping it in print for twenty years. The Boston Women's Health Book Collective liked the book so much they held two parties in Reitz's honor, one in Boston and one in New York. *Menopause: A Positive Approach* spoke directly to readers, many of whom wrote to Reitz to share their own menopause stories or to ask for her advice on how to start their own consciousness-raising groups.[33] Clearly, women hungered for a fresh take on menopause.

Women also wanted more information on estrogen. To learn about this hormone and its medical uses, women (and men) turned to another book published in 1977, *Women and the Crisis in Sex Hormones*, by Barbara Seaman and her husband, Gideon Seaman, a psychiatrist.[34] Seaman's research, like that of Reitz, was informed by both personal interest and a feminist commitment to women's health. Seaman became skeptical of estrogen as a young woman in 1959, when her aunt died at the age of 49 from endometrial cancer after taking Premarin to treat menopausal symptoms. This tragedy inspired Seaman to uncover evidence about estrogen not previously shared with the general public.[35] Whereas Reitz only came to the subjects of menopause and hormone replacement in her late forties after having worked at several different careers, Seaman began her investigation of estrogen products as a young woman in her twenties, and it remained the central focus of much of her life's work.

Barbara Ann Rosner Seaman was born in New York City in 1935, the eldest of three sisters. Her father was assistant commissioner of the city's Department of Social Services and her mother taught high school English. Seaman attended the New York public schools and then went to Oberlin College in Ohio. After graduating in 1956, she followed that familiar 1950s pattern: she got married, moved to the suburbs (Long Island, New York), and had three children. In 1963, when the children were still very young, the Seamans moved back into the city; Barbara worked as a journalist, contributing articles to women's magazines. She wrote monthly columns for *Brides' Magazine* and the *Ladies' Home Journal,* often on health care topics. Living on Park Avenue and working for the mainstream press, Seaman associated with equal opportunity feminists. She met Betty Friedan through the Society of Magazine Writers and attended the founding press conference of NOW. Unlike Reitz, who had always identified with radical feminists, Seaman began her career as a reformist feminist, hoping to promote changes within the existing system of health care.

In the 1960s Seaman began looking into oral contraceptives after hearing a number of disturbing stories about adverse effects experienced by women taking the pill. She knew that the pill contained estrogen, which raised suspicions in her mind because of her aunt's death. She realized that she needed more training to be able to decipher the medical literature on oral contraception and to ask the right questions in interviews with scientists and physicians, so she spent a year as a Fellow in the Advanced Science Writing Program at Columbia University. She then threw herself headlong into writing a book about the pill.

*The Doctors' Case against the Pill* came out in late 1969.[36] Based on the testimony of physicians, medical researchers, and women who had used oral contraceptives, the book presented a strong and to many readers a shocking case against the pill. Seaman challenged both the pharmaceutical industry and the medical profession, demanding that pill manufacturers and physicians share all available information about the health risks of oral contraception with patients, so that the women themselves could make informed decisions about birth control. Her charge that women received inadequate care from their physicians, particularly obstetrician-gynecologists, gave voice to the growing concern among feminists about women's health issues. When Seaman met the radical feminists of D.C. Women's Liberation at the Senate hearings on the pill in 1970, she underwent a transformative experience. She began to see the (mis)use of estrogen not simply as bad medical practice but rather as symptomatic of much more serious problems in the existing system of patriarchal medicine.

Seaman's next book, *Free and Female,* reflected this change in her feminist stance.[37] Published in 1972, the book explored women's sexuality and attempted to dispel the multitude of myths that surrounded that subject. Situated chronologically between Masters and Johnson's *Human Sexual Response* (1966) and Shere Hite's *The Hite Report: A Nationwide Study of Female Sexuality* (1976), *Free and Female* was based on Seaman's reading of published scientific and medical reports, her interviews with physicians and medical researchers, and her trademark source, the experiences and voices of women themselves. This time, however, she incorporated a more militant feminist analysis that borrowed from the ideas and language of her radical colleagues. In matters of sex and sexuality, Seaman argued, the personal was indeed political.

Seaman also included a critique of estrogen products (the pill, DES, and HRT) in her investigation of women and sex. She reproached doctors who advocated long-term hormone replacement therapy (naming Robert Wilson, specifically), and she exposed Ayerst's seemingly unethical practice of advertising Premarin to the general public via articles provided to magazines and newspapers by its "public relations service," the ambiguously named Information Center on the Mature Woman.[38] What occupied just a few pages in *Free and Female* came to fill an entire volume five years later in *Women and the Crisis in Sex Hormones*, the third book in what Seaman called her "estrogen trilogy."

In 1975, when the evidence linking estrogen with endometrial cancer came out, Seaman was active in the women's health movement as a founder of the National Women's Health Network and a member of the DES Action board. She had just turned down a job as health writer for the *New York Times* on principle, because the editor offered her a salary significantly less than that given to a recently hired food writer. (The editor reasoned that the restaurant critic would bring in advertising revenue, whereas Seaman's incendiary articles would probably scare away ads from drug companies.) The women's magazines refused to hire her on similar grounds. From a financial standpoint, Seaman needed to write a book that would sell well. Certainly, the time was ripe for a feminist approach to the estrogen issue. Although Seaman already had her own credentials as a medical writer, she decided to ask her husband to collaborate on this book for two reasons. First, she wanted to give advice to readers and she knew that only an author with a medical degree had license to make such recommendations (even though her husband was trained as a psychiatrist, not as a gynecologist or endocrinologist). Second, her marriage was faltering, and she hoped that such a joint production might bring the two of them back together.

*Women and the Crisis in Sex Hormones* addressed concerns about the three major estrogen products and the medical conditions for which these products were prescribed. The book began with the tragic history of DES and its initial synthesis by Charles Dodds in 1938. It then traced the widespread use of DES not only to prevent miscarriage but also as a postpartum milk suppressant, as a morning-after birth control pill, and in animal feed for livestock. This first section ended with the discovery in 1971 of vaginal cancer and other reproductive tract abnormalities in DES daughters and the reluctance of the federal government to forcefully regulate the use of DES. The next part picked up where *The Doctors' Case against the Pill* left off, with a review and update of the history of oral contraceptives and with detailed information on alternative methods of birth control, including three chapters on contraception for men. The final section on estrogen replacement therapy and menopause made up about a third of the book's contents. In these pages, Seaman revisited her familiar source base of female patients, research scientists, and "mop-up" doctors (the ones who treated patients with adverse effects, as opposed to the physicians who initially prescribed estrogen) in putting together a scathing exposé of estrogen replacement therapy in general and Premarin in particular, as its best-selling example. In simple, easy-to-understand language, Seaman described the misogynistic ways in which Ayerst advertised Premarin to doctors, the flawed ideas on which Wilson based his *Feminine Forever*, and the latest epidemiological results on endometrial cancer. She then explained the physiological changes of menopause and offered alternative therapies to treat menopausal symptoms such as hot flashes, insomnia, and depression, and to prevent age-related conditions such as osteoporosis. Ginseng and vitamin E ranked high on her list of "wholesome remedies," along with regular exercise, good nutrition, and vitamin and mineral supplements. At the end of the book, Seaman reiterated the tenets of the women's health movement. The "old system" in which patients "let the doctors do the worrying" was outdated; instead, patients must be "informed of the full risks and benefits of any treatment—and alternatives—as well as the probable outcome of getting no treatment at all." Women must "listen to what our own bodies tell us" and "respond to the news of medical miracles or fountains of youth with healthy skepticism."[39]

Much to Seaman's satisfaction and relief, *Women and the Crisis in Sex Hormones* was a hit. It was a featured alternate of the Book-of-the-Month Club, and it was serialized in *Ms., Family Circle, Working Woman,* and *Playgirl,* thus reaching a diverse readership. In its first six months, the book sold 60,000 copies; the Erie (PA) *Times-News* reported that it was the second "best-read" nonfiction book in Erie County, based on library requests.[40] Like *Menopause: A Positive Approach,* *Women and the Crisis in Sex Hormones* was hailed by the women's health movement for its honesty and comprehensiveness. It went through several paperback reprintings; the back cover of the 1982 Bantam paperback edition featured quotations from Betty Friedan and the authors of *Our Bodies, Ourselves.*[41]

Both Seaman and Reitz rejected the medicalization of menopause, particularly the use of estrogen as replacement therapy. Not only did these authors present their readers with alternative approaches to dealing with symptoms of menopause, they also offered an alternative source of information, more comprehensive than a magazine article, a promotional brochure, or a chat in the doctor's office. These books struck a responsive chord with women. After Barbara and

Gideon Seaman appeared on the Phil Donahue Show in September 1977 to promote their book, dozens of women wrote to ask more questions, to request a copy of the book, or to express their appreciation. One viewer from Pennsylvania wrote, "I saw you on Donahue this a.m. You're a refreshing breath of fresh air in our pill polluted society."[42] The women's health movement had encouraged women of all ages to become active participants in their health care; the books by Seaman and Reitz gave midlife women some information they needed to take steps in that direction.

Feminists, however, were not the only ones writing books on menopause, aging, and estrogen. Reitz and Seaman represented one end of the spectrum of opinion on the management of menopause. Physicians wrote their own books to encourage women to seek the counsel of medical specialists. Many of these books sought to reassure women that physicians and pharmaceuticals should continue to play a central role in guiding women through menopause. *No Pause at All*, published in 1976 by Louis Parrish, M.D. (authors with medical degrees were always clearly identified as such), is an excellent example of this genre. An advertisement for this book in the *New York Times* announced, "A renowned endocrinologist/psychiatrist shows you that there need be no pause at all in the normal functioning of your physical and psychological being. He explains all the tremendous strides medicine has made in the cure and prevention of middle age discomfort."[43]

Louis Parrish followed in the tradition of William H. Masters and Robert A. Wilson. He believed that "Estrogen is the closest thing to a 'fountain of youth' that science has yet discovered. Not only does it relieve and prevent the common symptoms of the climacteric change, it forestalls much of the degeneration we associate with 'old age.' When combined with a rigorous program of medical attention and personal care, it can keep a woman vital long beyond the time allotted by Mother Nature."[44] Parrish wrote his book as a first-person narrative, using case histories from his private practice—with patients' names changed and with lengthy dialogue provided—to illustrate the role of estrogen in the female body, the problems caused by its declining production in midlife, and the benefits of replacement therapy for older women. Thus he told the story of Miss Bigby, who suffered from both diabetes and menopause.[45] He explained that she was both insulin deficient and estrogen deficient and needed to replace both hormones to regain her health. By setting up the parallel between insulin and estrogen, Parrish made it clear that he considered menopause to be a hormone deficiency disease just like diabetes. He relied on the Maturation Index, popularized by Wilson in *Feminine Forever*, to confirm his diagnoses of estrogen deficiency.

Other case histories detailed the uses of estrogen replacement therapy for hot flashes, atrophic vaginitis, and osteoporosis. Parrish acknowledged the existence of the reports associating estrogen with endometrial cancer, but he rejected their conclusions. So long as estrogen was prescribed as cyclic therapy, to allow the endometrium to shed periodically, he saw no danger in its use.[46] He also advocated the use of tranquilizers for patients with depression, insomnia, and other psychological dysfunctions. However, he was adamant about patients following doctor's orders; he offered the cautionary tale of Mrs. De Marina, whose attempts to adjust her own tranquilizer doses wreaked havoc with her emotions and her relationships.[47]

*No Pause at All* included a chapter on skin care and another chapter on cosmetic surgery, diet, and exercise. Parrish lamented the FDA's limitation of the amount of estrogen allowed in over-the-counter skin creams, but noted that stronger preparations were available by prescription.[48] He also recommended certain kinds of cosmetic surgery, especially "smaller-scale procedures" such as eye lifts. Parrish recognized the challenges of growing older in a youth-obsessed culture. "Youth and beauty are seen as synonymous," he wrote; "therefore age and ugliness have become equated."[49] He sympathized with the plight of the postmenopausal woman in much the same way that Wilson did. Both physicians offered the miracles of medical science to individual women to counteract what others (feminists, for example) interpreted as collective problems inherent in the structure and institutions of American society and culture.

Parrish, like Masters, believed that men also underwent psychological and physiological changes at midlife. In the cases described in his chapter on the male climacteric, he attributed some patients' problems to hormone deficiencies, treatable with testosterone injections. For another patient, whose examination revealed a "normal" testosterone level, he prescribed weekly vitamin injections and what he cryptically referred to as "a psychic energizer that also had a mild tranquilizer effect" to treat the patient's fatigue, insomnia, irritability, indecision, headaches, digestive complaints, and loss of interest in sex.[50] Most likely, this drug was an anticonvulsant compound called diphenylhydantoin, or phenytoin (brand name Dilantin); the back cover of the book noted that Parrish was a consultant for the Dreyfus Medical Foundation, studying the effects of Dilantin on almost one thousand patients. The Dreyfus Medical Foundation had been set up by Jack Dreyfus, the founder of the Dreyfus Fund who was convinced that Dilantin had cured him of depression, to try to obtain FDA approval for this and other clinical indications (as of 2007, Dilantin was still approved solely for the treatment of seizures).

The biography on the back cover also mentioned that Parrish had served as medical director at Ayerst Laboratories (which may have helped to shape his enthusiasm for conjugated estrogens). Whereas Masters had promoted the use of an estrogen-testosterone combination for both male and female members of "the neutral gender" in the 1950s, Ayerst's aggressive promotion of Premarin in the 1960s made it the replacement hormone of choice for physicians like Parrish, who acknowledged the gender discrepancy in treating the climacteric in men and women: "Men may . . . find themselves with more than a twinge of jealousy over the advantage that estrogen replacement therapy offers to women."[51]

Parrish conceived his role to be that of an educator; it was his mission to teach the women of the world about "the revitalizing properties of estrogen replacement therapy." He lamented the slow pace of the dissemination of scientific and medical information and promised "to present the information in applicable form."[52] In this sense, he was doing exactly what the women's health movement requested, but the information he presented, and the paternalistic manner in which he presented it, could not have been further from feminists' intentions.

However, other physicians had imbibed the full meaning of the feminist campaign for improved women's health care and took it upon themselves to provide balanced information on controversial topics. *The Menopause Book,* published in 1977, promised "up-to-date helpful medical facts by eight women doctors" (three gynecologists, three psychiatrists, and two internists); advertising copy for this volume made much of the gender and expertise of its authors.[53] The content of the book was clearly informed by the principles of second wave feminism and the tenets of the women's health movement. "We have written this book for women who want intelligent and informed answers about menopause," the authors declared in the introduction, "for women who are no longer content to receive a pat on the head and a prescription for tranquilizers from their doctors when they complain of stress, anxiety, depression, or hot flashes . . . We also hope that it will help to change the outmoded child-parent relationship of patient to doctor. We would like to see the patient's role become a more active one, that of an informed and intelligent health services consumer."[54]

The first chapter began with a telling example of the low regard in which menopausal women were held in American society. It recounted the unfortunate episode in 1970 in which Dr. Edgar Berman, Hubert Humphrey's personal physician and a member of the Democratic Party's Committee on National Priorities, publicly ridiculed Congresswoman Patsy Mink's approval of a female president. "Suppose that we had a menopausal woman president who had to make the decision on the Bay of Pigs," he commented scornfully. When Mink asked that

Berman be removed from the committee, he replied that her response was a "typical example of an ordinarily controlled woman under the raging hormonal imbalance of the periodical lunar cycle."[55] The authors segued from this incident to the myriad problems faced by menopausal women based on confused and confusing social expectations for all women. One of their recommendations was to join or to start a "rap group," or consciousness-raising group, to discuss similar issues with other women in similar situations.[56] An entire chapter of the book consisted of seven in-depth interviews with menopausal women. The authors believed, as Rosetta Reitz did, that openly sharing personal experiences would help to bring menopause out of the closet and to make midlife women more comfortable about themselves. "Whatever your particular experience of menopause," they promised, "there is someone out there who knows what you mean."[57]

The chapter on estrogen fell somewhere in between the enthusiastic approval of Louis Parrish and the blanket condemnation of Barbara Seaman. The authors firmly pronounced that "estrogen has not proved to be the eternal fountain of youth," but they did condone its use in treating hot flashes and in treating (not preventing) osteoporosis.[58] They explained why and how studies of the side effects of estrogen were undertaken and the difficulties of making risk-benefit calculations for individuals based on epidemiological statistics. The physicians who withheld estrogen from all women were condemned as strongly as those who continued to prescribe it freely. The authors urged women to learn as much as they could so that they could participate in making the decision about estrogen therapy.

To facilitate the process of self-education, the final chapter of the book offered readers a resource directory of women's health centers, national and local women's organizations, and publications on a wide variety of health and lifestyle topics. The marriage of medical expertise with feminist sensibility in *The Menopause Book* acknowledged the enduring influence of the established system of medicine amid the growing appeal of alternative models of health care delivery. The authors, themselves practitioners of conventional medicine, recognized the value of well-informed patients and looked to the women's movement for guidance in providing that information.

Women scanning the shelves of bookstores and libraries in the late 1970s to read up on menopause and midlife health issues encountered a far greater selection of books than in previous decades, but the advice given by the diverse array of authors varied tremendously. Even celebrities weighed in with their own advice

manuals; their authority in this area presumably rested on their ability to maintain their good looks. In her book *Winning the Age Game,* model and television
personality Gloria Heidi "share[d] the secrets of her own Ageless Woman look"
in "her new, step-by-step, comprehensive guide."[59] Which recommendations
should a woman follow? The path to enlightenment as an informed medical
consumer was by no means clear.

Like the medical profession, the pharmaceutical industry also tried to maintain its influence in patient education. Ayerst Laboratories, for example, contem-
plated various strategies to improve its public image and revive Premarin sales
after the drug was linked to an increased risk of endometrial cancer. In December
1976 the president of Ayerst Laboratories received a letter suggesting ways to put
a more positive spin on estrogen replacement therapy from an executive vice-
president of the New York public relations firm Hill and Knowlton.[60] This firm
had also orchestrated a public relations campaign for the Tobacco Institute, starting in 1958, to refute claims that smoking cigarettes caused lung cancer.[61] The
Hill and Knowlton executive proposed a media blitz that included placement of
articles in women's magazines and general magazines; coordination with science
editors and syndicated columnists on newspapers' women's pages; development
of a film on menopause that could be shown in full at women's clubs and medical
conventions and in edited clips on television; and the cultivation of a spokes-
woman who could make appearances on local television shows. "If the spokes-
woman is properly prepared and rehearsed," the PR man wrote, "these appearances could avoid violating in any way the letter or spirit of regulations. In fact,
they should be portrayed as public service ventures." He also recommended that
at least two Ayerst executives be coached by Hill and Knowlton to make public
statements as needed. Ayerst did not follow up on Hill and Knowlton's suggestions to develop a widespread media campaign, but serious damage was done to
the drug company's reputation when the letter was intercepted (apparently by an
employee of one or the other of the two firms) and given to the feminist newspaper *Majority Report,* which published it in its entirety. Morton Mintz, the *Washington Post* reporter who had written critically about thalidomide and oral contraceptives in the 1960s, got hold of the story and co-wrote a trenchant exposé in
*The Progressive.*[62] Feminists and mainstream reporters agreed in their reproach of
both firms for demonstrating insensitivity toward women's health.

The provision of menopause information soon expanded beyond printed
sources. In the early 1980s local hospitals and Planned Parenthood affiliates
offered menopause seminars; some women started their own menopause discus-

sion groups to "counter the negative attitudes of doctors."[63] These support groups sprang up around the country, modeled, like Reitz's workshops, on the women's liberation consciousness-raising rap groups of the previous decade. In the absence of definitive medical evidence, women turned to each other to share their experiences of menopause and aging.

Feminist periodicals such as *Majority Report* also provided an alternative forum for women to read about menopause and postmenopause. As mentioned in chapter 4, *Prime Time*, the newsletter put out by the Over-45 Committee, solicited readers' experiences of menopause and published several of them in a "Menopause Speak-Out" in 1974.[64] The April 1976 issue prominently featured an article written by Rosetta Reitz as she was completing her book on menopause.[65] The journal *Women and Health*, launched in the winter of 1976 to "link academic research with social action," ran in its third issue a lengthy article on menopause by therapist Marilyn Grossman and sociologist Pauline Bart, author of the 1970 article "Portnoy's Mother's Complaint," on depression in midlife women.[66] One year later, it published a critical review of estrogen products.[67]

Older women's issues became increasingly visible within both the women's movement and federal programs on aging. In 1976, *Women: A Journal of Liberation* ran a special issue on aging. The opening editorial confessed, "This has been a difficult issue for us to produce and we feel we have changed a lot in our understanding of the aging process. At first, the issue of women and aging seemed like such a heavy topic, depressing; some of us on the staff couldn't relate to it personally, or were afraid to. We had many sessions of consciousness-raising."[68] In spite of their initial reluctance, the younger women came to appreciate the lived experiences of older women, expressed in this magazine through interviews, recollections, and poems. At the first National Women's Conference held in Houston in November 1977 (authorized, sponsored, and financed by Congress and the White House), almost two thousand delegates developed positions on a wide range of topics, such as child care, the equal rights amendment, reproductive freedom, women in elective and appointive offices, battered women, disabled women, and older women. The platform on older women called for the federal government, state governments, and both public and private social welfare groups and women's organizations to support efforts on housing, health care, public transportation, and other initiatives to improve the lives of aging citizens. "The image of older women is changing," announced the Plan of Action, the conference's official recommendations. "The effective use of the media is essential to furnishing information to the older woman so as to insure her informed participation in the decision-making process which continually affects the quality of her life and the life of her community."[69]

As women's organizations began to acknowledge the concerns of their older sisters, government agencies responsible for senior citizens' affairs also started to recognize the special plight of aging women. In 1978 the National Institute on Aging (NIA) held a three-day workshop in conjunction with the National Institute for Mental Health (NIMH) on "The Older Woman: Continuities and Discontinuities" to stimulate the interest of behavioral and social scientists in developing and pursuing research on the health needs of the growing population of women over the age of 65.[70] The director of the NIA, Dr. Robert N. Butler, sympathized with older women and chastised physicians for cavalier attitudes toward them: "Older women cannot count on the medical profession. Few doctors are interested in them. Their physical and emotional discomforts are often characterized as post-menopausal syndrome, until they have lived too long for this to be an even faintly reasonable diagnosis. After that they are assigned the category of senility."[71] At the end of the 1970s, the head of this research institute, one of the more than twenty that made up the National Institutes of Health (NIH), rejected the assumption that the medical problems associated with aging were directly caused by menopause.

In April 1981 the Federal Administration on Aging sponsored the first national conference on the health issues of older women. One hundred fifty health care professionals from twelve states met in Stony Brook, New York, to develop a proposal to present at the 1981 decennial White House Conference on Aging.[72] After fulfilling their mission, the participants decided to start a newsletter, called *Hot Flash*, to provide education on health-related issues for midlife and older women. Six months earlier the federal government had funded another meeting in Des Moines, where four hundred women between the ages of 50 and 70 met to ensure that their interests would be represented at the 1981 White House Conference. Half of the participants stayed for an additional day to start their own group, the Older Women's League. The founders planned to develop a national organization, with chapters in every state and a lobbying base in Washington.[73]

Thus by the start of the 1980s older women had begun to appear on the radar of advocacy groups and federal agencies. Health issues ranked high on the list of priorities, and thanks to the extensive media coverage of the potential adverse effects of estrogen use, hormone replacement therapy seemed no longer to be considered the panacea for all of aging women's ills. Among the recommendations made by the Stony Brook conference was a call for the development of treatments for osteoporosis other than estrogen replacement. The *New York Times* reported that "so concerned were many participants with the possible cancer-causing effects of estrogen treatment that the official recommendation will be that estrogen not be used at all."[74]

The widespread coverage of safety concerns about estrogen had a significant impact on the use of the drug, as both women and physicians became apprehensive about hormone therapy. The number of estrogen prescriptions written annually dropped by 50 percent in five years, from twenty-eight million in 1975 to fourteen million in 1980.[75] One survey of doctors found that while the majority still recommended estrogen to relieve the symptoms of menopause (such as frequent, severe hot flashes), they tended to prescribe smaller doses for shorter periods of time. Few advocated the use of high-dose estrogen for more than six months or low-dose estrogen for more than three years.[76] Many women, however, had a difficult time giving up estrogen, especially if they had been using it for many years. Those who tried to quit "cold turkey" often found that their menopausal symptoms returned with a vengeance. Perhaps for this reason, some of the female participants at Stony Brook took issue with the official recommendation against estrogen, preferring instead that "women should be informed of the risks of estrogens, and then be allowed to choose for themselves."[77]

This dissent also indicated acceptance of the notion of informed medical consumerism, one of the guiding principles of the women's health movement. Through its own efforts and through its influence on the popular media, the movement had triggered many women to re-evaluate their use of prescription medications and medical services, their relationships with their doctors, and their feelings about their bodies. After a decade of activism, health feminists may not have revolutionized the American system of medical care, but they had changed the vocabulary of the discourse about women's health issues.

Secure, perhaps, in the notion that they had advanced their goals of providing balanced information about menopause and estrogen, Rosetta Reitz and Barbara Seaman moved on to other endeavors in the 1980s. Reitz had indeed helped to bring menopause "out of the closet," a process that would accelerate in the following decades, especially as the women of the baby boom generation began to reach their middle years. In 1980 she founded a record label that reissued long-lost recordings of female jazz and blues singers, and she began to work on an anthology of biographies of these artists. Seaman continued to stay involved in the women's health movement, keeping a watchful eye on the status of estrogen. However, after two decades of writing about women and hormones, she longed for an entirely different project, so she turned her attention to writing a biography of the author Jacqueline Susann.[78] In spite of her efforts to keep her marriage together by co-authoring *Women and the Crisis in Sex Hormones,* Seaman parted ways with her husband, divorcing him in 1982.

Seaman and Reitz had contributed to the feminist critique of mainstream medicine which resulted in a dramatic change in the climate of opinion about women and health care in general and estrogen in particular by 1980. Menopausal and postmenopausal women encountered a wider range of information about hormone replacement therapy and were encouraged to participate in the decision whether or not to take it. In both alternative and mainstream publications, skepticism and caution replaced enthusiasm and conviction as the prevailing attitude toward estrogen.

Perhaps the most telling indication of this shift in attitude was the new position taken by the federal government in 1976. After more than three decades of a relatively laissez-faire approach to estrogen, the Food and Drug Administration initiated a course of action that swept doctors, pharmacists, pharmaceutical manufacturers, feminist and consumer activists, and women and their husbands into a national debate over estrogen replacement therapy and, more broadly, about medical intervention in menopause and aging, informed consent in medicine, and government regulation of medical practice.

# CHAPTER SEVEN

# Enter the FDA

## A Patient Package Insert for Estrogen

The disclosure of the estrogen-endometrial cancer link coincided with a general loss of confidence in medicine in the 1970s, as doctors and hospitals faced challenges to their political influence, economic power, and cultural authority from the women's movement, the consumers' movement, and the patients' rights movement.[1] In 1966, the Harris Survey found that 73 percent of Americans had "a great deal of confidence" in the medical profession; ten years later, that figure had dropped to just 42 percent.[2] In the recession economy of the seventies, the ever-rising cost of medical care drew criticism from consumer advocates. Patients became concerned about the abuse of power by doctors and hospitals, and they banded together to demand their rights in health care. And feminists protested a whole host of injustices perpetrated on women by the largely male medical establishment, including gender discrimination in medical school admissions, lack of informed consent in medical decision-making, and medicalization of too many aspects of women's lives, such as birth control, pregnancy, childbirth, and menopause.

Recall that the newly formed National Women's Health Network held a demonstration on the steps of the Food and Drug Administration in December 1975 to draw attention to estrogen-related deaths and to demand government action. This protest was timed to coordinate with the regularly scheduled quarterly meeting of the FDA's Bureau of Drugs' Obstetrics and Gynecology Advisory Committee in mid-December 1975. At that meeting, the committee debated the estrogen-endometrial cancer issue, as well as the relationship between Depo-Provera (an experimental injectable contraceptive) and cervical cancer. The next month, the United States Senate Subcommittee on Health (part of the Committee on Labor and Public Welfare) and the Subcommittee on Administrative Practice and Procedure (part of the Committee on the Judiciary) held a joint hearing on "Evaluation of the Increasing Use of Estrogens, Both as a Means of Contraception and as a Treatment for the Effects of Menopause," to address what Senator Ted Kennedy called this "subject of unusual importance and interest to the American people—particularly to this nation's women."[3] In the mid-1970s both contraceptive and noncontraceptive estrogens came under government and public scrutiny in response to medical and scientific evidence of their adverse health effects. The news media followed the controversy closely as it played out in the pages of medical journals, in the testimonies of witnesses on Capitol Hill, in the actions of federal regulators, and in the reactions of feminist and nonfeminist women alike.

The United States Senate had already been participating in the critical review of medicine and the pharmaceutical industry for several years when the estrogen-endometrial cancer link was exposed. From 1967 to 1977, the Select Committee on Small Business's Subcommittee on Monopoly held a thirty-three-part set of hearings investigating the American drug industry. Senators heard testimony on a wide range of topics, including drug pricing, testing, and advertising; competition between brand-name and generic drugs; the relationship between the medical profession and the drug industry; and the use and misuse of certain categories of drugs, such as psychotropics (barbiturates and tranquilizers) and combination antibiotics. As part of this investigation, the subcommittee turned its attention in 1970 to the safety of birth control pills; these hearings attracted widespread media coverage and public interest.[4]

The hearing on oral contraceptives and estrogens for postmenopausal use in January 1976 called in government officials, medical scientists, and industry representatives to testify on the use of these hormones. Robert N. Hoover, a medical researcher at the National Cancer Institute (and author of a study of the relationship between menopausal estrogens and breast cancer, to be published in the *New England Journal of Medicine* the following August), applauded the more rigorous methodology of the most recent studies showing an increased risk of cancer for estrogen users, especially when compared to earlier investigations that claimed estrogen protected against cancer. He dismissed these earlier studies, "conducted by gynecologists who, while I am certain were well intentioned, had little or no expertise on how to do an epidemiologic study," as "so heavily flawed in design, conduct and analysis as to make any conclusions reached from them meaningless."[5] For this epidemiologist, the practicing physician had no business trespassing in the world of biostatistical analysis.

Hoover acknowledged the efficacy of estrogen in treating hot flashes, but he averred that "[estrogen] is prescribed for many more women than have severe symptoms of the [menopausal] syndrome, and is used for durations much longer than is consistent with the treatment of this condition."[6] The commissioner of the FDA, Alexander M. Schmidt, concurred. In his testimony, he presented data that confirmed what he described as the overuse of estrogen in menopause: he compared the number of women estimated to go through menopause each year (between 1.3 and 1.5 million) with the amount of Premarin sold (enough to treat five million women per year) and concluded that "the menopausal syndrome [should] be defined more narrowly."[7] Senator Kennedy, head of the subcommittees holding the hearing, asked one witness, "Do you think we may very well be heading toward . . . a deadly manmade cancer epidemic?"[8] The hyperbole of his inquiry notwithstanding, Kennedy expressed the growing conviction that the medical-pharmaceutical complex might be endangering the American public.

In addition to addressing doubts about the safety of estrogen generated by the *New England Journal of Medicine* reports on the correlation with endometrial cancer, this group of legislators also expressed concerns about the prescription of estrogen to treat a wide variety of vaguely defined symptoms allegedly associated with menopause, such as nervousness, anxiety, irritability, and fatigue. Senator Kennedy described an advertisement for Premarin in *Medical World News* that said, "While you are calming her down with a tranquilizer, treat what may be her real problem with Premarin."[9] Noting the lack of evidence for estrogen's efficacy in treating psychological and emotional disturbances, Kennedy questioned the motives of manufacturers and physicians in "the pressing and pushing and forwarding of that particular product [Premarin]."[10] He called on the FDA, with support from Congress, to develop and maintain a system for continued surveillance of drugs after they earned approval to be marketed.[11]

Between the 1950s and the 1970s, the FDA had grown from a small, relatively insignificant government bureau under the radar of both politicians and the public into a highly conspicuous, politicized agency held accountable by multiple interest groups.[12] Its previously limited influence expanded greatly with the Kefauver-Harris amendments to the Federal Food, Drug, and Cosmetic Act, passed in 1962 in the wake of the thalidomide tragedy.[13] Thalidomide, a German-made sedative, had been prescribed to pregnant women in Europe and had caused thousands of babies to be born with missing or malformed limbs. The drug had not been approved for sale in the United States, thanks to the vigilance of one FDA medical reviewer, Dr. Frances O. Kelsey. The disaster pointed up the need for tighter drug laws, so the amendments gave the FDA increased authority over the manufacture and sale of prescription drugs. First, manufacturers had to demonstrate not only the safety of proposed new drugs but also their efficacy. Second, the FDA earned control over medical advertising (previously under the jurisdiction of the Federal

Trade Commission), which enabled it to require manufacturers to distribute a detailed package insert directed to physicians. These pamphlets contained instructions for using the medications, as well as information on indications, contraindications, efficacy, and side effects. However, physicians were not obliged to share any of this information with their patients.

The FDA's visibility increased not only because of its increased regulatory clout but also because of greater attention from what FDA Commissioner Schmidt called "the public voice."[14] Consumers began to demand a seat at the table, alongside industry representatives and government personnel. Whereas in the 1960s advisory committee meetings were closed to the public, pressure from feminist activists compelled the agency to open these meetings in the 1970s to interested lay people.[15] By 1975, the FDA could expect a fairly large crowd to attend meetings whose agendas listed topics of concern to the general public.

Thus, at the meeting of the FDA's Obstetrics and Gynecology Advisory Committee in December 1975, an invited panel of fourteen physicians and scientists deemed to be experts in the field discussed the validity of the epidemiological studies linking estrogen and endometrial cancer with an audience consisting of private citizens and representatives from the news media; pharmaceutical manufacturers, and health activist organizations, as well as the members of the advisory committee itself. Those who were skeptical about the correlation emphasized the alleged long-term benefits of estrogen. Dr. Charles Hammond, of Duke University, claimed that "women may live many years in a hormonally deficient state" which "may lead to a decreased quality of life, osteoporosis, and an increased rate of cardiovascular disease, and may be altered by estrogen therapy."[16] Those on the other side of the table called for more stringent package labeling for estrogen. Dr. Sidney Wolfe, a consumer activist with the Health Research Group, contended that more than three-quarters of all estrogen prescriptions (an estimated twenty-two million per year) were "unwarranted either because of their lack of efficacy or by suggesting benefits which are trivial in comparison to the risk of cancer."[17] He proposed that the estrogen product labeling (for physicians) be revised to include data on the risk of cancer and that physicians obtain patients' informed consent to use the drug. To the surprise of many, the FDA went even further and mandated direct *patient* labeling for estrogen.

The first patient labeling had been ordered for oral contraceptives in 1970. That year, at the end of the Senate hearings on the safety of the pill, FDA Commissioner Charles Edwards announced that his agency planned to require pill manufacturers to include a patient package insert in every package of birth control pills. This pamphlet, to be written by the FDA in lay language and directed to the patient, would outline the health risks associated with taking the medication.[18] Evidence pointed to a causal role for the pill in the development of thromboembolisms (abnormal blood clots), but this correlation remained under contention. The association between oral contraceptives and cancer was ambiguous, because of the long latency period for the development of cancer and the relatively recent availability of the pill. Dozens of metabolic reactions to the pill had been documented, but the implications of these effects were not at all clear. The first draft of the insert listed more than twenty-five symptoms and conditions that women might experience while taking birth control pills.[19]

Caving in to pressure from professional, industrial, and government interests, the FDA ultimately produced a much shorter, less detailed version of the patient package insert.[20] This abridged draft was just one hundred words in length; the original had been six times as long. When the FDA published the text of the proposed package insert in the *Federal Register* and invited all interested parties to respond with comments on the proposal, some eight hundred individuals and groups wrote letters.[21] More than half of the correspondents wrote to object to the shortened form of the insert and to call for reinstatement of the longer version. Some commented on the unequal distribution of power in the doctor-patient relationship. Others added their concern about the integrity of the pharmaceutical industry and its control over government agencies. Still others expressed the opinion that women had the right to full disclosure on medical matters.

Doctors wrote to the FDA to protest any kind of patient package insert for oral contraceptives. Their objections fell into two main categories: (1) it would interfere with the doctor-patient relationship, and (2) the government should not regulate what information the doctor must give to each patient. Drug companies also opposed the inclusion of an FDA-mandated warning in packages of oral contraceptives. The manufacturers pointed out that the patient label contradicted the intent of the federal Food, Drug, and Cosmetic Act, which clearly distinguished between prescription and over-the-counter drugs. They argued that the law designated the class of prescription drugs to be issued on a physician's prescription only and thus precluded the necessity of directing detailed information to the patient. The drug industry sided with the medical profession in preserving the sanctity of the doctor-patient relationship.

In June 1970 the FDA announced that the agency would require the distribution of a brief (hundred-word) insert with every prescription of birth control pills.[22] The insert informed the patient that she could request from her doctor a booklet with more complete information. Since the onus fell on the patient to ask the physician for the longer version, feminists and consumer advocates were disappointed by what they considered to be an insufficient provision of information. Initially the manufacturers mailed the inserts to physicians to hand out with every prescription for oral contraceptives. In the five years from 1970 until 1975, doctors distributed a total of only four million copies, although ten million American women were taking the pill each year.[23] Subsequently, the patient package inserts were included by the manufacturers in individual packages of birth control pills, which were distributed by pharmacists.

By the spring of 1976 the FDA had decided to require a patient package insert for all estrogen products, not just for oral contraceptives. The agency was irked by a letter sent out by Ayerst Laboratories to physicians after the meeting of the FDA's Advisory Committee on Obstetrics and Gynecology. This communication, which the *FDA Drug Bulletin* described as "irresponsible," alluded to the "controversy that has arisen" but insisted that "it would be simplistic indeed to attribute an apparent increase in the diagnosis of endometrial carcinoma, solely to estrogen therapy."[24] The letter went on to reassure physicians that it was still safe to prescribe Premarin for their patients. Doubtful that the pharmaceutical industry would self-police and provide balanced information to physicians, and convinced that patients had the right to full disclosure on the drugs they were given, the FDA decided to intervene.

In the 29 September 1976 issue of the *Federal Register,* FDA Commissioner Alexander M. Schmidt announced his agency's intention and invited interested persons to submit written comments on this proposal.[25] This notice was picked up by wire services, and articles were published in newspapers around the country. In response to the request for comments, almost four hundred individuals and organizations from forty states and the District of Columbia sent letters to the FDA.[26] Individual physicians, nurses, and pharmacists recorded their personal views on professional letterheads, national associations (such as the American Medical Association, the Congress of County Medical Societies, and the National Association of Chain Store Druggists) sent official responses on behalf of their memberships, and pharmaceutical manufacturers sent bulky dossiers addressing the proposed regulations. Many correspondents were not affiliated with the health professions; they were just average women (and some men) who wanted to express their opinions about the hazards and benefits of estrogen for menopausal and postmenopausal use.

About half of the correspondents supported the FDA proposal of a patient package insert for estrogen. Many of these letters reflected the influence of the women's movement and the consumer movement, employing the language of informed consent and the patient's "right to know." One included as an appendix Ralph Nader's statement on prescription drug information for consumers. Another letter came from the Nassau County Homemaker's Council in suburban New York, with 107 signatures listed, almost like a petition in favor of the patient package insert. Clearly, these individuals took seriously their roles as informed participatory citizens; they felt obligated to express their opinions on government actions that would have a direct impact on their personal lives.

Other advocates of the patient labeling for estrogen based their positions on personal experiences rather than convictions about patients' rights. They expressed skepticism about the time and attention physicians devoted to individual patient care and declared their shrinking confidence in their physicians' actions and recommendations. Said one disgruntled ex-Premarin user: "I've gotten to the point where my faith in doctors and pills is at an all time low ebb. So please, print the warnings already—my money doesn't grow on trees."[27] This writer's fulminations against what she perceived as the insensitivity of doctors to patients' needs were reiterated by others, who cast themselves as victims and their doctors as villains.

Doctors and druggists were overwhelmingly opposed to the patient package insert for estrogen. Some of their concerns were fiscal; the National Wholesale Druggists' Association objected to the insert because of "inflammatory effects of these proposals on drug prices and the serious distribution problems that may be created."[28] In general, physicians' complaints echoed those of their peers who had written to the FDA in 1970 to oppose the patient package insert for birth control pills: interference with the doctor-patient relationship and government regulation over the doctor's discretion. However, the letters in 1976 revealed a greater skepticism about the ability of female patients to handle drug information. One internist from Florida wrote three pages of single-spaced invective against the patient package insert, concluding: "Even now I am compelled to discharge from my care more and more women who are too suggestive and believe everything they read. I can no longer stand the strain necessary to educate frightened and brainwashed women."[29] Pharmacists supported the sanctity of the doctor-patient relationship and concurred that women would not benefit from estrogen labeling. A New York pharmacist wrote:

> I have been a licensed pharmacist for nearly half a century and your recent regulation concerning the distribution of literature to patients with their prescriptions on estrogen fills me with horror . . . What makes you think that patients who receive

estrogenic medication can intelligently comprehend the literature? We have scores of patients who cannot even speak English and most of our patients who receive estrogenic treatments are highly neurotic . . . This interference in the physician-patient relationship is very dangerous and intolerable.[30]

These health professionals believed that all medical decision-making authority should rest in the hands of physicians. Their denunciation of government regulators as meddlesome and detrimental to the practice of medicine was nothing new; indeed, the medical profession had worked hard to maintain its autonomy throughout the twentieth century. What was particularly striking, especially given the scope of the women's movement by the mid-1970s, was that their attack on the patient package insert consistently belittled the intellect of the female patient: either she would not understand the content of the warning label or she would react hysterically to its message.

Pharmaceutical manufacturers echoed the claim that the patient package insert would provide the patient with information that she should not have. The twenty-one-page response submitted by the senior attorney for Hoffmann-La Roche clearly identified the physician's right to withhold information from his patient: "In our opinion, the decision concerning what additional information is to be provided to the patient should be left entirely in the hands of those responsible for treating the patient."[31] At issue here was the control of access to medical information. Doctors and drug companies regarded this information as privileged knowledge, to be meted out in appropriate doses to patients.

This privileging of knowledge related to women's bodies was exactly what the women's health movement condemned. These feminists believed that women, and men, had the right to all available information regarding medical care, so that they could participate in informed decision-making. The FDA's proposals to grant women access first to information about the pill (albeit limited), and then to information about estrogens for menopausal and postmenopausal use, were important steps toward reducing paternalism in medicine and creating more of a partnership between the providers and the consumers of health care.

Indeed, one of the central tenets of the women's health movement was, and is, the patient's right to informed consent. The executive director of the National Women's Health Network, in a statement before the FDA, argued that "patient package inserts are a legitimate, reasonable, and inexpensive method to disseminate important consumer education . . . Most Americans do not realize what it means when a drug or medical device is approved as 'safe' by the FDA. There is no drug . . . that is without risk. Citizens need to know about these risks—and benefits—in order to make an informed decision."[32] For this organization, representing more than a hundred health groups and thousands of health consumers nationwide, patient package inserts meant much more than a consumer safety issue; they stood for a realignment of the doctor-patient relationship. "A license to practice medicine does not confer the right to make someone else's decision," rebuked the same executive director in a letter to a physician who opposed the patient package insert.[33] She rejected his assertion that "mass distribution might actually do harm in terms of unnecessary fear, iatrogenic hypochondriasis, and further erosion of patient compliance,"[34] and she chided him for "fail[ing] to recognize the growing sophistication of the consumer movement."

These new sophisticated consumers also wrote to the FDA about the estrogen patient package insert. Letters from individual women who had suffered adverse effects from taking estrogen most clearly expressed the feminist perspective. These women turned their personal experiences into a political issue. They blamed the overprescription of hormones on financially motivated physicians and pharmaceutical companies, as this angry letter reveals: "I took Premarin ten years and found that it was the reason for ten years of headaches which sent me to the hospital many times . . . I finally got tired of taking Premarin and found to my surprise that I no longer had headaches. It cost me a fortune in medical bills which labeling could have prevented. Why did I take it in the first place??? Because some doctor probably had stock in American Home Products."[35] Other correspondents looked past the immediate health effects of using estrogen to cast doubt on the medicalization of menopause and aging. One woman asked, "Why all this artificial medication? Why not let nature take its course? Aging women are not supposed to remain sixteen."[36] A man, who signed his letter "A concerned husband," complained about "a general practitioner who prescribed a twenty-three month supply of Premarin for my wife and told her she should expect to take the drug the rest of her life. His reason: 'It will help you guard against, prevent brittleness of your bones.' . . . It's high time certain doctors discontinue selling the *youth myth* to women and begin warning them of certain drug dangers."[37] For these concerned consumers, the solution appeared obvious: provide full disclosure about the effects of estrogen directly to patients and let women decide for themselves. Based on their own trials and tribulations, these estrogen opponents felt confident that many women, if given the whole story about estrogen and freed from the paternalistic advice of the physician, would choose not to take the medication.

However, not all women regarded estrogen with suspicion. Several correspondents took issue with the FDA's narrowed list of indications and offered their personal experiences with hormone replacement therapy as testimonials to the drug's benefits. These women misconstrued the plan to order patient package inserts for estrogen, fearing that the FDA was planning to take the hormone off the market, and they wrote to the FDA in defense of what they perceived to be a miracle drug. They were mainly concerned with the role of estrogen in maintaining their personal health. The few who did politicize the issue sided with their physicians and argued against the "intrusion" of government in the private practice of medicine: "I plead with the FDA to not restrict the doctors any further. I would find life completely without meaning and worthless. Probably would end up in a mental institution or worse—dead by my own hand. Estrogen is like a miracle for me and I am sure for many other women. Without it life is a veritable hell."[38] Others doubted the veracity of scientific research, favoring instead the advice of their trusted physicians: "To be fair to patients, they should be advised that some of the so-called 'studies' were not conclusive, and in fact were carried out in an unscientific manner. The media has blown these up all out of proportion to the point where many women are afraid to take estrogens . . . Estrogen therapy under supervision of a qualified physician (gynecologist) is good, and women should not be frightened away from it by half-baked 'studies.' "[39] This group of estrogen advocates did not participate in the larger cultural critique of medicine; rather, they believed medicine could and should be used to help women feel better and to combat the signs of aging.

Also, for many of these women, estrogen "worked." That is, they truly believed that estrogen not only relieved their hot flashes but also controlled nervousness, produced softer skin, and resulted in a more youthful appearance and demeanor: "I have been taking hormones for 6 yrs. And it makes *all* the difference in the world as to how I *feel, look,* and *react* . . . I tried going without them for 6 months and this is what happened. I became so nervous I could hardly put on my make-up. I lost all interest in sex, I gained weight, and I was so plain mean and short tempered with everyone, especially my husband, that he was about ready to divorce me. How can you place *all* these problems against the slight chance of cancer[?]"[40] This woman worked out her own risk-benefit analysis for estrogen therapy; she much preferred the short-term benefit of improved mood to the long-term risk of tumor development.

Even more significantly, the women who wrote to the FDA in defense of estrogen gave voice to something that remained unstated in the debates over estrogen among physicians and legislators: the ambivalence about aging felt by women living in a youth-centered culture. The woman who wrote to report, "I not only look, but feel much younger than my age (56) . . . I agree with the women who claim to feel 'sexier' and strongly believe that many middle-aged marriages who suffer from sexual doldrums because of the wife's unresponsiveness during this period in their lives, could be more harmonious,"[41] relied on estrogen to help her to preserve her position as a sexually satisfying wife. For such women, an important objective was to maintain their youthful appearance and demeanor so as to remain attractive and pleasing to men. They trusted their (largely male) physicians to provide them with a medical regimen that would help them to meet that goal, and they perceived the FDA's intent to regulate the use of estrogen as an unnecessary impediment.

The goal of the FDA was not to restrict the use of estrogen but rather to provide patients with enough information to help them participate in the decision whether or not to use long-term hormone therapy. The agency announced its final order on the estrogen patient package label in July 1977. Manufacturers were compelled to supply, and pharmacists to distribute, a lengthy leaflet that discussed the uses, dangers, and side effects of estrogen. The pamphlet warned the patient in explicit detail about the increased risk of endometrial cancer, as well as the potential risk of breast cancer, gall bladder disease, and abnormal blood clotting. In the wake of this mandate, the FDA decided to revise the oral contraceptive labeling requirements to be consistent with those of other estrogen products. The new 1978 version of the patient package insert for birth control pills repeated the information in the physician package insert in lay language in simplified form (as did the estrogen labeling) and represented what consumer and feminist groups had wanted all along.[42]

On 29 July, just seven days after the FDA published its ruling on estrogen labeling, the Pharmaceutical Manufacturers Association (PMA) filed suit in federal court against the FDA, challenging that agency's authority to require such an insert and contesting the regulation as an unconstitutional interference with the practice of medicine.[43] The pharmaceutical companies claimed that only Congress had the authority to mandate such labeling of drug products; furthermore, they argued, the mandatory nature of the regulation requiring patient package inserts to be dispensed with prescription drugs interfered with physicians' right to practice medicine. The American College of Obstetricians and Gynecologists, the National Association of Chain Drug Stores, the American Society of Internal Medicine, and several other medical associations joined the drug companies as plaintiffs; the Consumers Union, the Consumer Federation of America, the National Women's Health Network, and the Women's Equity Action League filed as intervenor-defendants in the case, alongside the FDA.

The immediate goal of the lawsuit was to stop the FDA from putting into effect the patient package insert for estrogen, but the U.S. District Court in Delaware denied the plaintiff's motion for a preliminary injunction, on 5 October 1977. A summary judgment was handed down in February 1980, in which the judge ruled against the PMA and in favor of the FDA, acknowledging the agency's regulatory authority in prescription drugs and declaring that "there is simply no constitutional basis for recognition of a right on the part of physicians to control patient access to information concerning the possible side effects of prescription drugs."[44] Nine months later the decision was upheld by the Third Circuit Court of Appeals.[45]

This ruling represented a limited victory for consumer advocates: physicians had lost their omnipotence in regulating the flow of information to patients, but the battle to gain full access to information about *all* prescription drugs had yet to be won. However, in 1980 feminists and consumer activists had reason to be hopeful that the requirement for patient package inserts would be extended to all medications. In March of 1975 the National Organization for Women, the Women's Equity Action League, and the Women's Legal Defense Fund joined Consumers Union and Consumer Action for Improved Food and Drugs in filing a petition that asked the FDA to require patient labeling for all prescription drugs.[46] In November 1975 (one year *before* the estrogen patient package insert proposal), the FDA published the petition in the *Federal Register* and invited comments from the public.[47] Of the 1,006 letters received, an overwhelming 940 approved of consumer drug information.[48] Encouraged by this show of public support, the FDA held two conferences (in 1976 and 1978) that convened doctors, pharmacists, drug manufacturers, consumer advocates, and the public. The first meeting explored the desirability and feasibility of the patient package insert; the second addressed more specific questions of the style and content of the proposed consumer insert. In 1979 the FDA commissioned the Rand Corporation to test prototypes, and the following year it decided to mandate a three-year pilot program for ten classes of drugs.[49] Although consumer advocates would have preferred patient information for all prescription drugs, this action seemed to be a step in the right direction.

Unfortunately, the timing of the FDA's action could not have been worse. Two months after the ruling was published, Ronald Reagan was elected president. In the first weeks of his administration, President Reagan ordered all federal agencies to review the necessity and cost-effectiveness of existing and proposed regulations. His newly appointed commissioner of the FDA, Arthur Hull Hayes Jr., stopped the pilot program before it ever began. In the new spirit of deregulation, privatization, and smaller government, the FDA revoked its earlier ruling on patient package inserts, leaving the initiative to inform patients about prescription drugs to the private sector.[50] Thus estrogen was one of only four prescription medications (along with oral contraceptives, progestins, and isoproteronol, an asthmatic inhalant) for which the FDA required manufacturers to provide information directly to patients.

Ultimately, the movement to extend patient labeling to all prescription drugs was not strong enough to overcome the revived forces of conservatism in the early 1980s. The backlash against expanded government included a contraction of the regulatory authority of the FDA. As evidenced by several of the letters written to the FDA protesting the estrogen labeling proposal, the feeling that government should keep its nose out of the private practice of medicine was shared by doctors and patients alike. Those women who opposed the patient package insert expressed a larger, and growing, antigovernment sentiment. However, their position was not so different from that of the women who supported the estrogen labeling. Both groups wanted the right to choose whether or not to take prescription medications. The labeling proponents wanted access to written information in order to participate in drug decision-making; the opponents wanted no government restrictions on drugs already on the market and in their medicine cabinets. The ambivalence over the estrogen patient package insert in the 1970s highlighted enduring uncertainties about the role of the government in regulating the practice of medicine.

How well did the estrogen patient package insert work? This is a difficult question to answer. Effectiveness depended on distribution. In 1979 the FDA undertook a study to investigate what percentage of patients actually received the information leaflet.[51] Unlike birth control pills, which (after 1975) came prepackaged in individual packets with the insert enclosed direct from the manufacturer, estrogen prescriptions were custom filled by pharmacists, who transferred a month or two's supply of the pills from bulk packaging to individual containers. It therefore fell to the pharmacist to give the leaflet to each patient when he or she dispensed the prescription. FDA researchers sent undercover agents, armed with prescriptions for estrogen, into drugstores in twenty cities across the country to find out how many pharmacists actually handed out the patient package information. In 39 percent of the stores, the "shoppers" (who included female "patients" and men posing as patients' husbands) automatically received the information along with their pills, and in 55 percent, he or she received the leaflet upon request. Thus more than half of estrogen prescriptions were filled without the patient package insert if the consumer did not specifically ask for it. As the authors of the study pointed out, "simply initiating a requirement does not guarantee regulatory compliance."[52] They hypothesized that many pharmacists might not have automatically dispensed the information because of resentment over government intrusion in their practice. This reaction was certainly consistent with the letters pharmacists wrote to the FDA, and the fact that the FDA was considering patient package inserts for all drugs may have compounded pharmacists' antagonism.

Pharmacists' opposition notwithstanding, many women did receive the patient package insert along with their estrogen prescription. Whether the insert convinced them not to take their pills is hard to say. It is true that the number of estrogen prescriptions dropped from twenty-eight million in 1975 to fourteen million in 1980, but that decrease cannot be fully attributed to the patient package insert (in effect as of September 1977), since the insert was distributed by the pharmacist at the time of sale, *after* the prescription had been written by the physician.[53] Indeed, this was a point of contention discussed by an FDA advisory committee reviewing the estrogen labeling in 1984: to be truly effective, some on the committee argued, the information should be presented to the patient *before* the prescription was even written, so that she could participate in making the decision about her medical treatment.[54] However, upon reading the patient package insert distributed with the initial estrogen prescription, a patient may have concluded that the risks outweighed the benefits and decided to discontinue the hormone replacement therapy (thus contributing to the decrease in prescriptions written and filled). The decline in estrogen prescriptions in the late 1970s was also fueled by the widespread media coverage of the news that estrogen use greatly increased a woman's chance of developing endometrial cancer.

The value of patient package inserts is suggested by contemporaneous evidence from the 1979 Rand Corporation study, which found that 70 percent of consumers read the drug information supplied to them, 50 percent kept them for future reference, and 20 to 30 percent reported reading the information more than once.[55] Patients in this study clearly appreciated the opportunity to learn more about the drugs they were taking. Some doctors may have resented patients' increased knowledge (and the questions this knowledge provoked), but this result was exactly what feminists and consumer advocates wanted: for patients to be informed and active participants in health care. Greater access to information about estrogen did not make the decision to take the drug any simpler, but it did help to change decision-making from a unilateral judgment by the physician to a more cooperative process involving both physician and patient.

However, there were some patients for whom the insert provided too much information. At the 1984 FDA advisory committee meeting convened to discuss revisions of the estrogen labeling for both physicians and patients, one committee member commented, "I do think that some women who need this are being frightened off by the current package insert . . . I had one woman who called me and said, 'Doctor, are you trying to kill me?' "[56] Although this remark prompted laughter from the committee, it reveals ambivalence, among both doctors and patients, about providing patients with comprehensive information on estrogens for menopausal and postmenopausal use. The woman who asked if her doctor was trying to kill her found much that was troubling in the estrogen patient labeling. Not only was she concerned about the potential adverse health effects of the drug itself, but she also now had reason to doubt the wisdom and intentions of her physician. Of course, she did not truly believe that he meant to do her any harm, but she did question his judgment about her medical treatment. No longer could she place complete trust in her physician, because she now had access to additional, previously guarded information. Some patients, and many physicians, regretted the passing away of the era of "Doctor Knows Best."

The estrogen patient package insert left more questions unresolved than it answered. Providing information to patients about the drug's usage, contraindications, side effects, and adverse reactions did not even begin to address the larger medical question of whether or not women should take hormones for ten or twenty or thirty years after menopause. Prescribed for long periods of time for postmenopausal women, estrogen was one of a class of preventive drugs with ambiguous results. In the late 1970s, no group—neither government regulators nor medical researchers nor practicing physicians nor patients—had enough information to ascertain clearly the benefits and risks of long-term hormone replacement therapy.

Estrogen labeling landed definitively on the side of consumer representatives and feminist activists in the debate over government regulation of medicine and pharmaceuticals. The estrogen patient package insert was another victory in the battle to reform the practice of prescribing drugs that had begun with the first patient package insert for birth control pills in 1970. These little leaflets armed patients with information necessary to participate in making knowledgeable decisions about their medications. In this way, "the people" appeared to gain some leverage against "the system" of traditional medical practice. The FDA agreed with consumer advocates that informed consent was a good thing, important enough to be mandated by government fiat.

In spite of the pharmaceutical manufacturers' legal challenge, the patient package insert had little immediate impact on the advertising, sale, and distribution of drugs. True, those companies that made estrogen products had to print and deliver the inserts, but in general it was business as usual. Since manufacturers advertised directly to the medical profession, they could still reference the authority and control of the physician in the prescription of estrogen. They may have curbed the therapeutic claims for their products, but they continued to appeal to the physician's expertise, leaving him (and, increasingly, her) to deal with the decision whether or not to include the patient in decision-making.

Although under no obligation from the government, several firms had by the mid-1990s voluntarily expanded the patient information program, developing package inserts for some thirty to forty prescription drugs besides those containing estrogen. In the previous decade, the pharmaceutical industry began advertising prescription drugs directly to consumers in magazines and newspapers. Although the FDA called for a moratorium on product-specific direct-to-consumer ads in 1983, the ban was lifted two years later. These ads were subject to the same FDA regulations as those in medical journals; both had to provide the official professional labeling. Thus consumers gained more access to information about prescription drugs not as a result of increased government regulation but rather as an unanticipated outcome of changes in the marketing of medicines.[57]

Physicians learned to live with the patient package inserts and the implicit notion of informed consent. Although some may have resented the challenge to their omnipotence, most realized that they were swept up in a much larger social trend in the 1970s, in which American citizens demanded both greater accountability from and greater participation in both the public and private sectors.[58] The estrogen patient package insert represented an important milestone in women's efforts to play a role in their health care during menopause and postmenopause. It also seemed to signal the demise of estrogen as an antidote to aging. Indeed, after the turn of events in the late 1970s, many observers believed that estrogen had peaked in 1975 and would henceforth play only a minor role in the management of menopause. They could not have been more wrong.

# CHAPTER EIGHT

# Resurrecting Estrogen, I

## Osteoporosis and Medical Science

In 1980 American pharmacists filled about fourteen million prescriptions for estrogen, just half the number dispensed in 1975, when the Premarin brand alone had been the second most frequently prescribed drug in the country.[1] By 1980 Premarin had dropped to number eighteen on the list of America's top prescription medications.[2] Then, after five years of steep decline, the annual number of estrogen prescriptions began to rise again; almost twenty million were dispensed in 1985 and almost thirty million in 1990.[3] Premarin became the fifth most frequently prescribed drug in 1988; four years later, in 1992, it reached number one and remained either the first or second most popular drug in America every year for the rest of the century.[4]

How did estrogen make such a stunning comeback? What transpired in the course of less than a decade to revive its status as one of the most widely used drugs in the country? Both scientific developments and cultural inclinations coalesced in the 1980s to create conditions suitable for the resurrection of long-term hormone replacement therapy. In this modified climate, individuals and groups with vested interests in estrogen succeeded in re-presenting the drug as a preventive health measure for postmenopausal women; many of these women in turn bought the new rationale for the prescription and use of estrogen. This chapter examines the scientific evidence for the reappraisal of hormone replacement therapy; the following chapter looks at the cultural context in which these changes took place.

It was not a new discovery that breathed life into what appeared to be a moribund therapeutic regimen. What did happen was that new data were collected that reinforced an existing hypothesis and extended its clinical implications. This hypothesis—that estrogen loss after menopause caused bone loss, which could lead to osteoporosis—had been around for decades. Recall that Fuller Albright and his colleagues at the Massachusetts General Hospital first identified the condition they called postmenopausal osteoporosis in 1940. In a speech before the Association of American Physicians, Albright presented data that suggested that estrogen therapy (in the form of intramuscular injections) resulted in a positive calcium and phosphorus balance (two minerals necessary for the production of bone tissue) in osteoporotic postmenopausal women. The notion that estrogen could be used to treat osteoporosis gained support as researchers confirmed Albright's findings in the 1950s and 1960s. By 1972, the Food and Drug Administration decided that enough evidence had been accumulated about the relationship between estrogen and osteoporosis to warrant a relabeling of the drug's indications. Not only were estrogen products considered to be effective in treating various menopausal symptoms (such as hot flashes and vaginal atrophy), but they also were deemed "probably effective" in treating postmenopausal osteoporosis. This finding was announced in the *Federal Register* in July 1972 and appeared soon after in both the products' package labeling for physicians and advertisements in medical journals.[5]

By the early 1980s, an important shift had occurred in the medical discourse about estrogen and osteoporosis. Discussion now focused not only on *treatment* but also on *prevention*, thanks to the publication of the results of two prospective studies on the effect of estrogen therapy on osteoporosis in postmenopausal women. These multiyear studies, one in the United States and the other in the United Kingdom, demonstrated that estrogen could forestall the development of postmenopausal osteoporosis. These findings were corroborated by several retrospective case-control studies of estrogen use and non-use among older women with bone fractures. Osteoporosis became a hot topic in clinical research: the number of articles with the keywords "osteoporosis and estrogen" published in English-language medical journals (as identified by PubMed, the National Library of Medicine's index of articles in biomedical journals) tripled from the 1970s to the 1980s (176 articles in 1970–79 to 560 articles in 1980–89). Similarly, articles with the keywords "postmenopausal osteoporosis" multiplied sevenfold between the two decades (from 45 in the 1970s to 313 in the 1980s).

Increased attention to the relationship between estrogen and osteoporosis among scientific researchers brought about a concomitant interest in the subject among government agencies, specifically the NIH and the FDA. In April 1984 the NIH convened a consensus group to develop a statement on osteoporosis. This group asserted that estrogen replacement was the most effective way to prevent bone loss.[6] Later that month, the FDA called a meeting of its advisory committee on fertility and maternal health drugs to discuss the use of estrogens for postmenopausal therapy. One of the tasks set before the group was to decide whether to advise the FDA to expand the indications for Premarin and other estrogens to include the prevention of osteoporosis. The committee voted to recommend estrogens both to treat existing osteoporosis and to prevent its development.[7]

Evidence from scientific studies and support from government agencies for the use of estrogen as a prophylactic against postmenopausal osteoporosis delighted executives at Ayerst Laboratories. The company launched a spirited campaign to educate both doctors and the public about the dangers of osteoporosis (and the merits of Premarin). Estrogen manufacturers were not the only ones publicizing osteoporosis prevention; as the role of calcium in building strong bones was elucidated, the makers of dairy products and calcium supplements joined the action. Osteoporosis received so much attention in the popular press that in December 1986 the *New Republic* cynically declared it to be "the disease of the week."[8]

Estrogen faced a major hurdle in its claim to be the best preventive against the development of osteoporosis: its putative causal relationship with endometrial cancer. So long as estrogen carried the taint of cancer, it could not be seriously considered as a health promoter. In order for advocates of estrogen to successfully promote its long-term use as a protective therapy, they had to find a way to nullify its carcinogenic effect. A second scientific development was needed to solve this problem before the indication for osteoporosis prevention could make a significant impact on the prescription and use of estrogen therapy. The solution, however, was readily at hand and had, in fact, been employed by some physicians for decades. It came in the form of the other female sex hormone, progesterone.

Progesterone is secreted by the corpus luteum (the "yellow body" that remains after an egg leaves the ovarian follicle) during the second half of the menstrual cycle and, in pregnant women, by the placenta. It was isolated and purified during the "heroic age of reproductive endocrinology" in the 1930s by workers on both sides of the Atlantic. One of these scientists was George Corner, who later became an influential mentor to William Masters; another was the German chemist Adolf Butenandt, who was one of the first to isolate estrogen in 1929. Corner and his collaborator at the University of Rochester, Willard Allen, called the hormone "progestin"; Butenandt adopted the name "luteosterone." At an international conference on the standardization of sex hormones in London in 1935, the men agreed on a compromise term, "progesterone," which combined syllables from the two names in contention.[9] The terms *progesterone, progestin,* and *progestogen* (or *progestagen*) may be used interchangeably, although typically progesterone refers to the natural molecule and progestin to one of the synthetic forms of the hormone.

It took two decades for progesterone to become a viable therapeutic agent. Although physicians were eager to experiment with the newly discovered hormone to prevent miscarriage or to treat infertility, it was extremely difficult to extract from natural sources. One ton of animal organs (Butenandt, for example, derived progesterone from the cholesterol in sheep wool and cattle brains and spinal cords) yielded just one gram of pure progesterone. The resulting product was inordinately expensive; that single gram of progesterone could cost one thousand dollars.[10]

Then, in the 1940s, a maverick chemist named Russell E. Marker figured out how to make progesterone for a fraction of the cost. Although he was a brilliant scientist, Marker's personal idiosyncrasies resulted in a checkered career. He abandoned a graduate program at the University of Maryland (before earning his doctorate) because he refused to take a required course in physical chemistry; he left a lucrative position at the prestigious Rockefeller Institute after a dispute over his choice of research project; and he walked away from a professorship at Penn State to undertake steroid production in Mexico. After having published 213 academic articles, he suddenly quit science in 1949.[11] But before he swore off chemistry, Marker succeeded in developing an innovative procedure for deriving progesterone from the roots of the wild Mexican yam. As the result of his efforts, the price of progesterone dropped to eighty dollars per gram, and, within a few years, to just five dollars a gram.[12] Although still pricey, progesterone could now be put to medical use. However, like the early forms of estrogen, progesterone had to be injected in order to be effective, and thus its clinical applications were limited.

Chemists at two pharmaceutical companies, G. D. Searle and Syntex, raced to synthesize a progestin that would be orally active. In 1951 Carl Djerassi, of Syntex, got there first. By tinkering with the chemical arrangement of testosterone, he produced a compound called norethisterone, a highly potent progestin that did not break down in the stomach. The following year, Frank Colton, of Searle, developed an alternative procedure to synthesize a related compound called norethynodrel. Colton received his patent in November 1955; Djerassi received his five months later. Both researchers hoped that their products would be put to use in treating various gynecological disorders. Neither anticipated the eventual ground-breaking application of synthetic progesterone as the world's first birth control pill. In May 1960 the FDA approved Searle's Enovid—consisting of norethynodrel plus a small amount of mestranol, a synthetic estrogen—for oral contraception.[13]

Indeed, what the progestin did was to trick a woman's body into thinking it was pregnant, thus preventing ovulation. (The estrogen helped to prevent breakthrough bleeding.) By taking the pill for twenty-one days and then stopping for seven days, women brought about menstrual bleeding each month, because the withdrawal of progesterone caused the uterine lining to shed. Although this bleeding did not represent true menstruation (because no ovum was expelled), it looked and felt like the real thing.

The endometrial shedding induced by progesterone withdrawal was exactly what was needed to counteract the proliferative effect of estrogen on the endometrium in menopausal and postmenopausal estrogen users. By adding a progestin for the last seven to ten days of the twenty to thirty days of estrogen taken each month (the replacement therapy regimen varied widely), physicians found that they could bring about a cyclic sloughing of the uterine lining, manifested by a few days of bleeding each month. Although the FDA had not approved the use of progestins for menopausal therapy, many doctors had already taken the step of prescribing them along with estrogen for their menopausal and postmenopausal patients. As with any drug, doctors were free to use their own discretion in prescribing for conditions not specifically indicated in the product labeling.[14] When the news broke of the association between estrogen and endometrial cancer, those who had been prescribing progestins as part of the menopausal replacement regimen congratulated themselves for their foresight.

For example, two physicians from King's College Hospital in London wrote to the *New England Journal of Medicine* to criticize what they called the "bizarre therapy" of prescribing continuous estrogen therapy, "a mistake that would not have been made by any final-year medical student." They noted that the regimen used in their menopause clinic—cyclical estrogens plus at least ten days of progestin—resulted in not one single case of carcinoma among a thousand patients receiving treatment for as long as twenty years.[15] These correspondents did not dispute the correlation between estrogen and a higher risk of endometrial cancer. Rather, they wrote to condemn the practice of giving postmenopausal women uninterrupted courses of estrogen, which led to abnormal growth of the uterine lining.

Several other investigators demonstrated that lower incidences of endometrial cancer could be achieved by prescribing both estrogen and progestin ("combined therapy") rather than estrogen alone ("unopposed therapy"). This body of work indicates an important shift in the medical consensus on menopausal therapeutics: from the late 1970s, more and more physicians began to prescribe both estrogen and progestin for their patients with uteruses (obviously, the threat of endometrial cancer did not exist for women with hysterectomies). This transition in practice was accompanied by a change in terminology, from "estrogen replacement therapy" (ERT) to "hormone replacement therapy" (HRT).

The results of studies undertaken by two gynecologists—one American, one British, working independently of each other—and published in a variety of journals between 1977 and 1981 are representative of the evidence in favor of the merit and necessity of combined estrogen-progestin therapy. In Texas, R. Don Gambrell of the Department of Obstetrics and Gynecology at Wilford Hall USAF Medical Center, Lackland Air Force Base, had access to thousands of postmenopausal women who used the medical center's outpatient services as beneficiaries of military personnel. In retrospective and prospective studies, he showed that the women who took progestin along with estrogen had a lower incidence of endometrial cancer than both women who took estrogen alone *and* women who took no hormones at all.[16] In London, M. I. Whitehead, of the Department of Obstetrics and Gynecology at King's College Hospital Medical School, performed microscopic and biochemical studies of endometrial biopsies taken from women on cyclical therapy (estrogen only) and sequential therapy (estrogen and progestin). His results, reported in the *Journal of the Royal Society of Medicine* and the *New England Journal of Medicine,* corroborated those of Gambrell: estrogen caused the potentially dangerous proliferation of the uterine lining, progestin counterbalanced the effect of the estrogen, and therefore the use of estrogen in postmenopausal women had to be coupled with progestin.[17] Both Gambrell's epidemiological surveys and Whitehead's laboratory analyses provided scientific justification for what had previously been an empirically based practice. With the evidence in place to support the contention that progestin negated the carcinogenic effect of estrogen, physicians could feel safe in prescribing hormone replacement therapy for the relief of menopausal symptoms. The progestin data also opened the door for reconsideration of long-term hormone replacement therapy to prevent osteoporosis.

In the early 1980s discussions about the prescription of hormone therapy for osteoporosis broadened from treatment to prevention. Not only was estrogen considered effective in halting (if not actually reversing) further bone loss in women with demonstrated osteoporosis, it was also recommended to prevent (or at least forestall) the disease from occurring. The discourse about osteoporosis was often cast in the language of public health and economics: with millions of older women potentially at risk for osteoporosis, the cost of medical, nursing, and hospital care for all of the resulting bone fractures and incapacitation could be astronomical. Viewed in this light, estrogen could save both lives and money. This approach appealed to many physicians; for members of a profession faced with increasing criticism for its rising costs, this rationale could be used to demonstrate efforts to deal with the crisis of medical expenses. The depiction of osteoporosis as both an individual threat and a communal public health concern allowed estrogen advocates to present hormone replacement therapy as the preventive health measure that would address the problem of brittle bones among the nation's aging female population.

A 1982 article in the *Journal of Family Practice* titled "Postmenopausal Osteoporosis and Estrogen Therapy: Who Should Be Treated?" began with the ominous statistics that three-fourths of the more than 200,000 hip fractures each year in the United States, costing over one billion dollars, could be attributed to osteoporosis. The author explicitly identified this crisis as a "public health issue."[18] In 1984 an article in the *Journal of Medicine* framed its discussion of postmenopausal osteoporosis in terms of an economic cost-benefit analysis; this author quoted the expense of caring for patients with complications resulting from osteoporotic hip fractures at more than two billion dollars a year. "Clearly," he asserted, "postmenopausal osteoporosis is a major public health problem in the United States today."[19] The following year, a review article on osteoporosis in *Medical World News* announced its estimate that "the toll in health care dollars is nearly $4 billion a year."[20] These authors made persuasive cases for the long-term use of hormone replacement therapy in women at risk for osteoporosis, by putting an exorbitant price tag on the medical and hospital expenses of fractures that could potentially be avoided.

However, the claim that estrogen could prevent osteoporosis had to be based on scientific evidence. In the late 1970s and early 1980s data from numerous epidemiological, clinical, and laboratory studies convinced both practicing physicians and government advisors of the validity of estrogen's protection against bone loss. As part of a continuing trend in biomedical research in the postwar era, these (and all other) investigations were held to much higher standards than were those of the previous generation. As mentioned in chapter 5, the design of experiments now had to meet certain methodological requirements, such as a statistically significant sample size and the inclusion of appropriately matched controls. Prospective studies were expected to be "double blind," which meant that neither investigator nor subject knew whether the subject was receiving estrogen or a placebo, so observations would not be biased. Nonetheless, adherence to these norms did not automatically ensure that the results and conclusions of these experiments were above reproach. On the contrary, investigations could be criticized for ignoring certain variables or for studying an inappropriate population or for not carrying on long enough, to give just a few examples. The challenges of experimenting with human subjects, with all their infinite variability, meant that no study could claim to be definitive. But as complementary evidence from several studies accumulated, many in the medical community were persuaded of the protective benefits of the estrogen in HRT against the progression of postmenopausal osteoporosis. Just as the combined weight of data, in spite of methodological concerns about individual studies, convinced medical investigators and clinicians that estrogen use increased the risk of endometrial cancer, so too did the findings on estrogen and osteoporosis begin to add up.

In 1979 the results of a ten-year prospective study of the relationship between hormone replacement therapy and osteoporosis were published in *Obstetrics and Gynecology*.[21] The research had begun in 1965, when a team of physicians at New York University Medical Center, led by Lila Nachtigall, an internist specializing in reproductive endocrinology, recruited women to participate in this randomized double-blind long-term prospective study of the effects of estrogen-progestin therapy.[22] All of the women were bedridden patients at the Goldwater Memorial Hospital, a hospital for chronic diseases on Roosevelt Island in the East River between Manhattan and Queens in New York City. The Goldwater opened in 1939 as the Welfare Island Hospital for Chronic Diseases, the first public hospital in the United States dedicated exclusively to treating chronic diseases. It was later renamed for a New York hospital commissioner, and after World War II it became a leading site for research into chronic disease, while continuing as a patient care facility. Just like the subjects of Masters's investigations at the St. Louis City Infirmary and Infirmary Hospital fifteen years earlier, Nachtigall's subjects were hospitalized for the duration of the study.

There were 168 women in the study, matched in 84 pairs with one of the pair receiving hormone therapy and the other receiving placebo. The pairs were matched according to age and disease (for example, multiple sclerosis or polio) to minimize confounding variables. In order to find these 84 matched pairs, Nachtigall and another gynecologist had to screen more than eight hundred postmenopausal patients (by taking each woman's medical history and giving her a thorough work-up), which took the better part of a year. Once enrolled in the study, the subjects received complete physical examinations annually, including bone density measurements (using an instrument that passed low-energy radiation from radioactive iodine through the middle finger bone in a new technique called single photon absorptiometry).[23] Neither the patients nor the physicians knew who received hormones and who received placebos, until the code was broken at the conclusion of the ten-year study period.

The differences in the "before" and "after" bone density measurements for the two groups of women (treated and control) were dramatic. Those who had taken placebo pills for ten years lost significant amounts of bone mass; at the end of the study, six new cases of osteoporosis had developed. Among the treated group, none developed the disease. When the treated group was divided into those who began hormone therapy within three years of menopause and those who started more than three years beyond menopause, the results were even more striking. While the women who started therapy more than three years after menopause had minimal bone loss (as compared to the control group), the women who initiated treatment within three years of menopause actually *gained* bone mass. The authors felt justified in concluding that "ERT virtually halts the osteoporotic process" and "when estrogens are therapeutically administered within 3 years of the menopause, the process can actually be reversed."[24] The fact that the study was prospective, controlled, randomized, and double-blind meant that its findings would be taken seriously by the medical research community.

The conclusions were also considered legitimate because the study was *not* funded by the pharmaceutical industry. Nachtigall had approached Ayerst and asked the company to make a placebo that looked exactly like a Premarin tablet. The company volunteered not only to supply the placebos and the drugs (which was, and is, standard practice in medical studies) but also to underwrite the entire study. Nachtigall consulted with her department chair at NYU, Gordon Douglas, who recognized the peril of accepting drug company money, namely, that the study would be criticized for having a pro-industry bias. Instead, Douglas offered to pick up the tab, so the entire study was paid for by the Department of Obstetrics and Gynecology at NYU's School of Medicine. Independent funding ensured that the study would be judged by its peers based on its scientific merit, and not on political grounds. Thus, when the *New England Journal of Medicine* rejected the paper, it was because of the editors' misgivings about the quality of the statistical analysis rather than any suspicion about partiality. Nachtigall corrected the statistics and submitted her work to *Obstetrics and Gynecology*, a relatively new journal in the field with a strong clinical focus. After peer review, the article was accepted immediately.

When Nachtigall and her colleagues began the study in the mid-1960s, estrogen was still riding the crest of the "feminine forever" wave. The study concluded in the mid-1970s, amid the flurry of publications linking estrogen use to endometrial cancer. Nachtigall had had the foresight to include progestin as part of the hormone replacement regimen for her subjects (the treated group took 2.5 mg of Premarin daily plus 10 mg of Provera for seven days each month); thus, at the end of the study, she also had ten years of data on combination therapy. She and her colleagues looked at other disease categories and found no increase in the incidence of endometrial cancer or breast cancer or heart disease among the treated group. The women who had taken hormones did have more cases of gallbladder disease, but overall there was no statistical difference in mortality between the treated group and the control group. Nachtigall reported these results in a companion article published in the same journal, *Obstetrics and Gynecology*, a few months after the first article appeared.[25] Although her conclusion was cautiously worded ("The study excludes only a high incidence of complications from estrogens"[26]), its implications were enormous. This prospective study appeared to demonstrate the relative safety of the long-term use of estrogens, when used in combination with progestins, for postmenopausal women.

The osteoporosis indication for long-term hormone therapy received further confirmation from another prospective study conducted by a research group in Glasgow.[27] Investigators followed 100 of the subjects, 58 of whom took estrogen (in the form of mestranol) and 42 of whom took placebos, for an average of nine years. Photon absorptiometry of the finger bones and measurements of height revealed that estrogen protected the treated group from bone loss in both the central and peripheral skeleton. The difference in height loss was especially remarkable: while 16 of the 42 women in the placebo group (38%) were shorter after the nine-year period, just two of the 58 estrogen users (4%) had shrunk. Although the women in this study underwent surgical menopause (because their ovaries had been removed), the authors maintained that bone loss in this group was no different from bone loss in women who experienced natural menopause. However, they were unwilling to go so far as to recommend preventive therapy for all postmenopausal women, "since the true incidence of osteoporosis is unknown."

Meanwhile, evidence continued to mount in support of the use of estrogen not only to inhibit bone loss but also to prevent fractures in older women. Half a dozen retrospective case-control studies appeared in the medical literature in the early 1980s, all with the same conclusion: estrogen users had roughly 50 percent fewer fractures of the hip and forearm as compared to age-matched controls.[28] The experimental design and findings of these reports looked very much like those on endometrial cancer from a few years earlier, except that estrogen users had a *decreased* risk of bone fracture as opposed to an *increased* risk of endometrial cancer.

Loss of height often served as proxy for potential spinal fractures, as both conditions resulted from decreased bone density. A prospective study of the effects of estrogen replacement conducted at Case Western Reserve School of Medicine in Cleveland corroborate the findings of the Glasgow prospective study, and it offered further support for the theory that estrogen could prevent spinal osteoporosis.[29] The Cleveland team followed 61 estrogen-treated and 63 control subjects and found that, after an average of eight and a half years of study, more than twice as many of the nonusers shrank by at least half an inch as compared with the estrogen users. These results, when added to those of the numerous other prospective and retrospective studies of the osteo-protective role of estrogen, made a persuasive case for the benefits of long-term hormone replacement therapy.

Curiously, there were no dissenting reports that disputed or discounted the effect of estrogen on bone. Indeed, there was virtually no debate at all in the medical press about the connection between estrogen loss (natural or surgical) and bone loss, nor about the prevention of bone loss (and subsequent fractures) by pharmaceutical estrogen. These relationships had been proposed four decades earlier by Fuller Albright and his colleagues; subsequent studies, particularly those undertaken in the late 1970s and early 1980s, only provided further scientific affirmation of both hypotheses. That estrogen could prevent the progression of osteoporosis, which was cast as a serious and widespread disease among older women, seemed to outweigh its potential to increase the incidence of endometrial cancer, which was portrayed as a relatively rare condition, treatable if caught early and avoidable in estrogen users if progestin was added to the replacement regimen. By the mid-1980s, osteoporosis had trumped endometrial cancer in the medical literature, and estrogen advocates worked hard to get the message across to their fellow physicians.

Only a small proportion of physicians engage in laboratory, epidemiological, or clinical research; the vast majority spend their time seeing patients. How, then, did practitioners learn about these latest developments? Many read the research studies published in medical journals, but since these reports often appeared in specialty publications (e.g., *Obstetrics and Gynecology*), they might not have reached other clinicians, such as general practitioners or internists, who saw older women in their practices. Many journals also published review articles that translated the research findings into recommendations for clinical practice, such as, for example, the 1982 *Journal of Family Practice* article mentioned earlier, "Postmenopausal

Osteoporosis and Estrogen Therapy: Who Should Be Treated?," which ran as part of the periodical's "Family Practice Grand Rounds" feature.[30] A similarly titled article appeared two years later in *Postgraduate Medicine*, along with a multiple-choice self-assessment test at the end.[31] The *Journal of the American Medical Association* commissioned its Council on Scientific Affairs to report on "Estrogen Replacement in the Menopause" to place recent developments and commentaries "in perspective for the practicing physician."[32] This 1983 article discussed the benefits and risks of estrogen therapy and concluded with a "relative risk assessment" to help guide physicians' decision-making. Of the three main indications for estrogen therapy (osteoporosis, vaginal atrophy, and hot flashes), osteoporosis was the first and most extensively discussed. All three of these articles declared that estrogen would prevent bone loss and reduce the incidence of fractures. However, all three stopped short of a universal recommendation for all postmenopausal women, citing the risk of side effects and the need for individualized evaluation of patients.

The panelists who developed the Consensus Development Conference Statement on Osteoporosis for the National Institutes of Health in April 1984 went further, recommending that "cyclic estrogen therapy should be given to women whose ovaries are removed before age 50 in whom there are no specific contraindications. Women who have had a natural menopause should be considered for cyclic estrogen replacement if they have no contraindications and if they understand the risks and agree to regular medical evaluations. *The duration of estrogen therapy need not be limited*" (emphasis added).[33] This recommendation was restricted to white women, since other racial groups had not been included in the efficacy studies reviewed by the panel. Although many retrospective studies had indeed focused mainly on white women, 30 percent of the population followed in Nachtigall's prospective study was black. By the mid-1980s most medical authorities considered Caucasians (and Asians) to be at greater risk than African Americans for developing osteoporosis, because several epidemiological studies had shown that both low bone densities and bone fractures were less common in black women than in white women.[34]

That no time restriction was placed on the duration of hormone replacement therapy was a stunning reversal from the position taken by the group of experts convened to develop a consensus statement on estrogen use and postmenopausal women just five years earlier. The authors of the 1979 statement urged that "unnecessary prolongation of therapy should be avoided."[35] In 1984, the compelling evidence that estrogen could slow postmenopausal bone loss and prevent osteoporosis, combined with the reassurance that the addition of progestin could reduce the risk of endometrial cancer, allowed government advisors to relax their concerns about long-term hormone replacement.

The 1984 Consensus Development Conference also evaluated other supplements and strategies for preventing osteoporosis. Along with estrogen, calcium was considered a "mainstay" of prevention and treatment of osteoporosis; exercise and nutrition were possibly "important adjuncts."[36] A number of studies published in medical journals suggested the potential bone-building benefits of weight-bearing exercise and adequate calcium intake.[37] Calcium also received a seal of approval from *The Medical Letter,* the nonprofit newsletter that reviewed medications and therapies, in 1982; this publication concluded not only that "daily calcium . . . may retard bone loss and reduce the incidence of fractures in women with postmenopausal osteoporosis" but also that "prophylactic supplementation with calcium beginning years before menopause might prevent osteoporosis in some women."[38] Dairy producers and the makers of calcium supplements were pleased to have these endorsements, and they quickly incorporated calcium's role in preventing bone loss into advertisements for their products.

Pharmaceutical manufacturers were even more thrilled by the positive reevaluation of estrogen by the NIH panel. The scientific studies that hailed estrogen as a preventive against osteoporosis gave the drug companies new hope for this old product. The advertisements published in medical journals in the wake of these studies and government reports provided yet another source of information for practicing physicians on the new indication for long-term hormone replacement.

Ayerst Laboratories had already begun to market Premarin as an antidote against osteoporosis in the late 1970s. One advertisement showed a photograph of an elderly woman walking with the aid of a cane. The caption warned, "Postmenopausal women can lose more than estrogen . . ."[39] The second page of the ad was illustrated with images of one woman clutching her back in pain, another showing early signs of a dowager's hump, and a third using a walker to get around. This ad never mentioned menopausal symptoms as an indication for Premarin, focusing only on the skeletal woes of postmenopausal women. Other Premarin advertisements added the "possibility of osteoporosis" to the list of indications for estrogen (along with sweats, hot flushes, and atrophic vaginitis) in the early 1980s.[40]

In the summer of 1984, after the release of the NIH Consensus Development Conference Statement on Osteoporosis, Ayerst launched a major campaign to promote Premarin as a treatment for osteoporosis in both the medical and popular presses. An ad that appeared in June in the *Journal of the American Medical Association* showed a photograph of a hunched-over women leaning on a cane superimposed on a graph that showed the rapid decrease in bone density in the four decades after menopause. Titled "The Disabling Course of Osteoporosis," the copy went on to advise, "The sooner treatment begins, the better." It also recommended that the physician evaluate the patient's diet, calcium intake, and physical exercise, in addition to writing a prescription for Premarin.[41] The following month in the same journal, Ayerst ran a rather unusual advertisement, not for Premarin directly but rather for another advertisement.[42] The three-page spread began with a page that was blank except for the words "THE AD YOU ARE ABOUT TO SEE WILL WARN YOUR PATIENTS ABOUT A SILENT EPIDEMIC." The second page described the problem, risk factors, and prevention of this silent epidemic, osteoporosis, in introducing Ayerst's plan to run an ad (featured on the third page) in popular consumer magazines. "Don't be surprised if many of your patients start asking you about the ad shown here," Ayerst told the audience of physicians. "We want to inform women about this 'silent epidemic.' And we want to urge them to see their doctors about it." The one-page ad depicted a middle-aged woman and her stooped elderly mother, and it explained osteoporosis in the first-person voice of the daughter. The text urged women to "see your doctor before the damage is done." It never mentioned estrogen (in the early 1980s, pharmaceutical manufacturers were not allowed to advertise prescription drugs directly to consumers); instead, it obliquely referenced a pharmaceutical solution: "Proper diet, calcium supplementation, regular exercise and cutting down on smoking and alcohol are important preventive measures. But these measures may not be enough. Only your doctor can give you an individualized program to prevent osteoporosis." The only clue that this ad was not a public service announcement was the notice at the bottom, in small print, that Ayerst was the sponsor of the message.

By 1985 Ayerst was running at least three different ads that featured osteoporosis as a widespread, debilitating disease whose progression could be slowed by Premarin. Even if physicians skipped over the dense research articles in medical journals, the bold headlines and dramatic images of the advertisements must have caught their eyes and alerted them to the recommendation to prescribe estrogen for patients at risk for osteoporosis.

The advertisements, the government reports, and the clinical studies all sought to address the questions that practicing physicians had about the prevention and treatment of osteoporosis. These questions can be simply reduced to "why," "what," "who," and "when": why osteoporosis must be treated, what therapies should be prescribed, who could be identified as an at-risk patient, and when treatment should be initiated. The first question, why, was answered by the frightening statistics of the number of women afflicted and the cost of their medical care. The public health threat, combined with the depiction of individual pain and suffering, made an obvious case for immediate attention to this disease. The response to the question of what therapies to use was usually plural: estrogen replacement and calcium supplementation were the first line of defense, backed up by regular exercise and a well-rounded diet. Although some researchers had investigated the use of other therapeutic agents (such as fluoride, calcitonin, and parathyroid hormone), the NIH panel refused to endorse these treatments because neither safety nor efficacy had been satisfactorily established for any of them.[43]

The identification of who was at risk for developing osteoporosis was based on epidemiological studies of women who presented at hospitals with fractures, particularly forearms and hips broken under nontraumatic conditions (that is, in circumstances in which someone with normal bone density would not have broken the bone). By the mid-1980s, risk had been associated with sex, race, menopause, and body weight-for-height. Thus, thin, postmenopausal, white women had the greatest risk of developing the disease. By 1986, prolonged inactivity, cigarette smoking, excessive alcohol consumption, and childlessness were also considered to predispose women to osteoporosis.[44] Clearly, these broadly defined categories included a large proportion of the population of older women as potential candidates for long-term hormone replacement. The NIH panel called for the development of screening technologies and methodologies "for determining the level of risk for osteoporosis in an individual, [and] to establish early diagnosis."[45] The use of photon absorptiometry to measure bone density as a screening test became a controversial topic later in the decade, as will be discussed in the next chapter.

Perhaps the most difficult question to answer was the fourth: when to begin treatment. Obviously, sufficient calcium, diet, and exercise were not the concern; these recommendations applied to all women (and men) as part of a healthy lifestyle. Most experts agreed that estrogen should be prescribed to treat existing osteoporosis; at issue was how estrogen should be utilized to prevent the disease from occurring.

This matter was taken up on several occasions in the late 1970s and early 1980s by two of the Food and Drug Administration's advisory committees. The FDA had paid close attention to estrogen replacement therapy since it had first proposed the patient package label in 1976. Issues surrounding the use of estrogen in postmenopausal women appeared frequently on the agendas of the scheduled meetings of these two advisory committees from 1977 through 1985. The Endocrinologic and Metabolic Drugs Advisory Committee focused on guidelines for the clinical evaluation of drugs (including estrogen) to treat osteoporosis; the Fertility and Maternal Health Drugs Advisory Committee spent several sessions reviewing both the physician and patient package labeling for estrogen.

The FDA considered the process of prescription drug labeling to be "dynamic"; federal regulations required periodic review, according to the following procedure. FDA staff members (in either the Division of Advertising and Labeling or the Division of Endocrinology and Metabolic Drug Products) drafted revisions, which were then reviewed by the advisory committee. The draft was then rewritten, re-reviewed, and when a version was agreed upon, published in the *Federal Register*. At that point, the public was invited to comment on the proposal, after which the FDA staff incorporated any changes into a final draft and then submitted it to the secretary of Health and Human Services for approval and to the Office of Management and Budget for an evaluation of its impact. After this official vetting, the order appeared in the *Federal Register* and the pharmaceutical companies were given a time frame (often up to a year) for printing and distributing the new labeling information. All in all, the process from the initial revision of the existing labeling to the production of the revised labeling could take almost four years. And if new data became available during that time period, several steps in the process might be repeated to incorporate that information.[46] In spite of its seemingly glacial pace, the review process was (and still is) considered an essential mechanism in relaying updated information to physicians in the field.

The advisory committees consist of a standing panel of physicians and scientists, who serve for a number of years. The meetings convene guest speakers—experts in the field—to report on their latest research findings to the committee. One or more members of the FDA staff are usually in attendance; since the meetings are open to the public, consumer advocates, feminist activists, and industry representatives are often in the audience as well. The discussions recorded in the transcripts of these sessions reveal the challenges faced by the advisors in evaluating data from clinical studies and translating them into recommendations for practicing physicians.

In February 1977 the members of the Endocrinologic and Metabolic Drugs Advisory Committee undertook an assessment of the safety and efficacy of estrogens in the treatment of osteoporosis. This latest evaluation, based on recently available information, was a first step in updating the FDA's classification of the hormone for this particular indication. Recall that Premarin (and several other estrogen products) had been classified as "probably effective" in treating selected cases of osteoporosis in 1972, during the FDA's review of all drugs introduced prior to the 1962 Kefauver-Harris amendments. The FDA was eager to have a final decision made one way or the other, that is, to reclassify the drug as either effective or lacking evidence for efficacy.

The group wrestled with a number of issues: the semantic distinction between loss of bone mass and osteoporosis, the appropriate clinical endpoint for making a diagnosis of osteoporosis, the size of the dosage and the duration of treatment, and the identification of populations (and individuals) at risk of developing osteoporosis. At the end of the day, they agreed that estrogen did prevent loss of bone, but they did not feel there was enough evidence to support the contention that estrogen prevented fractures. They identified two at-risk populations for whom the benefit of the prevention of further bone loss outweighed the risk of endometrial cancer: women who had their ovaries and uteruses removed surgically and postmenopausal women with demonstrated evidence of bone loss (as shown by radiography or the presence of a fracture). They were not yet willing to recommend preventive estrogen therapy for all postmenopausal women.[47]

However, seven years later, the members of the Fertility and Maternal Health Drugs Advisory Committee were ready to take that step. At a meeting convened in April 1984 to complete its review of the proposed revisions to the estrogen package labeling for both physicians and patients, the committee recommended estrogen not only for those postmenopausal patients with evidence of bone loss (to treat existing osteoporosis) but also as preventive therapy for *all* postmenopausal women. The group of physicians and researchers (and one consumer representative) discussed the implications of "opening the door to extremely wide estrogen use" and agreed that there was enough data to justify the blanket recommendation of hormone replacement to, in the words of one committee member, "prevent serious public health problems." A colleague concurred: "I think we've got an enormous amount of data that osteoporosis is one of the major causes of morbidity and mortality, and certainly an impairment of the quality of life in our older female population . . . I do think we have good evidence, consistent evidence that the estrogen will prevent it in a significant segment."[48] This conclusion was made in spite of a report heard earlier by the committee that the answer to the question of how long preventive hormone therapy should be used was unknown.[49]

A related issue addressed by this committee was the use of progestins in postmenopausal hormone therapy. By 1984 a number of studies hinted that progestins might promote adverse effects on carbohydrate and lipid metabolism, possible indicators of developing cardiovascular disease. At a meeting in February, the committee refrained from supporting a specific warning on the progestin labeling, because they were "uncertain about the uncertainties."[50] Better persuaded by the studies of Gambrell, Whitehead, and Nachtigall, which showed the positive effect of progestin on the postmenopausal uterus, the same committee voted two months later to recommend the addition of at least seven days of a progestin in hormone replacement therapy to prevent the development of endometrial cancer.[51] The only concession made to address the concerns about negative cardiovascular effects was the suggestion that "the choice of a progestin and dosage may be important in minimizing and eliminating a possible adverse effect of lipoproteins and arteriosclerosis."[52]

In these advisory committee meetings, both invited speakers and panel members recognized the tension between the relative weights of research studies and empirical experience in making clinical decisions. One presenter, who described himself as "a climacteric male physician, practicing climacteric medicine on behalf of climacteric women," made a case for considering the individual patient, noting that "a great deal of clinical judgment, and sometimes experience, is necessary, because if we rely entirely on the very good double-blind studies that we all would like to have, we are going to deprive many people of treatment which they should have."[53] Another physician, a woman appearing on behalf of the National Women's Health Network, asked that the physician labeling include references to articles in the medical literature so that doctors could see which studies had served as the bases upon which recommendations were made.[54] All of the participants acknowledged the importance of choosing their words carefully, because labeling (which also appeared verbatim in the *Physician's Desk Reference*) often served as the primary information source consulted by busy practicing physicians.

An excellent example of the significance of semantics played out in a discussion about the difference between treatment and prevention of osteoporosis at a meeting of the Endocrinologic and Metabolic Drugs Advisory Committee in June 1985. The committee chair raised the question of whether it was appropriate to call estrogen a treatment for osteoporosis if it *prevented* bone loss, as opposed to *increasing* bone mass.[55] One of the FDA representatives reasoned that the terms *treatment* and *prevention* really applied to the target groups for a given therapy, as opposed to the therapy itself. Thus treatment referred to a therapy prescribed for a woman with demonstrated osteoporosis, while prevention implied that the therapy was given to someone with no clinical evidence of osteoporosis but who was perhaps at risk for developing the disease.[56] When the committee chair then described a drug as "an important preventive treatment," the group broke into laughter.[57] He went on to make the point that precise wording was critical to fostering an understanding of osteoporosis therapies in the minds of both doctors and patients.

Although in the end the FDA chose not to use the term *prevention* in the revised labeling for estrogen, the wording implied that the hormone could be used by a large proportion of postmenopausal women. On 11 April 1986, the *Federal Register* announced the revised classification of estrogen as "effective for the treatment of postmenopausal osteoporosis."[58] The rationale for the widespread prescription of estrogen seemed logical: Loss of estrogen after menopause causes bone loss. Bone loss can lead to osteoporosis. Replacement estrogens can slow bone loss and thus, presumably, the progression of osteoporosis.

Indeed, by 1986 many physicians had already imbibed the message that estrogen could prevent bone loss in postmenopausal women, and osteoporosis had become one of the main indications for estrogen therapy. A survey of 234 physicians in southwestern Ohio showed that 96 percent believed estrogen would prevent osteoporosis.[59] Another survey of 103 gynecologists in the San Diego area in 1985 found that 73 percent of the doctors usually recommended estrogen replacement therapy for at least ten years after menopause, for the prevention of bone loss.[60] The new epidemiological evidence about osteoporosis justified the reexpansion of clinical indications for postmenopausal hormone therapy to include the long-term prevention of this debilitating disease.

By the mid-1980s hormone replacement therapy for women with intact uteruses had also come to mean the prescription of both estrogen and progestin, to avoid the development of endometrial cancer. Only those women whose uteruses had been removed took estrogen alone. The Ohio survey reported that 64 percent of this group of internists, family practitioners, and obstetrician-gynecologists (and 83% of the obstetrician-gynecologists alone) prescribed a progestin along with the estrogen. Of the San Diego gynecologists, all but five (95%) prescribed estrogen in conjunction with progestin.[61] Clearly, many physicians were convinced that scientific evidence supported two key developments in postmenopausal estrogen therapy: first, that the addition of a progestin conferred protection against endometrial cancer, and second, that long-term administration of combined hormone replacement therapy conferred protection against osteoporosis.

But doctors were only half of the consumer side of the equation for estrogen. Women, of course, were the ultimate users. In order for women to fill their estrogen prescriptions, they had to be persuaded to buy into the new and improved representation of hormone replacement therapy. Culture, as well as science, played a key role in this transformation.

# CHAPTER NINE

# Resurrecting Estrogen, II

## Osteoporosis and American Culture

Estrogen found its way back into women's medicine cabinets via their bones, or, rather, via the fear that their bones were in danger of crumbling. The campaign to increase public awareness of osteoporosis popularized the benefits of exercise, calcium, and estrogen in maintaining bone density. Although the sale of gym memberships and fitness clothing may have increased, the two industries that worked hardest to capitalize on "the silent epidemic" were dairy producers and estrogen manufacturers.

Ayerst Laboratories, the makers of Premarin, participated in the effort to educate the public about osteoporosis by hiring a public relations company in 1982 to coordinate an information campaign. The firm, Burson-Marsteller, began by conducting a survey, which found that 77 percent of American women had never heard of osteoporosis. To reach this vast uninformed audience, it placed articles in popular women's magazines and arranged for coverage on radio and television broadcasts. The message conveyed was simple and direct: osteoporosis was a serious women's health problem, and women at risk for the disease would be wise to see a doctor about prevention and treatment. Burson-Marsteller also worked to educate health care professionals, by developing seminar materials for the Nurses Association of the American College of Obstetricians and Gynecologists (NACOG), to train nurses to educate women in hospital outreach programs, church groups, women's clubs, and senior day care. A vice president of Burson-Marsteller justified its role (and, implicitly, that of its client, Ayerst) in health education as "a noble thing. The reality is that there is decreased funding for research and public education, resulting in the necessity for corporations to fill the gap."[1]

The dairy industry also climbed aboard the osteoporosis bandwagon, tailoring its educational programming and product advertising to address the importance of calcium in maintaining strong bones. The National Dairy Council, the indus-

try's nutrition education and research arm since 1915, developed and disseminated a flood of calcium-related materials for both health professionals and consumers, such as reference manuals, conferences, brochures (with titles such as
"Calcium: You Never Outgrow Your Need for It," "Sticks and Stones Can Break
Your Bones . . . And So Will Too Little Calcium," and "Like Mother, Like Daughter: A Woman's Guide to Bone Health"), films, slide presentations, and exhibits
for display at supermarkets, health fairs, shopping malls, hospital lobbies, and
airports. The council's public education campaign initially targeted 20- to 35-year-
olds, people decades away from the specter of osteoporosis, in an effort to decrease the occurrence of the disease (and, of course, to increase consumption of
milk, cheese, yogurt, and ice cream).[2]

Another organization, the National Dairy Promotion and Research Board,
spent millions of dollars advertising the dangers of osteoporosis and the need for
calcium to people of all ages. Its supermarket calcium consumer campaign—
which consisted of kits of leaflets, recipe cards, and audio and video tapes—
sought to reach four specific groups: children, teens, young adult women, and
older women. Its print ads were seen by an even larger number of people; in the
first six months of 1985, the board placed print ads that mentioned either osteoporosis or the role of calcium in bone health in twenty-eight national magazines. This board was set up by an act of Congress in 1983 "for dairy product
promotion, research and nutrition education as an effort to reduce milk supplies
and increase the consumption of dairy products," to be funded entirely by a
mandatory fifteen-cent assessment on every hundred pounds of milk produced
by the nation's dairy farmers. With a $60 million advertising budget at its dis-
posal, the National Dairy Promotion and Research Board could promote its products and, at the same time, contribute to the campaign to raise public awareness
of osteoporosis.[3]

To educate both the public and the medical community specifically about the
hidden problem of osteoporosis, a group spun off from the American Society for
Bone and Mineral Research in 1984 to form the nonprofit Osteoporosis Foundation. Headquartered in Chicago, the nine-member board of directors consisted of
six physicians, one professor of pharmacology, the executive director of the National Consumers League, and, as chairman, the former Democratic congress-
man from Florida, Paul G. Rogers, who had spent much of his twenty-four-year
career in Congress working on health care issues. The Osteoporosis Foundation
pledged an ambitious agenda—heightening public awareness, educating health
care professionals, establishing a resource registry for information about osteoporosis, supporting research and the training of researchers—predicated on ex-

tensive fund-raising. It saw itself as collaborative, not competitive, with other organizations devoted to bone diseases.[4]

As part of its public awareness initiative, the foundation planned to spread its message via television, radio, print media, events (such as golf tournaments and senior Olympics), cooperative ventures with groups with parallel interests (such as the American Association of Retired Persons and the Gray Panthers), and the formation of local chapters.[5] Members would be encouraged to act as advocates, by lobbying their state and federal legislators to increase appropriations for even more research and education on osteoporosis.

One of the strategies of this private nonprofit organization was to increase public support from the government. Osteoporosis advocates found a champion on Capitol Hill in Olympia Snowe, the Republican congresswoman from Maine. In June 1984 Snowe introduced a joint resolution to designate the first week in May as "National Osteoporosis Week," a popular strategy for increasing public visibility of a disease or health care issue.[6] Although no action was taken that year, Snowe reintroduced the resolution in January 1985, at the beginning of the next session of Congress. With 238 co-sponsors in the House and a companion resolution introduced in the Senate in February, the bill easily made its way through committees in both houses and onto President Reagan's desk, where he signed it into law on 20 May 1985.[7] Thereafter, the week of May 11–17 was officially observed as "National Osteoporosis Awareness Week."

The Senate turned its attention to osteoporosis the following month, with a one-day hearing before the Subcommittee on Aging of the Committee on Labor and Human Resources, chaired by Charles Grassley, the Republican senator who had sponsored the Senate version of Snowe's resolution. In his opening remarks, Senator Grassley described both the individual and the public health ramifications of osteoporosis, namely, the pain of debilitation (afflicting "in the neighborhood of 15 million Americans") and the great expense of medical care for resulting fractures ("$4 to $7 billion annually"), and announced, "Our purpose this morning is to help make the general public more aware of this problem and, more importantly, what to do about it."[8] Representative Snowe, invited to give a statement because of her interest in the subject, concurred that it was essential to educate women about the pro-active measures they could take to avoid the disease.

A procession of medical experts testified before the Senate panel that the keys to preventing osteoporosis were a calcium-rich diet, weight-bearing exercise, and, for those women considered to be at greatest risk for the disease, hormone replacement therapy after menopause. There was a great deal of discussion about the best way to identify high-risk individuals, comparing the advantages and disadvantages of different diagnostic methods. Most participants agreed that education—on identifying the risk factors for developing osteoporosis and making the lifestyle changes that could prevent it—was vital for both consumers and providers of health care.

Both calcium and estrogen received endorsement as effective measures to maintain strong bones. The senators were very receptive to the testimony of the National Dairy Council spokesperson who described that organization's efforts to promote both osteoporosis awareness and increased calcium consumption. In her opening remarks Senator Paula Hawkins, who co-chaired the hearing, commented on the irony that "a nation that produces so much food that the Government is forced to purchase the surplus has a virtual epidemic among its citizens due, in part, to bad dietary practices."[9] One witness, a professor of medical physics at the University of Wisconsin, pooh-poohed the preventive effect of calcium, but his derision of what he called the "calcium myth" put him at odds with all of the other experts.[10] Most of the physicians who testified recommended "selective and judicious use" of hormone replacement therapy for high-risk women, namely, those who had undergone surgical menopause or who had several of the risk factors for the disease.[11] The immediate past president of the American College of Obstetricians and Gynecologists, Luella Klein, identified women at risk as "those who have a small bone structure, are underweight, *who lack estrogen after menopause,* have a calcium deficiency, do not exercise, are Caucasian or of the oriental race, smoke, consume large amounts of alcohol, and have a family history of the disease" (emphasis added). Since estrogen production decreases dramatically in most women after menopause, the "lack of estrogen" factor meant that potentially all postmenopausal women might be at risk for osteoporosis. Klein went on to quantify the relative dangers of osteoporosis (which might be prevented by estrogen) and endometrial cancer (which might be triggered by estrogen) in terms of annual fatalities. She noted that fifty thousand women died each year because of osteoporosis as compared to the three thousand who died from endometrial cancer, adding that endometrial cancer had the highest cure rate of any gynecological cancer when detected early (presumably, the cancer was cured by removal of the diseased uterus). Although her warning that "the most serious health hazard of menopause is a condition called osteoporosis" may have seemed overly dramatic, she was merely echoing a growing consensus among gynecologists at that time.[12] This consensus also held that, while calcium and exercise were excellent preventives to osteoporosis, for those older women who may not have benefited from a lifetime of good nutrition and physical conditioning, estrogen was the only effective therapy on the market.

The Senate hearing not only encapsulated current medical and scientific thought on osteoporosis, it also reflected the social, cultural, and political contexts in which this information was translated for and evaluated by nonscientists. The senators and expert witnesses concurred in their depiction of osteoporosis as a serious problem for both individual Americans, in terms of pain and suffering, and the greater public health, in terms of economics. However, in spite of the dire calculations of the number of people afflicted and the cost of medical and nursing care, the disease was also portrayed as potentially preventable, *if* women were alerted to the importance of calcium, exercise, and, for those at highest risk, estrogen. Since prevention was obviously preferable to treatment, the message expressed at this hearing, and relayed to the public via the media, was the importance of pro-active health care and informed medical consumerism. Academic researchers had imbibed the notion from the women's health movement that patients ought to participate more actively in their health care, and they incorporated this idea into a medicalized view of postmenopause, which was then promulgated to a wider audience.

The medicalized approach to older women's bodies conflicted with the very different picture of health care envisioned by feminist activists. In written testimony submitted for the record of the Senate hearing, the National Women's Health Network (NWHN) applauded the increased public attention being paid to osteoporosis, but it warned, "There is a danger that existing treatment for osteoporosis may be mass marketed and that women will once again be guinea pigs for medical treatments which could prove to be fatal."[13] These feminists worried that pharmaceutical sponsorship of medical research would lead to an overemphasis on estrogen replacement at the expense of diet and exercise. In the same month as the hearing, the Network's bimonthly newsletter reported with outrage the osteoporosis public relations campaign conducted by Burson-Marsteller and financed by Ayerst Laboratories.[14] The article quoted a NWHN board member's response: "It is chilling to think a public relations firm that has learned to influence us to buy products we don't need or that may be harmful to our health is gearing up to educate health workers to convince us to buy a program of hormone therapy beginning before menopause."[15] Both feminist activists and women's physicians (most commonly, obstetrician-gynecologists) agreed that osteoporosis was an important problem and that this predominantly female medical concern deserved greater attention. Their central disagreement lay in whether or not pharmaceuticals should play a role in the prevention of the disease.

These tensions between medicalized and nonmedicalized approaches to osteoporosis, specifically, and to aging, more generally, must be considered in the light of contemporary cultural phenomena in America in the 1980s. Factors such as the enduring cultural authority of medicine, the revival of health conscious-ness among the middle and upper classes, and the ongoing obsession with youth and concomitant aversion to aging shaped the production and consumption of information about the use of estrogen after menopause and influenced the revival of long-term hormone replacement therapy.

The relative roles of lifestyle habits and pharmaceuticals in the prevention and treatment of osteoporosis seemed to hinge on the ascertainment of the progres-sion of the disease in individual women. Hardly anyone disputed the recommen-dation that women of all ages exercise regularly and consume sufficient amounts of calcium. Although some skeptics (notably, the National Women's Health Net-work and the Center for Medical Consumers) doubted the motives of the dairy industry and calcium supplement manufacturers,[16] calcium was never perceived as a potentially dangerous antidote to osteoporosis. Estrogen, as we have seen, faced exactly that charge. Its benefits in preventing osteoporosis had to be demon-strated to greatly outweigh its risks as a carcinogen (as well as the uncertain effects of the progestin in combined therapy). In the first half of the 1980s, estrogen was advised only for women at high risk for developing osteoporosis. The question remained: how could those high-risk individuals be identified?

By 1985 a long list of risk factors had been compiled, but these qualitative criteria (as listed above, in Luella Klein's testimony) lacked the persuasiveness of hard data. To obtain a quantitative measurement, some researchers advocated bone density testing, using either single photon absorptiometry (SPA, which measured the bones of the forearm) or the newer technique of double photon absorptiometry (DPA, which measured the bones of the spine). SPA became commercially available in the early 1970s (this was the technology used by Nachti-gall in her study); DPA came onto the market around 1980.[17] By the mid-1980s, five companies manufactured SPA machines and four produced DPA machines.[18] More than five hundred screening centers were in operation around the country, about 65–75 percent affiliated with hospitals and the rest owned and operated directly by physicians.[19] The idea was to bring in women who had several of the risk factors; if their bones proved to be below normal density, then they would be encouraged to begin hormone replacement therapy.

Some critics of bone density screening questioned the utility of the tests, since the measurements could not predict the incidence of fractures. They also protested the notion that the tests were in some way preventive, since the measurement indicated the actual presence of bone loss, not its potential to occur. In other words, early detection was being sold as prevention, which critics read as false advertis-ing.[20] Advocates maintained that measuring bone density was an important diag-nostic, analogous to measuring blood pressure and cholesterol: even though blood pressure levels could not predict the occurrence of strokes and cholesterol tests could not predict the timing of the development of coronary events, those tests did alert individuals to take steps to forestall disease. If changes in diet and exercise did not yield corresponding improvement in bone density, they argued, then estrogen therapy was a wise course of action.[21] Critics retorted that the bone density tests were intended only to facilitate decision-making about HRT, since *all* women were advised to exercise and take calcium. In this view, the only purpose of osteoporosis screening was to increase medical intervention, which was anathema in the eyes of the women's health movement.[22] Proponents, on the other hand, defended the use of testing to make decisions about HRT; from their perspective, the technology enabled physicians to make better-informed clinical judgments.

The concept of screening for disease was not new. In the 1910s, the National Association for the Study and Prevention of Tuberculosis urged people living with tubercular relatives to have a chest X-ray to discover early signs of the disease.[23] Two decades later, serological blood tests became widely used to test for the presence of syphilis. In the 1940s, the Pap smear was introduced into gynecologi-cal practice for the detection of cervical cancer. The first two of these examples were designed to target specific populations, namely, those individuals exposed to contagious disease. The Pap smear, by contrast, sought to identify a chronic disease, which could potentially afflict any female. Thus the Pap smear's target population included *all* women. And whereas chest X-rays and serological blood tests were initially administered by public health agencies, the Pap smear was offered as part of the annual checkup by gynecologists. However, the idea behind all three tests was similar: to identify those individuals within the at-risk popula-tion who might have the disease so that they could be treated.[24]

Indeed, screening for disease and annual checkups were gradually integrated into medical practice in the twentieth century as part of a movement toward preventive health care. As chronic diseases (diabetes, cancer, heart disease, obe-sity) replaced infectious diseases (those that could be avoided by improved sanita-tion or treated by antibiotics or prevented by vaccines) as the leading causes of death among Americans, physicians claimed as legitimate areas for medical in-volvement both early detection and instruction in lifestyle management to avoid or control chronic disease. At the same time, a countervailing trend developed in part as a reaction to the incursion of medicine into everyday life in the years after World War II.[25] The ideology of the modern health movement, while tracing its earliest roots back to the antiprofessionalism of the 1830s, was based on a de-medicalized view of health. An important component of this philosophy was the conviction that individuals ought to be personally responsible for their health and well-being. This principle was incorporated, for example, into the vegetarian lifestyle promoted by health reformer Sylvester Graham (1794–1851), inventor of the Graham cracker. A century and a half later, in the wake of the social movements of the 1960s, which asked citizens to question authority in public and private life, many Americans again rebelled against professional control of medicine, opting instead to take charge of their own health care. The widespread appeal of this approach was evidenced by the plethora of self-help books published in the 1970s (Rosetta Reitz's *Menopause: A Positive Approach* is one example of this genre). These advice manuals encouraged readers to improve their health and well-being through diet, exercise, and spiritual enlightenment. The burgeoning of this trend led one *New York Times* writer to pronounce self-help as one of the major developments in science and medicine during the decade of the 1970s.[26]

Interestingly, instead of resisting this new form of individual health conscious-ness and control, organized medicine embraced it and integrated its rhetoric into a medicalized version of preventive health care. In a prescient article written in 1980, political scientist Robert Crawford hypothesized that "a strong case can be made that the new health movements and consciousness may ultimately extend medical jurisdiction, even though they are presently developing in relative auton-omy from it."[27] In the decade that followed, the medical profession did indeed adopt language compatible to the ideology of healthism in the management of postmenopause and the use of estrogen. Since the modern version of healthism advocated a lifestyle that entailed pro-active control of one's health and well-being, the decision to take hormone replacement therapy could be portrayed as a positive step toward staving off the progression of osteoporosis. Both the pharmaceutical industry and the medical profession incorporated into the marketing and pre-scription of estrogen therapy the message that women could and should take charge of their health.

In spite of challenges to its control of health and healing, medical science retained its cultural authority in late twentieth century America. Although survey data indicated that medicine experienced a loss of confidence among Americans between the 1960s and 1970s, the level of confidence remained consistent from the late 1970s through the 1980s.[28] Furthermore, medicine was not the only profession being called into question. In fact, the General Social Survey con-ducted in 1986 by the National Opinion Research Center found that Americans had the highest level of confidence in medicine, followed by the scientific community; respondents ranked their trust in these professions higher than the military, the Supreme Court, education, organized religion, major companies, banks and financial institutions, the executive branch of the federal government, the press, Congress, television, and organized labor. In another nationwide poll (the Harris Survey), conducted the same year, only the military received a higher vote of confidence than medicine.[29] Although Americans expressed an absolute decline in their willingness to believe wholeheartedly in their social institutions, including medicine, in the 1980s they retained a relative faith in medicine as compared to almost all other establishments.

Americans' continued, and growing, concern with health-related issues certainly created a receptive environment for the public campaign against osteoporosis. Another perceptive author observed, in a 1977 article entitled "The Medicalization and Demedicalization of American Society," that "the shifts in emphasis from illness to health, from therapeutic to preventive medicine, and from the dominance and autonomy of the doctor to patients' rights and greater control of the medical profession do not alter the fact that health, illness, and medicine are central preoccupations in society."[30] This interest was reflected in the increase in coverage of health and medicine topics in newspapers and magazines.[31] Osteoporosis took its place in the limelight after a series of other "media diseases" in the 1970s and 1980s, such as hypoglycemia, anorexia, toxic shock syndrome, herpes, and premenstrual syndrome.[32] And, by dint of their obvious associations with osteoporosis, menopause and estrogen also received renewed attention from the popular media.

Newspaper and magazine articles on estrogen and menopause in the early 1980s initially expressed the same cautious and ambivalent stance as did those written at the end of the previous decade; the shock of the endometrial cancer link was still a fresh memory, and osteoporosis had not yet registered on journalists' radar screens. "I cannot remember a time when there has been more confusion about the proper treatment of a condition than there is now about the use of estrogens for the discomfort of menopause," wrote one physician in a column for *McCall's*.[33] Another physician quoted by *Newsweek* averred, "The real dilemma of menopause . . . is ERT."[34] The *Newsweek* article illustrated this dilemma by discussing the role of estrogen both in slowing the progression of osteoporosis and in advancing the development of cancer. It also drew attention to menopause support groups started by women to "advise one another" and to "counter the negative attitudes of doctors," evidence of the spreading influence of the women's health movement.[35]

However, as early as 1982, popular articles began to report that "the medical pendulum is apparently swinging back in *favor* of replacement therapy."[36] With the publication of the medical studies on the addition of progestin (to slough off the endometrium) and the effect of estrogen on osteoporosis, the physicians (usually academics affiliated with medical schools) consulted and interviewed by journalists were more inclined to tout the benefits of hormone replacement. This change in attitude was reflected in the tone and content of the articles and their titles: "Breezing through Menopause," "Hot Flash: Good News about Estrogen," "Good News on Estrogen for Older Women."[37] These articles relied heavily on medical experts, either to provide information or to write the articles themselves (often excerpted from forthcoming books). Although a few journalists gave passing notice to women's efforts to avail themselves of nonmedical approaches to menopause (e.g., support groups, diet and exercise), most fell back on the medical model of menopause: not as a disease but rather a phase best navigated under the supervision of a physician, who would prescribe estrogen if necessary. The cultural authority of medicine in the realm of older women's health went unquestioned; rather, women were reassured that it was medically, and socially, acceptable to seek professional treatment.

Estrogen received a further boost from the publicity surrounding osteoporosis. While the first articles on this bone disease tended to emphasize diet and exercise as preventive remedies, estrogen began to receive greater endorsement as an effective, and safe, therapy. *People* magazine ran its first article on osteoporosis in 1985, an interview with Dr. Morris Notelovitz, who had recently published a book called *Stand Tall! Every Woman's Guide to Preventing Osteoporosis*, in which he promoted the importance of calcium in the diet and, to a lesser extent, exercise.[38] Two years later, the same publication (which reached more than thirty million readers each week) interviewed Dr. B. Lawrence Riggs, an endocrinologist at the Mayo Clinic, who warned against relying on calcium intake to prevent bone loss. His prescription: estrogen replacement therapy. Riggs railed in particular against advertisements for calcium supplements that, he claimed, were "way out ahead of the scientific evidence."[39]

Riggs also presented a mixed message as to the value of public education on osteoporosis. On the one hand, he declared that even more women were susceptible to the disease, announcing that "half of all elderly women eventually have bone fractures because of osteoporosis." That ominous statistic alone would seem to indicate the importance of awareness, prevention, and early detection. Certainly, the public awareness campaign had been extraordinarily successful: whereas only 15 percent of the population could identify osteoporosis in 1982, in five years that figure had leapt to 85 percent.[40] However, he bemoaned the "young women coming to see me in their 30s who are afraid that if they don't have their five glasses of milk a day they're going to have a hip fracture in two or three years."[41] A little knowledge, he seemed to imply, could be dangerous in fomenting unnecessary hysteria.

For the food and calcium supplement industry, this fear was a boon for business. Manufacturers rushed a wide variety of calcium-enriched foods onto the market. The Coca-Cola Company, for example, came out with calcium-fortified Tab (a diet cola) to appeal to those 30-year-old women who were trying to maximize their calcium intake. Dairy producers developed calcium-fortified cheeses, yogurts, and even milk, in an attempt perhaps to increase the calcium to calorie ratio, since women tended to perceive dairy foods as fattening.[42] Pharmaceutical companies saw sales of their over-the-counter calcium products soar.[43] Tums, already a popular antacid, earned a whole new set of customers eager to get their calcium in tablet form.

However, as the decade progressed, estrogen definitely trumped calcium as the solution to the problem of osteoporosis. The *New York Times* reported the results of a Danish study published in the *New England Journal of Medicine* that found women who took large daily calcium supplements still lost bone.[44] Jane Brody, the *New York Times* health reporter, reinforced this conclusion. "While calcium wars rage in pharmacies, health food stores, and supermarkets throughout the country," she wrote, "research in osteoporosis is revealing that only a minority of cases of this bone-wasting disorder result from diets deficient in calcium."[45] One week later, in her "Personal Health" column, Brody revealed that "the decline of estrogens after menopause" was the most important cause of osteoporosis, which meant "a return to favor of long term E.R.T., at least for the large proportion of post-menopausal women who face rapid bone loss."[46] Recall that just ten years earlier Brody had asked, "Why Has Estrogen Fallen on Such Difficult Times?"

Other commentators writing in a variety of mass-market publications hailed the return of estrogen. "Medical consensus about estrogen replacement therapy has gone from love to hate and back to love again in two short decades," announced the *Washington Post* in 1985.[47] "Estrogen therapy is revived," proclaimed the *Commercial Appeal* of Memphis, Tennessee. *Redbook*'s article on estrogen claimed that "the facts can change your life,"[48] and *U.S. News & World Report* hyped hormone replacement as the "prescription for a healthy old age."[49] All of these articles made it clear that, in the late 1980s, a prescription for estrogen was no longer intended merely as a short-term remedy for menopausal symptoms;

rather, long-term hormone replacement was back in style, as a lifelong commit-ment to ward off disease.

The *Redbook* article promised more than good health to long-term estrogen users; it also suggested that estrogen could improve the appearance of their skin. Slowly but surely, anti-aging rhetoric crept back into the rationale for hormone replacement therapy. An article in *Ladies' Home Journal* set forth a manifesto for "the anti-aging lifestyle" that would attract many adherents in the coming years: "This is the best of all times for women. We have come into our own in an age of scientific breakthrough and medical miracles. But women, not surprisingly, al-ways want more . . . not only longer lives, but health and beauty, too. With science as our handmaiden, we may yet achieve that Eden. Plastic surgeons now rival sculptors with their results; cosmetic companies compete with pharmaceutical companies to develop more effective products."[50] Just a dozen years before, Susan Sontag had bemoaned the gender inequity in the greater obligation of women to maintain their youthful appearances. Now, in 1984, according to the (female) *Ladies' Home Journal* writer, this duty had become an opportunity to take advan-tage of the fruits of medical science, generously brought onto the market by American industry. Beauty and health were intertwined, with the ultimate goal of looking and feeling young. The observation made by historian Dorothy Porter that "the competition to be healthy in the twentieth century has been historically bound to the competition to be beautiful" held true.[51] While 1970s cultural critics like Sontag questioned the reasons for and fairness of this trend, particularly in terms of gender disparities, social commentators in the next decade voiced no concern about the social implications of the combined efforts of medicine and commerce to help women maintain a youthful façade.

What happened to the vigorous efforts to make room for older women in American society? With the oldest president ever to serve in the White House, the 1980s might have been an opportune time for improving the status of senior citizens. But Ronald Reagan personified America's obsession with youth (or the illusion thereof). With his full head of dark hair, his actor's good looks, and his younger second wife by his side, President Reagan defied the physical signs of aging. Moreover, his conservative administration fostered an anti-feminist back-lash that stalled much of the progress made by women in the previous decade.[52] Social activism on behalf of older women faded in the light of the commercial and cultural promotion of youth. And medicine was enlisted in the fight against aging, to forestall its effects on both physiology and appearance.

Pharmaceutical companies, armed with the latest scientific evidence on es-trogen, also promoted the medicalization of aging in the name of disease prevention and the maintenance of a youthful vitality. CIBA Pharmaceutical's educational pamphlet "MIDLIFE—NO CRISIS" offered this explanation of estrogen replacement therapy to prospective patients: "Basically, ERT replaces what nature has taken away so that the body can continue to experience the healthy, protective benefits present in the premenopausal years."[53] Most of the informational pamphlets produced by drug manufacturers in the 1980s (and strategically placed in physicians' waiting rooms) endorsed the medical management of menopause and the use of hormone replacement therapy, while paying lip service to the feminist ideal of patient education and involvement in health care. A booklet put out by Mead Johnson Laboratories, entitled "For the Woman Approaching Menopause," advised: "Menopausal women should keep up to date with the latest thinking about menopause from reputable professional sources. Avoid forming your judgments concerning management of your menopause on the basis of casual chats with friends, hearsay, and sensationalist media reports."[54] The message here was that there were both good and bad sources of information; medical experts were "right," nonprofessionals were "wrong," and women should steer clear of the latter. It was acceptable for women to pursue their own education in health matters, so long as they sought information from within the medical-pharmaceutical community.

Some physicians offered advice to women on menopause, aging, osteoporosis, and estrogen in books written for the mass market, just as they had in previous decades; lay authors also contributed volumes on these topics to bookstore and library shelves. Books written in the 1980s bore witness to the cultural trends identified in this chapter: the enduring authority of medical science, the individual pro-active approach to staying healthy, and the preoccupation with avoiding the appearance of aging. Taken together, this literature characterizes the evolution of medical and popular thought during the 1980s, and it is worth examining three examples in some detail to get a sense of just how rapidly estrogen regained its status as a long-term therapy.[55]

*Stand Tall! Every Woman's Guide to Preventing Osteoporosis* came out at the end of 1982, co-authored by Morris Notelovitz, a South African-trained obstetrician-gynecologist who founded and directed the Center for Climacteric Studies at the University of Florida, and his colleague Marsha Ware.[56] Over the next two years, articles excerpted from or based on this book appeared in *Prevention, Vogue,* the *New York Times, Your Good Health, Family Circle, Shape, Health, Glamour, Newsweek,* and *Consumer Reports. Stand Tall!* helped to create the buzz about osteoporosis in the early 1980s.[57]

Notelovitz and Ware used scare tactics to get women to take action against bone loss. In the text of the frontispiece, they listed several questions "For Every Woman Over 30," including "Are you thin? Are you petite? Do you have a fair complexion? Are you a compulsive dieter? Do you hardly ever exercise? Is your diet heavy in protein? Do you take antacids? Are you a smoker? Do you avoid milk and other dairy products? Do you suffer from excessive stress?" They then warned, "If you've answered 'yes' to even a few of these questions, you may be headed toward chronic disability for the last third of your life."[58] Women who continued turning the pages read in the preface that "more common in older women than heart attacks, strokes, diabetes, rheumatoid arthritis, or breast cancer, osteoporosis presents one of the greatest health threats to your later years both in terms of the quality and the length of your life."[59] Later on, in their discussion of the emotional costs of osteoporosis, the authors solemnly declared, "You may feel less attractive and much older than your peers. You must face the fact that *never again* will you look the way you used to. The physical changes in shape and posture will make it difficult for you to find clothes that fit properly and attractively."[60] Whereas other sections told of physical disability, this description played on women's fears of the outward signs of aging.

But the book was not all doom and gloom. At the outset, the authors promised readers the information necessary to enable them to work cooperatively with their physicians to prevent osteoporosis. "Designing an individualized program of prevention involves decisions to be made *together by an informed woman and her physician* . . . The information you need is here in *Stand Tall!* Read it, think about it, and go to your physician with questions . . . Become an informed health-care consumer, and take an active role in making these years healthy ones!"[61] Notelovitz and Ware adopted the language of informed consumerism championed by the women's health movement, while at the same time staying true to the authority of the medical profession in matters of health care.

The authors touted nutrition and exercise as the best ways to fight the disease. Consuming adequate calcium, avoiding excessive amounts of protein, cutting down on tobacco and alcohol, and performing weight-bearing and cardiovascular exercises were pitched as common sense, low-cost preventive health measures. They pointed out that these efforts might not be enough for some women, in which case they suggested hormone therapy as "the final option."[62]

In their discussion of the pros and cons of hormone therapy, Notelovitz and Ware emphasized the former over the latter. They acknowledged the controversial history of estrogen, then reassured readers that "hormone therapy can prevent osteoporosis and can also be amazingly safe if used properly by healthy women."[63] In discussing the possibility of endometrial cancer, the authors minimized the

absolute risk by pointing out that "even if estrogen therapy increases that incidence to 8 per 1,000 per year [the expected incidence being 1 per 1,000], this
means that of 1,000 women receiving estrogen therapy, 992 will *not* develop
endometrial cancer."[64] To put the relative dangers of osteoporosis and cancer in
perspective, they noted that hip fractures caused 40,000 deaths annually as compared to 2,300 deaths from endometrial cancer, not all of which could be attributed
to estrogen therapy.[65] It is interesting to contemplate the number of women who
might have been seduced by the ease of taking a daily pill to prevent osteoporosis,
rather than undertaking an arduous campaign of daily exercise and mindful
eating. Still, Notelovitz and Ware advocated first and foremost the importance of
establishing good habits at an early age; they recommended hormone therapy only
for those older women who had effectively passed the point of no return with
respect to bone loss. Their advice mirrored the cautious optimism of academic
medicine toward the use of estrogen in the early 1980s.

Whereas *Stand Tall!* emphasized the perils of the age-related disease of osteoporosis, *Women Coming of Age*, by Jane Fonda, focused on strategies to stave off
a wide range of physical and aesthetic effects of aging. Published in 1984, *Women
Coming of Age* followed on the heels of Fonda's enormously successful *Workout
Book*, which topped Publisher's Weekly's nonfiction best seller list in 1982, remained on the list (at number five) in 1983, and sold more than 1,500,000 copies
by 1984.[66] Fonda traded on her good looks and her star quality to sell women a
nutrition and fitness program to "help them meet time and nature head on with
confidence, good health and vigor."[67] Illustrated with photographs of the actress
skiing, jogging, aerobicizing, and preparing her signature protein shake, the book
offered a compelling reinterpretation of what middle age could look like. The
paradox she presented—aging women should be accepted in the same way that
aging men are accepted ("My pal Redford gets furrows and character lines. I get
wrinkles and crow's feet," she protested. "It ain't fair!"[68]) *and* women should fight
the signs of aging—went unquestioned. Fonda was not the only one to straddle
the two horns of this dilemma; much of the popular discourse about women and
aging in the 1980s (and in subsequent decades) communicated this same quan-
dary. Her approach epitomized the health consciousness of the 1980s, as well as
the determination of many midlife women to control the visible signs of growing
older.

Informed by the rhetoric and ideology of second wave feminism and the
women's health movement, Fonda promised both information about and instructions for the care of the midlife body. She envisioned herself and her age cohort as
being "at the heart of a major redefinition of 'middle age,'" and she began by

debunking what she considered to be the "mythology of aging."[69] Aging, she declared, was "not a straight downhill dive after thirty . . . not a loss of intelligence or the beginning of senility . . . not a disease."[70] That said, she implicitly acknowledged that aging could bring about debilitation and decrepitude, because much of the rest of the book was devoted to a diet and exercise program to "prevent, slow down, and in some cases even reverse the degenerative changes associated with both aging and disease."[71]

One of these potential degenerative changes was, of course, osteoporosis. Fonda spent several pages explaining the causes of osteoporosis and how to prevent bone loss. Not surprisingly, she recommended a calcium-rich diet and frequent exercise. She noted that estrogen therapy was sometimes prescribed for women at high risk, and she allocated one full chapter to "the hormone therapy decision," in which she presented the risks and benefits of estrogen. Although Fonda described herself as "extremely conservative when it comes to the administration of any drug or hormone," ultimately, she concluded, each individual woman had to make her own decision, based on the latest information and in consultation with a physician.[72] In this way, Fonda's advice matched that of Notelovitz; both authors advocated a synthesis of informed consumerism and medical expertise in making choices about estrogen.

For Fonda, like Notelovitz, estrogen was a measure of last resort. She hoped to convince women that her "prime time workout" (the details of which occupied almost one-third of the book's pages) would provide the antidote to physical decline in middle age. She clearly struck a chord with her readers, who helped *Women Coming of Age* climb to the number seven spot on the nonfiction best seller list for 1984.[73]

Jane Fonda's celebrity appeal probably accounted for a significant portion of the interest in her book; the photograph of her beautiful smiling face on the cover must have attracted many prospective readers. By contrast, the cover of *Estrogen: The Facts Can Change Your Life*, published two years later in 1986, was all business: it featured the title, several more lines of text describing the contents, and the M.D. suffix displayed prominently after the name of the lead author, Lila Nachtigall. The message of this book signaled the dramatic about-face on the merits of estrogen for postmenopausal women.

*Estrogen: The Facts Can Change Your Life* sought to allay women's anxieties about the safety of hormone replacement therapy and to convince them of both its benefits and its necessity. Drawing on her ten-year prospective research study (see chapter 8), Nachtigall assured her readers that "ERT is safe when used correctly in the new medically approved way," by which she meant in low doses and in conjunction with a progestin.[74] She described the main reasons for taking estrogen: severe menopausal symptoms, osteoporosis, sexual difficulties, recurring urinary infections, early menopause, and instant (surgical) menopause. Although the first of these conditions could potentially be treated with short-term therapy, the other problems implied a long-term commitment to prescription hormones. According to Nachtigall, women at high risk for osteoporosis should begin taking estrogen soon after menopause and continue taking it for the rest of their lives.[75] However, if osteoporosis did not seem to be a good enough reason to join the estrogen bandwagon, Nachtigall provided another compelling motivation. "Virtually every woman," she warned, "will eventually have to give up sexual intercourse unless she starts taking estrogen."[76] This dire pronouncement surely sent many heterosexual women running to their physicians with requests for estrogen prescriptions.

Nachtigall also touted the "fringe benefits" of estrogen: younger skin, stronger muscles, and firmer breasts.[77] "Women on estrogen replacement therapy definitely tend to look more youthful than their years," she averred.[78] Although she cautioned women not to take estrogen solely for cosmetic purposes nor to believe that estrogen could actually "stop the clock," the news that estrogen could make faces and breasts—those outward signs of femininity and sex appeal—look younger must have acted as a further incentive for women considering estrogen. *Estrogen: The Facts Can Change Your Life* promised a pharmaceutical fix, in contrast to Jane Fonda's diet and exercise regime, for fighting the appearance of growing old.

However, it would be wrong to paint Nachtigall as merely a pill pusher. Her advice was based on her research, which had been widely accepted by academic medicine. She did not proselytize estrogen for everyone; rather, she identified specific cases in which women would, she believed, benefit from its use. Nachtigall also advocated exercise and nutrition as ways to maintain strong bones, but she recognized that many older women had not had the benefit of a lifetime of good health habits; they were her prime candidates for long-term estrogen replacement therapy to halt the loss of bone. Nachtigall also respected the role of the patient in the doctor-patient relationship. She advised her readers to "shop around for a doctor . . . remember you are not married to this person and you don't need a divorce if the match doesn't work out."[79] Yet, along with patient empowerment, Nachtigall, like Morris Notelovitz, remained committed to the medical management of menopause and postmenopause. "Take charge of your own body," she wrote. "And make your own decisions—with the help, of course, of a competent and knowledgeable physician."[80]

*Estrogen: The Facts Can Change Your Life* did very well. The first edition, which was also reprinted in paperback, sold around 100,000 copies. Two updated editions were published (in 1995 and 2000) and translated into seven languages, which accounted for additional sales of another 100,000. The success of this book is all the more surprising when contrasted with the failure of Nachtigall's first book for a lay audience, published in 1977. A comparison of the fates of these two volumes shows how greatly opinions about estrogen (and, more broadly, the medical management of menopause) had changed in the space of less than a decade.

The 1977 book was called *The Lila Nachtigall Report*, because the editor hoped to cash in on the success of Shere Hite's eponymous study of female sexuality, *The Hite Report*, which had been a best seller the year before.[81] *The Lila Nachtigall Report* was more comprehensive than *Estrogen: The Facts Can Change Your Life* as an account of menopause, aging, and the therapeutic role of estrogen. In fact, the second book seemed to recycle and abridge much of the material from the first. In that earlier work, Nachtigall took care to explain thoroughly the physiology of menopause. She came across as very open-minded about alternative remedies, such as vitamin E, for menopausal symptoms, and she stressed the importance of informed participation in health care, telling readers, "You must educate yourself about your body and have a good idea of how you want it treated."[82]

However, Nachtigall was an academic physician who believed in the power of the randomized controlled trial and in the ability of modern endocrinology to treat hormonal problems. She displayed her confidence in estrogen as a worthy therapy, and she disputed the findings of the studies of estrogen and endometrial cancer (because of their retrospective nature) and the publicity those studies received. "If the conclusions of the three studies had not been jumped upon with such fervor by the media, which, of course, can't be blamed for searching out newsworthy stories," she observed, "and then taken up with enthusiasm by well-meaning politicians, a more balanced view might have been possible."[83] Her faith in estrogen stemmed, of course, from her recently completed study; although the articles had not yet been published in the medical literature, she featured the results of her work prominently in this book. However, her advocacy of estrogen in 1977 was cautious; she presented arguments both for and against estrogen, as well as chapters on its possible side effects and "when it's wrong for you." Only about a third of the book discussed estrogen; almost half was given over to what to expect at menopause. By contrast, almost two-thirds of the chapters in the 1986 book focused on the positive aspects of estrogen therapy, with only brief mentions of its potential downside.

In her first book, Nachtigall demonstrated her sensitivity to the plight of the older woman in American society. She acknowledged the multifaceted etiology of menopausal symptoms: "To this female, fairly feminist physician's mind, there are both physical and psychological aspects which reinforce each other into a complex interchange that makes them hard to sort out."[84] She recognized the social pressure and validated the individual desire felt by some women to continue to look young. "Whatever the reasons [for long-term estrogen replacement therapy], it doesn't matter whether anyone else considers them unimportant, archaic, anti-feminist, or anything else . . . it's your decision."[85]

It is interesting to contemplate why *The Lila Nachtigall Report* sold so poorly. Nachtigall contends that her editor neglected to advertise the book properly, and the lack of a strong public relations campaign may indeed have rendered the book invisible.[86] But it is likely that other cultural factors came into play. Recall that 1977 was the year in which Rosetta Reitz published *Menopause: A Positive Approach* and Barbara and Gideon Seaman published *Women and the Crisis in Sex Hormones*. These two books reflected the contemporary critical stance against the organized medical profession and the pharmaceutical industry and the concurrent fascination with so-called natural approaches to health care. In this climate, *The Lila Nachtigall Report* didn't stand a chance. Written by a doctor who was promoting the use of a drug that was at the nadir of its popularity, castigated by scientists and feminists alike, the book was out of step with the times.

Yet, just nine years later, its thesis, shortened and simplified in *Estrogen: The Facts Can Change Your Life*, found a receptive audience. By 1986 the weight of medical opinion had shifted back to support the use of estrogen and this new positive reevaluation of an old drug had been successfully communicated to the public. Nachtigall had helped to turn the tide of medical opinion with her clinical studies of estrogen and osteoporosis in the late 1970s, but she had to wait until the next decade—when medical authority was less contested, when preventive drug therapy became more accepted, and when personal efforts to retain youth trumped social initiatives to change the status of aging women—for her message to be heard.

Lila Nachtigall considered herself to be an advocate for women. Indeed, a review of her career should earn kudos from the staunchest feminists. She entered medical school in 1956 at New York University, where she was one of four women in a class of 160.[87] Trained as an internist, she also qualified in obstetrics and gynecology so that she could work in the budding field of reproductive endocrinology. Nachtigall married during medical school and had three daughters while working full-time in the 1960s; in this way, she was among the early pioneers of professional women who juggled families and careers. Both her clinical research and her medical practice centered exclusively on women's health, on matters such as diabetic and other high-risk pregnancies, infertility, menstrual disorders, menopause, and osteoporosis. She spent twenty years as an attending gynecologist at Goldwater Hospital, the public chronic disease facility in New York City, providing care for debilitated women. It might be expected that she would be embraced by the women's health movement as a leader in women's health advocacy.

Nothing could be further from what actually happened. Although she was not individually singled out for reproach, Nachtigall fit the description of what radical health feminists derisively described as "white-coated gods, antiseptically direct-ing our lives."[88] Activists who claimed the mantle of health feminism in the 1970s and 1980s protested the incursion of the medical profession into natural events of the life course such as menopause. They were most critical of those physicians who continued to prescribe hormones, especially after the association between estrogen and endometrial cancer had been made.

The question that arises is what did it mean to be a feminist in the area of health care? Both physicians and activists believed they were working on behalf of women's best interests, but they came to radically different conclusions about the best course of action for midlife and older women to take. In order to more fully explore the multiplicity of conceptions and definitions of "feminist" health care for postmenopausal women, we must turn to a consideration of older women's health activism in the 1980s, as well as the tensions and contradictions experi-enced and articulated by older women in the face of competing models of meno-pause and aging.

# CHAPTER TEN

# Skeptics and Believers

## Varieties of Women's Responses

As early as 1979 the Boston Women's Health Book Collective had plans to incorporate information on aging into subsequent editions of *Our Bodies, Ourselves*. The members acknowledged that, as young and middle-aged women, they had no firsthand knowledge of what it meant and felt like to grow old, but they were seeking input from women with that direct experience.[1] Initially, the Collective planned to add a single chapter, called "Women Growing Older," to *Our Bodies, Ourselves*, but, as collaborators Paula Brown Doress and Diana Laskin Siegal recalled, "the task of squeezing the health and living issues of four or five decades of life into one chapter seemed insurmountable." They realized that the second half of life deserved a book of its own, which led to the publication in 1987 of *Ourselves, Growing Older: Women Aging with Knowledge and Power*.[2]

The authors anticipated that the generation of women who had learned about their bodies from *Our Bodies, Ourselves* would turn to this sequel as they grew older. "Taking control of our own lives and of our bodies is the most basic feminist principle there is," proclaimed Tish Sommers, one of the founders of OWL, the Older Women's League, in the foreword. "This book moves us in that direction, for it breaks down that formidable barrier in our minds—the fear of growing old."[3] But she also hoped it would have a bigger social and cultural impact. By exposing the combined impact of sexism and ageism, Sommers wrote, the book "opens up new ways to free ourselves of both of these, not only in our own minds but also in the society around us."[4]

*Ourselves, Growing Older* closely followed its predecessor in tone and approach. Forty-five women collaborated in the writing and editing, with input from another three hundred participants. The book was divided into three sections. The first discussed personal issues such as diet, exercise, and appearance. The second addressed sexuality, relationships, housing and living arrangements, and financial matters. The third, and longest, section tackled medical problems, from dental health and urinary incontinence, to osteoporosis and cancer, to dying and death. The presentation of all this compiled information—more than four hundred pages, plus sixty-two pages of additional resources—was infused with the spirit of self-help and self-empowerment pioneered in *Our Bodies, Ourselves*. The book also echoed the trenchant critique of the existing health care system first articulated in the earlier volume, and it offered specific strategies for older women to avoid abuse and neglect at the hands of medical providers. *Ourselves, Growing Older* encouraged women to become active participants in their medical care and health maintenance as they aged, but above all, it sought to change women's own attitudes toward the second half of life.

The topic of menopause merited its own chapter in *Ourselves, Growing Older*. Not only did the authors object to the characterization of menopause as a disease, they also rejected the use of the term *symptoms* to describe the changes experienced by women at this time of life, preferring instead to speak of the "signs" of menopause, namely, hot flashes, vaginal changes, and—the only universal sign of menopause—the cessation of periods.[5] After a lengthy description of hot flashes, the authors commented, "Physicians and researchers would gain much by listening carefully both to the descriptions that women give of their sensations of hot flashes and to the techniques women have found to minimize and relieve their discomforts. We hope that a better understanding of the mechanism of hot flashes will lead to new approaches and self-help techniques for relief rather than just to new drugs."[6] Clearly, these feminists were neither anti-science nor anti-medicine; they encouraged scientists and doctors to study the physiology of the menopausal hot flash. What they opposed was the incursion of the pharmaceutical industry into medical research and practice, with the assumption that research must result in a commercial medication for doctors to prescribe to patients.

The authors also dismissed as a myth the notion that decreased hormone levels after menopause inevitably caused sexual intercourse to become intolerably painful for women.[7] They advised women to try over-the-counter lubricants first and to resort to low-dose estrogen cream (not oral estrogen) if the other products didn't work.[8] Contrary to the ominous warning in Lila Nachtigall's book *Estrogen: The Facts Can Change Your Life*, this one reassured older women that they could continue to have sex for the rest of their lives without establishing a pharmacological dependence on estrogen. Both Nachtigall and the authors of *Ourselves, Growing Older* wanted to validate and facilitate the sex lives of older women, but they advocated very different means to achieve the same end.

The authors of *Ourselves, Growing Older* also agreed with Nachtigall that osteoporosis was a "vital women's health issue," but they disagreed with her over the role of estrogen in the prevention of this disease. Whereas Nachtigall advocated estrogen as the best way to forestall bone loss, the authors of *Ourselves, Growing Older* recommended nonmedical remedies such as diet and exercise. They expressed concern over the numerous disadvantages associated with estrogen use (such as the increased risk of cancer, the resumption of periods, and the greater expense) and skepticism over the motivations of the drug companies in representing estrogen as a therapy for osteoporosis in the 1980s.[9] Here, too, the feminist activist-writers and the female physician-author differed little in their shared concern for women's health but greatly in their approach to addressing the particular problem of osteoporosis. For Nachtigall, medicine and pharmacy offered a potent antidote; for the *Ourselves, Growing Older* group, the power of the medical-pharmaceutical complex obscured the dissemination of information about alternative ways to reduce bone loss. This disparity reveals the divide between what became known as the medical model and the feminist model of menopause and aging.

By the early 1980s feminist scholars in the fields of anthropology and sociology had begun to engage in critical discourse about the medicalization of menopause and aging.[10] What they discovered, however, was that the distance between the medical and feminist models of menopause was not as great as feminist activists claimed it to be. In 1979 anthropologist Judith Posner compared the presentation of menopause in gynecology textbooks from the 1960s and 1970s with the chapter on menopause in the 1976 edition of *Our Bodies, Ourselves.*[11] She found that the feminist authors and the physician authors shared assumptions about the psychological nature of menopause. While the gynecologists often considered women's complaints to be "all in their heads," she noted that the feminists also tended to stress social-psychological causes over physiological factors, which resulted in a similar dismissal of women's lived experiences as being somehow imagined. In 1982 social scientist Patricia Kaufert compared two guides to menopause written for lay audiences and published in 1977: *A Doctor Discusses Menopause and Estrogens,* written by M. Edward Davis, M.D., and published by Budlong Press, which specialized in booklets on health topics, and *Health Resources Guide: Menopause,* written in part by Rosetta Reitz (author of the 1977 book *Menopause: A Positive Approach*) and published by the newly formed National Women's Health Network (NWHN).[12] Kaufert found that the guides differed significantly in their ideas about who should control the diagnosis and treatment of menopausal experiences, doctors or women themselves. However, she observed that both models based their dissimilar advice on a similar assumption, namely, that menopausal women should strive to maintain both their physical attractiveness and their sexuality. Davis's recommendation to take estrogen pills and Reitz's recommendation to use safflower oil on one's skin assumed that women shared the desideratum of looking young. Although the medical view regarded sex as legitimate only within the context of marriage and the feminist view embraced homosexuality and masturbation, both models were, in Kaufert's words, "equally class- and culture-bound."[13]

It is important to realize that these social scientists were working with models of menopause articulated in the 1960s and 1970s. This version of the medical model still smacked of Wilsonian paternalism, and this early feminist response was the first attempt by activists to counter the "feminine forever" mentality. By the mid-1980s, both models had softened a bit, but they retained their basic tenets. That is, the medical model still expected physicians to take charge of the management of menopause, but it allowed for input from the patient. And the feminist model still held that menopause was a natural part of aging, but it recognized that some women really did need relief from distressing physical symptoms. However, as the 1980s progressed and more and more scientific data came out in support of the use of hormone replacement therapy, the gulf widened between the two approaches on this issue. Whereas physicians enthusiastically advocated the long-term use of estrogen to forestall the diseases of aging, feminist health activists objected to the medicalization of the remaining decades of a woman's life after menopause and posited instead that aging could be "managed" with diet, exercise, and spiritual, emotional, and intellectual fulfillment.

To promote the dissemination of nonmedical information about menopause and aging, a couple of newsletters began publication in the early 1980s. In 1981 *Hot Flash* was founded as the official publication of the National Action Force of Midlife and Older Women. Consistent with the approach and objectives that would be taken up by the authors of *Ourselves, Growing Older*, the newsletter was a collaborative effort of "a voluntary group of 14 women ranging in age from 30 to 85, dedicated to bringing you, our members, the latest information of concern to midlife and older women."[14] In 1984 Montreal author Janine O'Leary Cobb began a newsletter called *A Friend Indeed for women in the prime of life* to provide "balanced, woman-centered health information and knowledge about menopause and mid-life."[15] Featuring articles on topics such as menopause, estrogen replacement therapy, osteoporosis, breast cancer, and sexuality, both newsletters offered subscribers a holistic approach to health in the second half of life.

Some women preferred to talk about menopause instead of just reading about it, so they joined self-help groups or attended seminars. As the notion that women should seek out health education began to take hold, the provision of menopause information expanded beyond printed sources. In the early 1980s local hospitals and Planned Parenthood affiliates offered menopause seminars run by nurses and health educators; Planned Parenthood of Central Missouri sponsored the development and publication of a teacher-training manual in order to involve more lay people in facilitating these sessions.[16] Some women started their own menopause discussion groups to "counter the negative attitudes of doctors."[17] These support groups sprang up around the country, modeled, like the workshops convened by Rosetta Reitz, on the women's liberation consciousness-raising rap groups of the previous decade. Self-help groups also provided midlife and older women with a forum in which to describe broader concerns about aging. As one participant commented, the groups "provide[d] an opportunity for older women to plug into the feminist movement."[18]

Feminist-run seminars and support groups had to compete with physician-run menopause clinics that also arose in the 1980s. These clinics were often affiliated with university medical centers, where women could get medical advice and treatment from academic gynecologists and endocrinologists. By 1987 Yale, UCLA, Baylor, George Washington University, the University of California, San Diego, and the University of Illinois had set up clinics, as had well-known hospitals such as Brigham and Women's in Boston, Northwestern Memorial in Chicago, and Mt. Sinai in Cleveland.[19] These clinics, in contrast to the groups and sessions sponsored by feminist health workers, operated within the medical model of menopause, and as estrogen regained favor within academic medicine, those physicians associated with the clinics advocated its long-term use.

The impact of the messages conveyed in feminist newsletters and scholarship was limited because they reached a relatively small number of women. Feminist scholars published their research in low-circulation professional journals, read mostly by other academics. The newsletters—started in the days before desktop publishing really took off—had at most a few thousand subscribers. *Ourselves, Growing Older* fared better. The first edition sold 100,000 copies (from 1987 until 1994) and the second edition (from its publication in 1994 through 2004) sold 87,000; these sales figures approached those for Lila Nachtigall's *Estrogen: The Facts Can Change Your Life* in the same seventeen-year time period, but they never reached the numbers necessary to make the national best seller lists.[20] It was especially difficult for alternative models and critical theories to compete with the medicalization of menopause and aging and the widespread dissemination of this model in the mainstream popular media, whose audiences measured in the millions.

For example, the recommendation to use hormone replacement therapy to maintain one's sex life after menopause reached more than twenty-one million Americans who read the article "Does Sex End at Menopause?" in *McCall's* November 1987 issue.[21] Although the article (presented as a sidebar to the feature article "Menopause: A Complete Medical Report") began with the statement, "In general, a woman's potential to experience sexual pleasure is not diminished by menopause," it went on to catalog the numerous problems that might arise as a result of decreased estrogen production: decreased clitoral sensitivity, loss of ability to orgasm, painful intercourse as a result of vaginal dryness, and urinary incontinence during sexual excitement. The solution: estrogen replacement therapy, which "can dramatically remedy these losses." The article ended by advising women to seek help at a menopause clinic, where they would be more likely to receive a prescription for estrogen, rather than a suggestion to purchase an over-the-counter lubricant.

Recall that estrogen received an increasingly positive presentation in several popular magazines as the 1980s progressed. Women's magazines like *Redbook* and *Ladies' Home Journal* reached fourteen million and nineteen million readers each month, respectively, while twenty million people read *Newsweek* and thirty million leafed through the pages of *People* every week.[22] Articles that touted the osteoporosis-preventing and anti-aging properties of estrogen were far more widely distributed—and thus had a far greater influence—than anything written in *Ourselves, Growing Older* or other feminist publications.

However, the increase in estrogen use after 1980 can be attributed not merely to the scale of the circulation of the medical model of menopause and aging, but also to the substance and style of the message contained within. What seemed to be persuasive medical evidence of the long-term benefits of hormone replacement therapy was presented as the scientifically proven best course of action for postmenopausal women. Moreover, this message was conveyed without the paternalistic overtones of the previous decade. In the cultural context of the 1980s (as discussed in the previous chapter), physicians and other mainstream authors relied on women's interest in maintaining their health (and youth) along with the cultural authority afforded to medical science to make the convincing case that taking estrogen was the right thing to do for well-informed, pro-active women.

Of course women did not have to subscribe exclusively to either the medical model or the feminist model. And individual attitudes toward menopause, aging, and the use of replacement hormones could change over time, depending on one's personal circumstances. The question of what women thought, and when they thought it, is a very difficult one to answer. The editors of the *Saturday*

*Evening Post* tried to address this issue with a tear-out survey in the December 1986 issue. The responses to their questionnaire—more than forty-five hundred altogether—represent an extraordinary snapshot of American women's experiences of and opinions about menopause, postmenopause, and hormone replacement therapy in the mid-1980s.

The *Saturday Evening Post*'s "Menopausal and Postmenopausal Women's Survey" appeared in December 1986 along with two feature articles on estrogen. The first, "Might Estrogen Prevent Memory Loss?," presented the text of an address given by a Rockefeller University geriatrician at a seminar on hormones and the brain. The second, "Is an Estrogen Skin Patch for You?," consisted of an interview with the aforementioned geriatrician, Dr. Howard Fillit, conducted by the magazine's publisher and editor, Cory SerVaas, M.D. The monthly food column offered several calcium-rich recipes under the heading "Bone Boosters"; it was accompanied by a large advertisement for the Citracal brand of calcium citrate, with a notice that *Post* subscribers could receive a free hundred-tablet bottle of this calcium supplement when they renewed their subscriptions or ordered a gift subscription for someone else. Other articles in this holiday season issue included four pieces about Christmas, a short story ("The Christmas Visitation"), and reports on Dr. Heimlich ("The Man behind the Maneuver"), air bags, and Lowell Davis ("Ozark Artist").

The *Post* liked to trace its ancestry to 1728, with Benjamin Franklin as the first editor and publisher. Two hundred and fifty-eight years later, the monthly periodical described itself as "a magazine of content that matters for readers who care." In 1986 it was read by almost five million adults. The average reader was a 42-year-old female homemaker, with at least a high school education and a household income well above the national median.[23] Given the demographics of the *Post*'s audience, articles on estrogen and aging were bound to be of interest. It is worth considering the two articles in some detail, because their content and tone shaped the design of the survey and the responses it elicited.

The first page of "Might Estrogen Prevent Memory Loss" featured a large photograph of the author, Dr. Howard M. Fillit, with the caption, "Dr. Fillit has a primary goal: 'Not to make people immortal, but to improve the quality of life for the elderly.' "[24] Fillit's objective, along with his characterization of older women as victims of their declining ovaries, was strikingly reminiscent of Dr. William H. Masters's position three decades earlier. Fillit began his article with a recitation of the alleged differences between males and females ("men have better spatial abilities . . . women are better at verbal communication . . . men are more oriented to objects . . . women are more concerned with people . . . little boys, for example, seem to be always running around, while little girls are usually more sedate and calm"), attributing these variances to the effects of hormones on the brain.[25] Having established the critical role of estrogen in behavioral differentiation, he went on to catalog the physiological and psychological effects of decreased estrogen production in postmenopausal women. In addition to hot flashes (experienced, according to Fillit, by up to 80 percent of women), vaginal atrophy (and painful intercourse), urogenital atrophy (and incontinence), and osteoporosis, menopause also brought on emotional disturbances (depression, anxiety, irritability, memory loss, and difficulty concentrating). "Although grief associated with loss of reproductive capacity or the perceived onset of aging may contribute in psychodynamically predisposed women, psychological symptoms may be the direct result of the loss of estrogen's effect on the brain."[26] With this statement, Fillit perpetuated the notion that a woman's personal identity was tied to youth and reproduction and biologized the cause for what feminists would have judged a socio-cultural issue.

A biological problem could be remedied by a medical solution. Fillit asserted that "data from a number of well-controlled studies indicate that estrogen replacement therapy (ERT) relieves many of the psychologic symptoms during menopause, including depression and memory loss," although he did not reference any specific research.[27] He did cite two studies that he thought demonstrated a beneficial role for estrogen in the treatment of Alzheimer's disease: one dated from 1968, almost twenty years earlier, and the other was conducted with a total of eight subjects at Rockefeller University. Although Fillit cautioned that "we do not recommend the widespread clinical use of estrogen for Alzheimer's disease until further research has proved without doubt that such therapy is both effective and safe," he offered statistical evidence of estrogen's safety (he compared the 1 in 20,000 risk of death from uterine cancer among estrogen-progesterone users to the 1 in 60 risk of death from osteoporotic hip fractures among nonusers) along with his conviction that "the risks of long-term postmenopausal ERT are probably far outweighed by the benefits."[28]

*Post* readers had every reason to take Fillit seriously; he was presented as a top-notch medical researcher at one of the nation's premier scientific institutions. The editorial introduction to the article described Rockefeller University as "staffed with Nobelists and eager scientists engaged in research at the very frontiers of biological discovery" and Fillit as the author of "numerous papers in the fields of immunology and Alzheimer's disease during his ten years at Rockefeller University."[29] The second article, "Is An Estrogen Skin Patch For You?," further solidified Fillit's image as a distinguished authority on the subject of estrogen; the transcribed conversation between Fillit and *Post* editor Dr. Cory SerVaas consisted of leading questions from SerVaas and lengthy responses from Fillit, in which he pointed up the multiple benefits of estrogen for menopausal and postmeno-pausal women. To the very first query, "In your opinion, what percentage of postmenopausal women in the United States should probably be on estrogen," Fillit replied, "Probably most of the women should be on replacement therapy."[30] The following six pages of reassurance about the safety of estrogen and pro-gesterone, optimism about the positive health consequences of long-term hor-mone use, and enthusiasm for the newly approved skin patch delivery system must have convinced many middle-aged and older female readers to consider, or reconsider, going on estrogen.

We can uncover some of these readers' reactions within the survey forms they filled out and returned to the *Post*'s editorial offices in Indianapolis. The survey, designed by Fillit himself, reflected the author's assumption that there was a direct connection between the loss of estrogen at menopause and negative mood and cognitive changes, in addition to the generally accepted effects of hot flashes, vaginal atrophy, and osteoporosis. The front side of the survey asked eight multi-part questions, most of which could be answered yes or no, about cognitive abilities, mood disturbances, hot flashes, and the use of estrogen. The back side was divided into three sections: the top third asked five more yes/no questions about long-term replacement therapy, osteoporosis, height loss, and broken bones; the middle third was addressed to the *Post*, with space for a stamp and return address; and the bottom third showed a cartoon character of a physician pointing to a list of six risk factors for osteoporosis (calcium and estrogen deficien-cies, cigarette smoking, postmenopausal white females, excessive alcohol con-sumption, prolonged inactivity, and childlessness), which might very well have influenced women's answers to the questions about estrogen and osteoporosis.

Within two weeks, 1,822 women had returned their completed surveys. The *Post* tabulated these results and published them in the next issue of the magazine, as part of an article entitled "More about Estrogen Skin Patches," along with interviews with women who switched from estrogen pills to the patch and advice from Lila Nachtigall, who ran a clinical trial of the transdermal patch.[31] Clearly biased in favor of estrogen and the medical management of menopause and postmenopause, this article typified popular magazine coverage of the subject in the mid-1980s. The article did note that Nachtigall advised women to try non-prescription remedies before opting for hormone therapy, but it also reprinted six paragraphs verbatim from *Estrogen: The Facts Can Change Your Life* on the need for women over forty to consult regularly with physicians: general practitioners, gynecologists, and, if necessary, reproductive endocrinologists. Menopause, according to Nachtigall, ought to be medically supervised, and the physician-editor of the *Saturday Evening Post* concurred.

The results of the survey supported the success of the current medical model; of course the respondents were a self-selected group and not at all representative of the larger population of midlife and older women in America. Why would a woman bother to fill out and mail in the survey? Altruism, perhaps, since the survey was introduced with the notice, "Your willingness to participate will help discover better ways of treating the very real problems that some women suffer during and after menopause."[32] Since most of the respondents simply checked the yes or no spaces on the forms, it is hard to know their motivation for participation. The results indicated that many of them had experienced discomfort during or after menopause. Of the 1,822 women, 78 percent reported experience with hot flashes, and more than half said they suffered anxiety (55%), irritability (55%), or depression (51%). Two-thirds had taken estrogen at one time or another for menopausal or postmenopausal reasons. Although only 7 percent reported a vertebral fracture and 2 percent a broken hip, 29 percent claimed to have lost height and almost half (45%) planned to take long-term estrogen replacement therapy to prevent osteoporosis. Interestingly, the original survey was called "Menopausal and Postmenopausal Women's Survey," but the report the following month was titled "Preliminary Results of *Estrogen* Study" (emphasis added). The one-page summary listed the raw data: the number who replied yes or no or who gave no answer to each of the questions. Although the heading "preliminary" hinted that a more complete final report might follow, no such analysis ever appeared in the *Post.*

However, the responses kept flooding in. After the speedy arrival of the first 1,822 completed surveys in the first two weeks of November, another 1,800 came in by the end of December, 800 more by the end of February, and 200 stragglers dribbled in over the next eighteen months. Altogether, the *Post* recorded the receipt of 4,570 replies.[33] The survey form asked respondents to provide name, address, telephone number, height, weight, age, and race, and most of them complied. Every state in the union was represented, plus the District of Columbia, Puerto Rico, and Canada. A handful came from Americans living abroad, on every continent but Australia. The women who replied ranged in age from 26 to 94, with the large majority between the ages of 50 and 70. Most of the younger respondents had undergone surgical menopause, as a result of hysterectomy and oophorectomy; a significant proportion of older women had also had their reproductive organs removed. Except for a few who identified themselves as being of Asian descent, all of the respondents were Caucasian. Not a single person identified herself as black or African American, Hispanic or Latina.

In addition to checking the yes or no spaces on the survey form, many women provided additional information by elaborating in the margins or on extra sheets of paper, since the form allotted no space for comments. These unsolicited annotations reveal a wide spectrum of attitudes and practices concerning hormone replacement therapy. Even within the relatively narrow demographic slice of America represented by the *Saturday Evening Post* readers who took the time to fill out and return the survey—that is, mainly middle-class, middle-aged white women —there was a great range of positions, not only on hormone use but also on the roles of doctors and patients in the management of menopause and postmenopause, the importance of the dissemination of health information for lay audiences, and conceptions of female aging. Although the women who added personal remarks to their survey forms made up only a small fraction of the total number of respondents, their scribbled notes and neatly typed letters provide us with a glimpse of the ups and downs encountered by midlife and older women and the strategies they developed to cope with these challenges. The fact that thousands of women chose to complete the surveys demonstrates that women were interested in sharing their experiences of menopause and aging and learning about those of others; the added comments on hundreds of these forms further uncover the influences on women's interpretations of their experiences. These influences came from the contemporary pro-estrogen climate in the medical community, the competing anti-estrogen perspective among feminist health activists, and media coverage of both positions.

Many respondents added comments to illustrate the physiological and psychological details of their menopausal years, but a few women wrote to say that they had sailed easily through "the change." "I am sending reply," wrote one 70-year-old from Parsons, Kansas, "because I feel most of those replying had problems and I feel there are lots of us who did not."[34] Others located their experiences within the context of family changes. A 52-year-old from Camp Hill, Pennsylvania, described herself as "victimized by the menopause." In one year, her father died, her youngest son moved out "under angry circumstances," and her husband lost his job and then took up with a younger woman, because, she wrote, "he believes I deliberately lost interest in him (especially sexually)."[35] She felt that her initial ignorance about menopause and its attendant physical and emotional changes meant that she learned too late that medical help might have saved her marriage and her relationship with her child.

For some women, the information they gleaned from the article prompted them to consult their physicians about a prescription for estrogen and, specifically, the patch, not only to relieve vaginal dryness and hot flashes, but for the alleged long-term benefits as well. A 70-year-old from Columbus, Georgia, announced, "After reading your articles I went to doctor and got Rx for & am wearing skin patch for memory loss & osteoporosis."[36] Fillit's claims for the power of long-term estrogen also resonated with much younger women. Wrote a 42-year-old from Portland, Oregon, "I'm so grateful for your articles this month. I will certainly discuss it with my Dr with mag in hand! I'm terrified of this memory problem & osteoporosis."[37]

A repeated theme in the margin notes was gratitude to the *Saturday Evening Post* for publishing the articles and the survey. "I *appreciate* this article as I've been searching for more information and help," said one 44-year-old from Yadkinville, North Carolina.[38] A 58-year-old from Muleshoe, Texas, penned, "I'm so glad someone is interested."[39] The articles helped women to feel connected to others going through similar experiences. "Thank you, thank you, thank you!" gushed a 49-year-old from La Canada, California. "At last—this very real problem is being addressed. It's so good to know I'm not alone or imagining symptoms."[40] Some felt empowered by the information they read and emboldened to face up to their physicians. A 57-year-old from Glencoe, Minnesota, ended her five-page handwritten letter, "I thank you for your information. It was terrific! And helped me feel like I could talk intelligently at least to my Dr."[41] It is especially interesting that this woman identified herself as a registered nurse. A 52-year-old from Pittsburgh, Pennsylvania, added to her survey form: "Note: After reading your two articles, I phoned my gynecologist who prescribed Estraderm & Provera: vaginal atrophy & dryness are gone, PTL [Praise The Lord]! Thank you & keep up the good work."[42] A thousand miles west, a 57-year-old in Prairie Village, Kansas, reported, "After reading this article, I saw my gynecologist and requested the skin patch . . . Thank you very much for this information!"[43]

However, many women reported resistance from their physicians. "A lot of doctors I went to in search of an answer *do not recognize* ERT & think it's all psychosomatic," complained a 43-year-old from Satartia, Mississippi.[44] A 52-year-old from Huntsville, Alabama, wrapped up her three-page letter with the comment, "Doctors need to listen to what's happening to women and not write it off as 'nerves' or 'tired housewife syndrome' or any of the other excuses they find convenient to use."[45] And a 48-year-old from Issaquah, Washington, extrapolated from her personal struggle to draw conclusions about American medicine more broadly:

If my own experiences and those of acquaintances are any indicator, there are many physicians who would chose [*sic*] to dismiss all women's health complaints as "nerves". The passage through menopause taught me a great deal about my own resourcefullness [*sic*] and even more about the current state of medicine in America . . . Information in my case had to be dug out of libraries and medical texts, because each question asked of a physician seemed to be answered with some type of discounting statement, ranging from:

"How would you know about that?" to

"You're not the first person to go through menopause."[46]

This woman and several others reported that they had to plead with physicians to get a prescription for estrogen. Some were successful, others were not. All articulated their frustration with not being taking seriously or not being included in the medical decision-making process.

But for every respondent who expressed dissatisfaction or anger with an ill-informed or dismissive physician, there was another who was perfectly content to defer to her doctor's expertise. In response to the survey question, "Do you plan to take estrogen for *long-term* replacement therapy to prevent osteoporosis?," many answered along the lines of: "if the doctor advises it," "if doctor thinks I need it," "only if prescribed by doctor," "depends on my doctor."[47] Women conveyed a wide range of opinions about their relationships with their physicians; what is interesting—and probably reflective of changes brought about by the women's health movement and the consumers' movement—was the resolve of some women to control the course of their own health care. According to one determined 58-year-old from Ocala, Florida, "My Dr. won't prescribe it. I'm going to change Dr's."[48]

Several respondents who had not previously taken estrogen or who had avoided or stopped taking it because of the association with endometrial cancer in the 1970s commented that the *Post* articles had compelled them to reconsider. A 54-year-old schoolteacher from Fort Worth, Texas, remarked, "My Dr. has been trying to convince me the past year, but your article mentioning memory loss did it. I'm terrified of being 'forgetful'."[49] A 63-year-old from Pana, Illinois, reported that she took the Ogen brand of estrogen for ten years, until she was diagnosed with endometrial cancer. In response to the question about taking long-term HRT to prevent osteoporosis, she wrote, "I would like to take estrogen again. Do you have any advice for me?"[50] And a 72-year-old from North Little Rock, Arkansas, reported optimistically, "I quit taking them [Premarin pills] during the cancer scare . . . I have an appt. with my doctor for annual checkup and hope to have a discussion with him regarding *another* try at estrogen & progesterone."[51] The positive portrayal of estrogen as an agent capable of forestalling the dreaded diseases of aging struck a responsive chord with many readers.

Other women, however, adamantly refused to take estrogen under any circumstances. For some, the cancer risk was unacceptable, as for the 46-year-old from Hudson, New York, who announced, "I chose (choose) not to take medications [for menopause] because of 'cancerous' side effects."[52] Some doubted the effectiveness or necessity of hormones. "PROVE IT HELPS," challenged a 53-year-old from Wynne, Arkansas, "then I'll think about it!!!"[53] A 51-year-old disbeliever from Lakewood, New Jersey, scoffed, "I consider all talk of estrogen and extra calcium *quackery!!!* "[54] Some rejected hormones in favor of "healthy living," such as the 57-year-old from Bozeman, Montana, whose reply to the question of long-term HRT to prevent osteoporosis was "Definitely *No!* P.S. I exercise daily, eat nutritious food, downhill ski twice a week and have a very robust family history."[55] For some, it was mind over matter. "*Attitude*, in my opinion, prevents menopausal (post or pre) problems," declared a 61-year-old from Coronado, California. "I would *NEVER* use estrogen."[56] Others turned to a higher power, as did the 57-year-old from Chico, California, who professed to "have done a great deal of praying."[57] These estrogen resisters may have imbibed the ideas of the women's health movement, although none of them made any explicit reference to feminist ideology. It may have simply been personal conviction that led them to adopt what appeared to be the feminist model over the medical model of menopause.

By contrast, a significant number of women wrote glowing testimonials to the benefits of taking estrogen, celebrating its effects on their hair, skin, mood, and personality. "I've been taking estrogen for 30 years," noted a long-term user from Bay St. Louis, Mississippi, "[and] my complexion is *very good* for 73!!" A 62-year-old from Amarillo, Texas, attested, "I was taken off estrogen & was a monster, so have remained on it since 1961 [25 years]."[58] "Without estrogen therapy, I would have been and would continue to be almost a basket case," swore a 69-year-old from Woodsfield, Ohio. "Hurrah for estrogen and a doctor willing to use this treatment."[59] A 71-year-old who had been taking Premarin for 26 years reported, "I have vitality, energy, and a feeling of well-being."[60] And a 79-year-old from Lake Grove, Oregon, who had taken estrogen for 30 years boasted, "I am *much* younger *looking & feeling* than my friends who did not take estrogen."[61] The comments on these surveys looked very much like the letters women wrote to the Food and Drug Administration in defense of estrogen ten years earlier. Based on their personal experiences, these respondents sang the praises of what was, to them, a miracle drug.

In general, the estrogen advocates were older than the estrogen resisters in this 1986 survey. Most of those who commented favorably on the long-term use of estrogen were in their sixties and seventies, whereas most of the women who wrote about their rejection of estrogen were in their forties and fifties. The age of 60 seemed to be a significant dividing line between the two groups. If we assume that the average age of natural menopause was 50 (excluding women who had surgical menopause), then the women over 60 had entered menopause prior to 1976, in the "feminine forever" years, and the women under 60 had reached menopause after 1976, during a much more skeptical period, after the scientific association of estrogen use with an increased risk of endometrial cancer and the start of the feminist critique of the medicalization of menopause.

Above all, what we learn from the 1986 *Saturday Evening Post* survey is that the menopausal and postmenopausal experiences of American women—even when taken from a relatively homogenous population subgroup—could not be reduced to a few simple archetypes. Some were informed by the medical model and some by the feminist model; some incorporated parts of each paradigm into their health care decision-making, and some were guided wholly by their own principles. Women interpreted menopause as one thread within the fabric of their lives, intertwined with other experiences, such as getting divorced or going back to school or taking on a new job or caring for elderly parents or dealing with financial problems. What is clear, however, is that a majority (of this sample, at least) did turn to medicine for relief from physical discomfort or psychological distress during and after menopause. And more often than not, the treatment prescribed by physicians was hormone replacement. In spite of the so-called "cancer scare" of the previous decade, many women seemed to acknowledge the potential short-term and long-term benefits of estrogen, thanks, in part, to upbeat articles in mainstream magazines that brought news from the world of medical science of the rehabilitation of estrogen.

The impression that more menopausal and postmenopausal women were taking estrogen as the 1980s progressed was corroborated by the findings of a series of quantitative public health studies that looked at prescription figures from pharmaceutical marketing databases to determine the changes in hormone replacement therapy use patterns over time. One of the databases, the National Prescription Audit (NPA), estimated the total number of new and refilled prescriptions dispensed by retail pharmacies, based on information collected from a sampling of these stores across the nation. Another, the National Disease and Therapeutic Index (NDTI), provided data on disease patterns and treatments, drawn from reports submitted by more than two thousand doctors on the purpose and outcome of each patient visit during one forty-eight-hour period every three months. Thus the NPA data yielded the annual number of estrogen and progesterone prescriptions filled by drugstores, while the NDTI information indicated the reasons for which the hormones were prescribed to patients. The NDTI also revealed the ages of the patients, the specialties of the doctors, and the rates of hormone use according to geographical region.

After bottoming out at 14 million prescriptions in 1980, the oral estrogen market reversed its trend, growing to 17.8 million in 1983, 20 million in 1986, and 31.7 million in 1992.[62] The estrogen skin patch came onto the market in 1986; in 1987, the first full year it was available, 1.5 million prescriptions were dispensed.[63] Five years later, that number had jumped to 4.7 million. The number of progestin (synthetic progesterone) prescriptions also increased, along with the prevailing medical wisdom of prescribing it along with estrogen to encourage the endometrial lining to shed on a monthly basis (to avoid the development of cancer). In 1980, about 2 million prescriptions for progestins were dispensed; that number rose to 3.2 million in 1983, 5 million in 1986, and 11.3 million in 1992.[64] The reason for the rise in progestin use was clearly the new indication for combined hormone therapy. In 1979, 18 percent of progestin was prescribed for menopausal indications; by 1986, that figure had more than tripled, so that menopausal reasons accounted for 59 percent of all progestin prescriptions.[65] Another statistic confirms the increased use of progestins by menopausal and postmenopausal women. In 1980, 11 percent of Provera (Upjohn's brand of medroxyprogesterone) use was accompanied by Premarin; only three years later, that figure had risen to 30 percent.[66]

Most women receiving prescriptions for estrogen were between the ages of 40 and 59, accounting for 60–63 percent of the total in the 1980s.[67] The percentage of this age group accounting for progestin use increased significantly, from 42 percent in 1984 to 62 percent in 1992.[68] The percentage of women over the age of 65 on estrogen also grew, from 10 percent in 1982 to 15 percent by 1986 to 19 percent by 1992.[69] This increase reflected the trend to prescribe long-term hormone replacement to prevent osteoporosis in postmenopausal women. Although doctors most commonly prescribed calcium supplements for osteoporosis in the 1980s, the prescription of estrogen for that indication rose throughout the decade, according to the NDTI.[70] Perhaps not surprisingly, the majority of the physicians who prescribed estrogen specialized in obstetrics and gynecology (53% in 1984, 60% in 1992). Most of the rest practiced general or internal medicine (38% in 1984, 33% in 1992).[71]

What is surprising, and not easily accounted for, is the geographical distribution of estrogen use across the country. The NDTI measured "rates of total oral estrogen mentions per 100 women ages 45–64 by region," which can be roughly translated as the percentage of women in that age group who received prescriptions for estrogen.[72] In 1974, before estrogen was linked to endometrial cancer, the rate of use varied, as follows: East 30, Midwest 38, South 34, West 45. In 1979, at the nadir of estrogen usage, the rates had dropped in every region, but westerners still prescribed and used estrogen more than the others did: East 11, Midwest 14, South 15, West 18. By 1985, the difference according to geographic region was enormous: East 15, Midwest 27, South 26, West 43. It is by no means clear why the medical profession in the western states prescribed estrogen three times as often as did doctors in the eastern states. We might postulate that New England's Puritan heritage made its people more suspicious of a drug that could improve women's sex lives, and that Californians' freewheeling approach to sex made them more receptive to hormone replacement therapy, but these would be idle speculations based on hackneyed stereotypes. Of course, just because doctors prescribed estrogen does not mean that women went to the pharmacy to have the prescriptions filled, and even if women did fill the prescriptions, we cannot be sure that they actually took the pills. In fact, one study of twenty-five hundred postmenopausal women between the ages of 45 and 55 found that 20 percent stopped therapy within nine months, 10 percent took their pills only sporadically, and 20–30 percent never even had their prescriptions filled.[73] However, the NDTI data indicate that, for whatever reasons, women in the West would be much more likely to take estrogen during or after menopause as compared with their peers elsewhere in the United States.

Another interesting contrast showed up in estrogen use between the United States and other Western nations. In the early 1990s researchers at the U.S. Food and Drug Administration calculated that approximately one in six women over the age of 50 was taking estrogen (17%); Wyeth-Ayerst, the manufacturer of Premarin (Ayerst had merged with Wyeth in 1987), placed the proportion closer to one in four (25%).[74] By contrast, Australian researchers estimated that no more than five to ten percent of women (between 1 in 10 and 1 in 20) in Australia, the United Kingdom, and Scandinavia used hormone replacement therapy in the early 1990s.[75] In 1984 a pair of sociologists compared prescribing practices in the United States and the United Kingdom and found that the American doctors (who tended to be specialists in obstetrics and gynecology) were more willing to prescribe estrogen for their older female patients than were the British physicians (most of whom were general practitioners).[76] It is unclear whether these international differences in estrogen prescription and use figures reflected discrepancies in medical training and professional practices, or disparities in the politics and economics of the two systems of health care, or variances in women's attitudes toward aging.

The statistics derived from the pharmaceutical data clearly demonstrate that estrogen use was on the rise among American women in the 1980s. The percentage of women using HRT seemed too high to those (the feminist authors, for example, discussed at the beginning of this chapter) who believed that menopause and aging were not diseases and therefore did not require medication. On the other hand, the same fraction seemed too low (after all, 75–83 percent of women over 50 were *not* taking estrogen) to those who touted its long-term benefits. The feminist model certainly influenced some women's health care decisions. But for the women who did choose to fill their estrogen prescriptions and take their pills or wear their patches, the medical model offered powerful justification not only for the safety and efficacy of hormone replacement therapy to relieve immediate menopausal symptoms but also for the logic of long-term therapy to forestall osteoporosis. As the 1980s turned into the 1990s, the authority and persuasiveness of this model was further reinforced when medical science provided yet another compelling reason for menopausal and postmenopausal women to consider taking estrogen for the rest of their lives: namely, to protect their hearts.

# CHAPTER ELEVEN

# Weighing the Benefits and Risks of HRT

## Estrogen, Heart Disease, and Breast Cancer

On a summer morning in 1991, three dozen people gathered in the Parklawn Building on the campus of the Food and Drug Administration in Rockville, Maryland, for a two-day workshop on the current status of combined hormone replacement therapy. This meeting was part of the regular quarterly schedule of the FDA's standing advisory committee on fertility and maternal health drugs. What was unusual was that this single subject occupied the entire two-day session; ordinarily the agenda addressed several topics of women's health care. The participants included the thirteen members of the advisory committee (nine outside consultants and four FDA employees) plus about twenty physicians, scientists, and representatives of various professional societies and health care advocacy groups scheduled to testify, some of whom had been invited and others who appeared on their own (or their association's) initiative. The next day, another fourteen individuals (eight medical researchers and six pharmaceutical company representatives) offered testimony to the committee. The mandate to the committee was twofold. One goal was of immediate practical interest, especially to the drug companies: to decide whether to recommend FDA approval of a combination drug product for menopause and postmenopause (containing both an estrogen and a progestin). The second objective—to determine whether enough information existed to support the recommendation of *long-term* hormone replacement therapy—was ostensibly intended to justify the result of the first. But in a larger sense, it reflected the phoenixlike rise of estrogen in the past decade. No longer limited to the short-term relief of hot flashes during menopause, estrogen and its new partner, progestin, were once again being considered as a lifelong therapy for the prevention of the diseases of aging.

The stimulus for this reevaluation of hormone replacement therapy was the growing body of evidence suggesting that estrogen use reduced the risk of car-

diovascular disease in postmenopausal women. In the 1980s, while some researchers were building the case for estrogen as a preventive treatment against
the development of osteoporosis, others were investigating the cardio-protective
effects of estrogen. By 1991, more than twenty case-control and cohort studies
reported a relationship between estrogen use and lower levels of heart disease.
Because heart disease was the number one cause of death for American women,
this correlation had the potential to tip the scales decisively in favor of the widespread prescription of long-term HRT. The pool of potential estrogen users would
be greatly expanded, because the number of women at risk for cardiovascular
disease was much larger than the number of women at risk for osteoporosis. The
ability to reduce the effect of both of these "silent killers" raised the profile of
estrogen as an important preventive medication, one that women could take to
preserve their health, by protecting their bones *and* their hearts.

These long-term benefits of estrogen also changed the way some people
thought about menopause. According to one FDA investigator speaking during
the workshop, "menopause has now developed the concept of being a 'deficiency
disease.'" She went on to pursue the comparison with another endocrinopathy:
"If women can actually have their cardiovasculature [sic] protected and their
bones protected by long-term use of estrogens, some would argue, are we not
dealing with a deficiency disease of estrogen akin to that of hypothyroidism?
Women or men whose thyroids stop working are prescribed thyroid hormone
replacement for life. Why should not women, if estrogen deficiency is analogous?"[1] Could this really be true? Had some in the medical community returned
to the characterization of menopause as a disease, promulgated by men like
Robert Wilson and Allan Barnes thirty years earlier?

While the image of menopause as estrogen deficiency circulated widely in the
1990s, the simple remedy of estrogen replacement was complicated by the risk of
cancer, not only of the endometrium but also, and more ominously, of the breast.
Breast cancer came to the attention of the American public in 1974, when Betty
Ford, the wife of President Gerald Ford, and Happy Rockefeller, the wife of vice-
president-designate Nelson Rockefeller, were diagnosed within a month of each
other. In the 1980s, advocacy groups such as the Susan J. Komen Foundation,
with its enormously popular Race for the Cure, held annually in dozens of cities
around the country, worked to raise awareness of the disease, as well as money for
research.[2] Although heart disease claimed many more lives than breast cancer
each year, breast cancer evoked much greater fear among American women.
Perhaps because of the breast's symbolic and material associations with femininity, sexuality, and motherhood, the threat of cancer—with its long history as the

"dread disease"—in this particular body part held special concern. The studies that suggested a correlation between estrogen use and breast cancer tarnished the glow of long-term hormone replacement.

The addition of progestin to inhibit the development of endometrial carcinoma only further confused the picture. Since physicians had just recently begun to prescribe progestin along with estrogen, the long-term effects of this dual hormone regimen were not yet known. Would combined hormone replacement confer the same benefits—and risks—as unopposed estrogen replacement?

In order to answer this question, and the others on their agenda, the advisory committee members had to adjudicate the value of the available scientific evidence, as well as its role in medical decision-making. This matter became even more problematic when the data obtained from research studies appeared to be ambiguous or when the studies themselves did not measure up to accepted standards of scientific inquiry. Nonetheless, the committee gave its approval to combined estrogen-progestin products for long-term hormone replacement therapy, in spite of the paucity of studies on long-term combined hormone use and the problems inherent in the studies of long-term estrogen-only use.[3] Even as the participants questioned the design of the research studies presented to them, they gave credence to the evidence supporting estrogen's reduction of the risk of cardiovascular disease, estrogen's protective effect against osteoporosis, and progestin's reversal of the risk of endometrial cancer. They also privileged these data over what they considered to be inadequate data on the relationship between hormone therapy and breast cancer. At the end of their two-day powwow at FDA headquarters, they recommended that physicians consider virtually *all* women over the age of 50 as candidates for long-term hormone replacement therapy (estrogen alone for women without uteruses; estrogen and progestin for women with uteruses).[4] The studies done on estrogen, heart disease, and breast cancer, as the basis for the rationale for widespread long-term hormone replacement, warrant a closer look.

The association between estrogen and lower rates of heart disease, like the one between estrogen and lower rates of bone loss, was not new. Researchers had identified this relationship decades earlier, in the 1950s. Observations that premenopausal women rarely had heart attacks, but that they were more frequent in oophorectomized women and postmenopausal women, led to the conclusion that endogenous estrogen somehow protected women from the development of cardiovascular disease. The conclusion led, in turn, to the experimental use of exogenous estrogen in *men*, to prevent the development of atherosclerosis. As noted

in chapter 2, the positive improvements in the patients' cholesterol and lipoprotein profiles levels could not overcome the concomitant negative experiences of
breast development, impotence, and loss of libido, so estrogen therapy for men
never moved past the trial stage. However, since estrogen was already in use as a
short-term treatment for menopausal symptoms, the fact that it could prevent the
development of atherosclerosis contributed to the increasingly popular rationale
for its long-term use for women in the 1960s.

The 1960s also saw the advent of another pharmaceutical application of female sex hormones: the birth control pill. Just one year after the FDA gave the
green light for the sale of oral contraceptives, reports began to trickle in—first
from Britain and then within the United States—of women who had become ill or
died from vascular disorders while taking the pill. By 1968, scientific studies had
confirmed a clear correlation between oral contraceptive use and increased morbidity and mortality from thromboembolic phenomena (inflammation of the
veins, the formation of blood clots in veins or arteries, or obstruction of these
vessels by blood clots). It appeared that the synthetic estrogen component of the
pill was primarily responsible for causing thromboses.[5] Manufacturers rushed to
market a second generation of birth control pills, with smaller amounts of estrogen (the dosage of the synthetic progesterone component was also reduced).
Then, in the 1970s, investigators suggested that women on the pill were also
more likely than their nonuser counterparts to die from myocardial infarction,
hypertension, cerebrovascular disease, and other cardiovascular diseases.[6] Furthermore, these complications seemed to be attributable to the progestin in the
pill.[7] These adverse effects of oral contraceptive use were extremely rare; millions
of women accepted the tiny risk of cardiovascular disease in return for the large
benefit of freedom from unwanted pregnancy. But the data from oral contraceptive studies raised concerns about the parallel use of estrogen and progestin
(albeit different formulations) in menopausal and postmenopausal women.

So investigators turned their attention to the potential cardiovascular effects of
hormone replacement therapy. Given the results of the oral contraceptive studies,
many expected to find similar results, namely, that exogenous estrogen, with or
without progestin, would have adverse effects on a woman's circulatory system.
In the fifteen years from 1970 to 1985, nineteen studies of postmenopausal
estrogen use and cardiovascular disease were published in the English-language
medical literature. Some of these measured risk factors, such as blood pressure
and plasma triglyceride levels; others looked at actual disease endpoints, such as
myocardial infarction or angina pectoris. Only two studies found an increased
risk of cardiovascular disease among estrogen users. Thirteen found a decreased

risk (although in five of these studies, the results were not statistically significant), and four uncovered no effect at all.[8]

One of these investigations was part of Lila Nachtigall's ten-year prospective study conducted at Goldwater Memorial Hospital and published in *Obstetrics and Gynecology* in 1979. When Nachtigall designed the study in the mid-1960s, one of her collaborators (who happened also to be her husband, an internist) suggested that she measure the subjects' cholesterol levels, in addition to their bone densities, to evaluate the possible cardiovascular outcomes of hormone replacement therapy along with its effects on osteoporosis.[9] Of the patients who began the study with high cholesterol, those on HRT decreased their levels within the first six months and maintained lower levels throughout the study, while those in the control group persisted with elevated levels. One woman in the treated group and three in the untreated group suffered myocardial infarction. Neither of these results was statistically significant, but the authors did note in the published report that "an increased risk of heart disease was not found," implying that this null result was good news for long-term hormone replacement, particularly since its benefits in preventing osteoporosis had been clearly demonstrated.[10]

By the early 1980s, concerns that postmenopausal estrogens would have similar negative cardiovascular consequences to those caused by oral contraceptives had been allayed, for the most part. Instead, the bulk of the evidence seemed to confirm the cardio-protective effect of estrogen that had been postulated three decades earlier. However, the debate was rekindled in October 1985, when the *New England Journal of Medicine* published two contradictory research reports back-to-back in the same issue. These large-scale observational studies came to exactly opposite conclusions: the Framingham Heart Study found that postmenopausal estrogen users (as compared to nonusers) had a *higher* relative risk (1.76) for cardiovascular disease,[11] while the Nurses' Health Study found that current estrogen users (as compared to women who had never taken estrogen) had a *lower* relative risk (0.3). Even those women not currently on estrogen, but who had taken it in the past, had a significantly lower relative risk (0.5).[12] How could such drastically different results be reconciled?

In short, they could not. The editors of the *Journal* asked their statistical consultant, John C. Bailar III, M.D., a biostatistician at the Harvard School of Public Health who had spent much of his earlier career at the National Institutes of Health, to write an accompanying editorial to evaluate the relative merit of the two studies and to put their findings into perspective.[13] Bailar reviewed each investigation's methodology and found no reason to favor one over the other. Both followed large cohorts of women for several years. The Framingham study was an offshoot of the famous epidemiological study of residents of the town of Framingham, Massachusetts, begun in 1948. Investigators identified 1,234 participants who were postmenopausal and between the ages of 50 and 83 at their twelfth biennial examination between 1970 and 1972. The women were monitored for eight subsequent years with physical examinations and reviews of their hospital and physician records and, where necessary, death certificates, to document all possible cardiovascular "events." Twenty-four percent reported use of hormones at or before the beginning of the study (32% of those 50–59, 24% of those 60–69, and 14% of those 70–83); the follow-up did not ask if they continued to take hormones. The Nurses' Health Study identified its subject pool by sending a questionnaire to all female married nurses in eleven states who were between the ages of 30 and 55 in 1976. Follow-up questionnaires were sent in 1978 and 1980 to the 121,964 women who responded to the first; 23,608 of these were postmenopausal in 1976 and 8,709 became menopausal by 1978. Thus the investigators followed a total of 32,317 postmenopausal women for four years, obtaining permission to review their medical records and death certificates (plus autopsy reports) to confirm the occurrence of nonfatal myocardial infarction (heart attack) and fatal coronary heart disease. In this group (which was younger overall that the Framingham group), 53 percent had ever used hormones, and 35 percent were currently using them.

Satisfied that the differences in technical design could not account for such diametrically opposed conclusions, Bailar raised three other explanations and immediately dismissed two of them as highly improbable: some sort of catastrophic error in calculation or some sort of incommensurability between the two groups of women included as subjects. He conceded that the third possible explanation held some merit, namely, that the results of these observational studies depended on far more variability than accounted for in the analysis. These unknown variables could act as the proverbial wrench in the works, distorting the analysis of the known variables. "Statisticians themselves," Bailar noted, "have often warned against the uncritical and inappropriate use of powerful analytic tools in the presence of bias, correlated errors, and other problems."[14] In making this point, he acknowledged the danger of overreliance on sophisticated statistics in epidemiological research, particularly in observational studies where so many variables could not be controlled. Bailar concluded that the results of the two studies left him (and probably most other readers) completely ambivalent: "I simply cannot tell from present evidence whether these hormones add to the risk of various cardiovascular diseases, diminish the risk, or leave it unchanged, and must resort to the investigator's great cop-out: More research is needed."[15]

And more research was indeed forthcoming, as was a flurry of commentary on the conflicting *New England Journal of Medicine* reports. The *Journal* published some of these letters to the editor in its 10 July 1986 issue, along with responses from Bailar and both sets of authors. Several correspondents picked up on the premises within Bailar's possible explanations for the discrepancy between the two studies. One writer pointed out that the difference in the ages and menopausal statuses of the women studied may have skewed the results. All of the Framingham women were over 50, 60 percent were over 60, and 18 percent had undergone surgical menopause. The nurses, by contrast, were all under 55, 30 percent were under 45, and 60 percent of those using estrogen had had surgical menopause. He postulated that younger women may respond more positively than older women to estrogen, so older women with natural menopause should be cautioned against estrogen use.[16]

Others noted the biases inherent in self-selection and what was known as the "healthy user" effect. In the case of self-selection, the nurses' medical knowledge may have influenced their decisions to use estrogens at higher rates than the Framingham women, who had no medical training. According to a British correspondent, "the difference in [the studies'] outcomes can be readily understood if the factors prompting the decision to use estrogens differed in the two populations and hence were associated differently with the risk of cardiovascular disease."[17] Self-selection was also related to the healthy user effect. A team of three authors wrote about the results from their own study of postmenopausal estrogen use. They found that estrogen users had not only a lower relative risk of cardiovascular disease but also a lower risk of death from accidents, homicide, and suicide, as compared to never-users. Instead of making the preposterous assumption that estrogen use protected women from accidents, these investigators believed that the women who chose to use estrogen in their study (and, by extension, those in the Nurses' Health Study) were simply healthier in ways that had not been calculated. Of course, this proposal implied the contradictory conclusion that the estrogen users in Framingham must have been less healthy than their non-estrogen-using neighbors.

A possible solution to this conundrum was offered by an observer from the Netherlands, who noted that the Framingham subjects were taking their hormone pills in the late 1960s and early 1970s, a time of great optimism about the long-term benefits of estrogen, whereas the nurses responded to questions about their estrogen use in the late 1970s, a time of deep skepticism toward hormones, in particular, and medicine, in general. He hypothesized that Framingham women with other risk factors for cardiovascular disease (such as a family history of the disease) might have been *more* inclined to take estrogen, which might have translated into their higher rates of cardiovascular disease because of preexisting conditions.[18] According to this line of reasoning, if women at risk for heart disease shied away from postmenopausal hormone use in the 1970s and early 1980s (perhaps because of the association between oral contraceptive hormones and cardiovascular disease), then rates of heart disease would appear higher among nonusers (because that group would contain more high-risk candidates) and estrogen would appear to be cardio-protective, as in fact was the case in several investigations, including the Nurses' Health Study.

In his response, Bailar acknowledged the contributions of the various correspondents to what he saw as "three intertwined debates: about the technical merits of these two studies, about the effects of postmenopausal hormone use on cardiovascular disease, and about methods of inference from imperfect, nonexperimental data."[19] The last of these debates was the most far-reaching, because it dealt with issues common to all observational studies. Whether retrospective or prospective, case-control and cohort studies did not adhere to the so-called gold standard of clinical research: the double-blind, randomized, controlled trial.[20] While many researchers agreed that only a randomized, controlled trial could definitively answer the question of the relationship between postmenopausal hormone use and cardiovascular disease, they were daunted by both the ethics and logistics of designing and implementing such a study. In the mid-1980s conclusions had to be drawn from the existing and ongoing observational investigations.

The late 1980s saw a series of published reports in the medical literature that only added to the ambiguity of whether or not estrogen had a cardio-protective effect in postmenopausal women.[21] Authors of review articles tried to make sense of these conflicting research reports. Two prominent female researchers reviewed nineteen studies of the effects of estrogen replacement on the risk of heart disease and concluded that "unopposed estrogen replacement therapy appears to have a beneficial effect on lipoproteins and blood pressures and to be highly protective for the subsequent development of fatal and non-fatal coronary disease in women." However, they acknowledged that the "healthy user" effect might produce a selection bias in these observational studies. They also identified "the major unanswered question at this time": whether the addition of progestin to postmenopausal hormone replacement therapy influenced the cardio-protective effect of the estrogen component.[22] Evaluation of the long-term effects of the increasingly popular combined estrogen-progestin regimen would become a major issue in the next decade. Nonetheless, a growing consensus within the medical community favored the long-term use of hormone replacement to protect older women against the development of cardiovascular disease. According to one physician-researcher, writing at the end of 1990, "increasing epidemiologic and experimental evidence has suggested a hitherto unappreciated and immense CHD [coronary heart disease] health benefit . . . Although questions remain, we should not let our imperfect knowledge dissuade us from more widespread prescribing of hormone replacement therapy."[23]

This consensus gained strength in the early 1990s with the publication of several meta-analyses of the existing medical literature. Meta-analysis describes a set of statistical techniques that combines all of the results from many research studies on a particular topic—in this case, noncontraceptive estrogen and cardiovascular disease—to come up with a grand summary in an attempt to resolve seemingly contradictory findings. Subject to its own problems, such as literature search bias (in which not all of the relevant studies are included) and inability to control for other biases inherent in the individual studies (flawed studies will lead to flawed analysis; in other words, garbage in, garbage out), meta-analysis can be an effective tool for cutting through the bewildering morass of study designs, objectives, terminologies, and results.

These meta-analyses pooled the data from twenty or so individual studies to come up with an overall estimate that postmenopausal women who took estrogen reduced their risk of heart disease by 50 percent.[24] The authors of each article made the provisos that (1) the effect of progestin on the risk of cardiovascular disease was still uncertain, (2) the healthy user effect could have skewed the results of the research studies, and (3) only large-scale randomized controlled trials could definitely ascertain the relationship between hormone use and the prevention of heart disease. Furthermore, they acknowledged that cardiovascular disease was not the only concern when considering long-term hormone replacement. Women and their physicians had to integrate the information about heart disease into the existing data on endometrial cancer and osteoporosis. And complicating matters even further was the uncertainty about the relationship between HRT and breast cancer.

As with heart disease and osteoporosis, estrogen's association with breast cancer had a long history. Because estrogen exerted its primary effects on the breasts and the reproductive tract, these organs were considered most susceptible to tumor growth in the presence of the hormone. In 1932 the French scientist Antoine Lacassagne demonstrated that estrogen could trigger the development of cancer in the mammary glands of mice; carcinoma of the breast was in fact the first type of cancer shown to be influenced by a specific hormone.[25] This and subsequent experimental work on animals prompted both of the co-discoverers of estrogen, Edgar Allen and Edward Doisy, to caution in the early 1940s that overuse of pharmaceutical estrogens might lead to more cancers in women. As discussed in chapter 2, these words of warning did little to dampen enthusiasm for long-term hormone replacement in the following decades. In 1962 Robert Wilson published a paper in the *Journal of the American Medical Association* suggesting that prolonged estrogen use might *prevent* the development of breast cancer.[26]

Within a few years, however, ambiguous data about the relationship between oral contraceptives and breast cancer translated into anxieties about the carcinogenicity of menopausal estrogens. Although there was no hard evidence linking the pill with breast cancer in the 1960s, critics averred that not enough time had elapsed to ascertain whether or not the pill caused cancer. The endocrinologist Roy Hertz warned in 1966, "Our inadequate knowledge concerning the relationship of estrogens to cancer in women is comparable with what was known about the association between lung cancer and cigarette smoking before extensive epidemiological studies delineated this overwhelmingly significant statistical relationship."[27] Then, in 1976, the association between estrogen and endometrial cancer rekindled concerns about estrogen's effects on the breast. By 1990, no fewer than thirty epidemiological studies had been published on hormone replacement therapy and breast cancer.[28]

According to epidemiologist Barbara Hulka, the quality of these studies improved over the years, as investigators incorporated greater numbers of subjects and employed enhanced analytical strategies.[29] Those conducted prior to 1985 produced inconsistent results: some showed no association, while others hinted at an increased risk of breast cancer among long-term estrogen users.[30] None of the studies in her review indicated any protective effect for estrogen against cancer. The studies published after 1985, investigating both American and European populations, mostly agreed that long-term use of hormone replacement therapy increased a woman's risk of developing breast cancer. Interestingly, the European studies showed a greater increase in risk after a shorter duration of use. Hulka speculated that this discrepancy might be attributed to the use of different estrogen compounds in the United States and Europe (conjugated estrogens—Premarin—in the United States, ethinyl estradiol in Europe) and/or the more frequent prescription of progestin along with estrogen in Europe.[31] Although more physicians had begun to prescribe combined hormone therapy for American women since the late 1970s, the studies in the 1980s could not yet incorporate this relatively recent revision in the regimen into their analyses. The disturbing news from the European investigations that progestin might increase the risk and accelerate the development of cancer, combined with confirmation of the increased risk of breast cancer among long-term estrogen users in America, added another dimension to the risk-benefit calculus for postmenopausal hormone replacement.

The plot thickened with the publication of two meta-analyses of the literature in two major medical journals, the *Archives of Internal Medicine* and the *Journal of the American Medical Association*, in early 1991.[32] The first analysis, by two researchers at Vanderbilt University School of Medicine in Nashville, found that the risk of breast cancer might depend on the dosage of estrogen. Women who took 0.625 milligrams per day of conjugated estrogens had a risk of breast cancer 1.08 times greater than that of nonusers. Studies of those who took 1.25 milligrams per day or more reported higher relative risks of breast cancer, up to 2.0 times higher for users than for nonusers. Duration of use seemed to increase the risk for women on higher doses, but not for those on the lower dose. A history of benign breast disease also seemed to increase a woman's risk of breast cancer while she was taking estrogen; among this subgroup, estrogen users had a relative risk of 1.16, which the Vanderbilt researchers found too low to be troubling. Putting all of this data together, they concluded, "There is considerable and consistent evidence that a daily dosage of 0.625 mg of conjugated estrogens for several years does not appreciably increase the risk of breast cancer . . . Also, ERT consisting of 0.625 mg/d of conjugated estrogens is not contraindicated because of breast cancer risk in women with a history of benign breast disease."[33] For readers of this article, the take-home message was that the fear of breast cancer should not deter any physician from prescribing—or any woman from taking—low-dose estrogen for long periods of time.

This message was reinforced by another article in the same issue of the *Archives*, which reported the results of a seven-year prospective study of almost nine thousand postmenopausal women living in the Leisure World retirement community in southern California.[34] Taking death as its endpoint, this study found that women who had used estrogen lived longer than those who did not, and the longer they used estrogen, the better their chances of survival. Ever-users had a 20 percent reduction in mortality, while current users who had taken estrogen for at least fifteen years had a 40 percent lower mortality, as compared with neverusers. Specifically, the estrogen users died less often from heart disease and stroke. They also had fewer breast cancer deaths, but this figure was not statistically significant.

In spite of what seemed to be a green light for widespread estrogen use, the author of an accompanying editorial remained cautious. "The absence of reports resulting from a clinical trial that randomly assigned estrogen or placebo," he pronounced, "makes such a recommendation unjustified as a national public health policy."[35] Because the women who took estrogen in the Leisure World study were self-selected, and because the meta-analysis was similarly based on nonrandomized studies, both fell short of the "gold standard." Randomized double-blind controlled trials would, he hoped, provide the data necessary to evaluate estrogen as a "medical intervention to prevent disease."[36]

Three months later, the results of the second meta-analysis, published in the *Journal of the American Medical Association,* justified this cautious stance. Performed by a team of researchers at the Centers for Disease Control and Emory University's School of Public Health in Atlanta, this review came to a different and more sobering conclusion. It combined the results of sixteen studies to calculate the risk of breast cancer according to number of years of use. There was no increase in risk for the first five years of estrogen use, but after fifteen years women had a 1.3 times higher risk of breast cancer. The authors noted that this small increase in risk could translate into a large number of actual breast cancers. They estimated that long-term estrogen use could be responsible for at least 4,708 new cases and 1,468 deaths from breast cancer each year. They also acknowledged the consensus that hormone replacement decreased the risk of osteoporosis and heart disease, but insisted that the breast cancer risk had to be taken into consideration. "Although the overall benefit of estrogen replacement after menopause may outweigh the risks for most women," they wrote, "our analysis supports a small but statistically significant increase in breast cancer risk due to long-term estrogen use."[37] Refraining from making any explicit recommendation, the authors of this article left it up to doctors to figure out the risks and benefits for their patients.

In the next several months, more reports were forthcoming in the medical press, none of which helped to clarify the relationship between postmenopausal estrogen use and breast cancer.[38] The only thing about the relationship between estrogen and breast cancer that was clear in the early 1990s was that the data were unclear. In October 1992, the titles of two review articles, one in *Maturitas,* the journal of the European Menopause Society, and the other in the widely distributed *Journal of the American Medical Association,* announced that the issue had attained the status of a full-blown controversy. Both articles reviewed the existing literature—both the individual studies and the meta-analyses that had been published—in an attempt to offer physicians some applicable guidelines for clinical practice. The American author, Janet B. Henrich, writing from Yale University's

School of Medicine, concluded, "There is no compelling evidence that, overall, women who have ever used postmenopausal estrogens are at increased risk of breast cancer."[39] What about the several studies that reported an increased risk for current users of estrogen? She hypothesized that the immediate risk might be due to one of two factors: either estrogen-dependent tumors have some unique physiology (which had not yet been demonstrated) or women on estrogen see their doctors more regularly and therefore have tumors caught at an earlier stage (so-called detection bias). She also offered two notes of caution: first, to be wary of the results of European studies because of different kinds of estrogen and perhaps differences in the people studied, and second, to approach meta-analyses with a healthy dose of skepticism (because they depended so much on the quality of the data compiled). Like all other commentators, she called for more studies, specifically randomized controlled trials. Although her advice to physicians seemed to imply that they should continue to prescribe hormone replacement therapy, her comment that "the information used currently to guide clinical decisions regarding estrogen replacement therapy is based on antiquated medical knowledge and practice" must have left many readers to worry whether or not they were doing the best for their patients.[40]

The *Maturitas* article, penned by Regine Sitruk-Ware, a Swiss endocrinologist who also worked for the pharmaceutical company Ciba-Geigy, delved into the question of the effect of progestins on breast tissue and the development of tumors.[41] The data on the relationship between combined hormone therapy and breast cancer came mainly from Denmark and Sweden; although Hulka had interpreted the results of these studies as showing an increased risk of breast cancer, Sitruk-Ware argued that they did not demonstrate a statistically significant increase in risk.[42] The only American to look at the long-term effects of combined estrogen-progestin therapy was Lila Nachtigall; her study found a lower incidence of breast cancer in the hormone-treated group as compared to the placebo-treated group. (Because of the study's small sample size, Nachtigall and her colleagues could only say that there was no increase in risk of breast cancer for estrogen-progestin users; they refrained from asserting that hormone replacement therapy actually protected women against breast cancer.)[43] The Swiss author maintained that evidence did not suggest any major difference between estrogen-alone and estrogen-progestin therapy; thus the figure of 1.3 as the relative risk of breast cancer for estrogen-alone (as determined by the meta-analyses) probably held as well for the combined regimen. Given the evidence that estrogen (and, potentially, estrogen and progestin) reduced the risk of cardiovascular disease by half, and the fact that that many more women were at risk of death from heart disease as compared to breast cancer, she concluded, "There is no scientific argument to change the presently recommended schedule of adding progestins at least 12 days per month of ERT in women with an intact uterus."[44] Although Sitruk-Ware's conclusion (which may have been influenced by her position at the pharmaceutical house of Ciba-Geigy, the manufacturer of the Estraderm patch) was stated with greater certainty than that of Henrich (an American academic physician with no obvious ties to a drug company) in the *Journal of the American Medical Association* article, both (female) authors called for more "scientific" evidence from randomized controlled trials before making significant changes to the current clinical practice of prescribing long-term hormone replacement for the prevention of disease.

When the members of the FDA's Fertility and Maternal Health Advisory Committee sat down for their meeting on hormone replacement therapy in June 1991, they must have experienced a feeling of déjà vu, since they had covered similar material in two recent meetings. In February 1990, the committee discussed the association between estrogen (and progestin) and breast and endometrial cancer; four months later, they tackled the question of whether Premarin had a significant enough cardio-protective effect to warrant a change in its labeling. By the time they convened for their two-day workshop the following year, they were well versed in the existing research on the risks and benefits of postmenopausal hormone replacement.

The participants in this series of meetings recognized the limitations of the evidence before them, namely, that data had been collected in observational studies, which were potentially subject to a whole host of biases. They echoed the consensus that randomized controlled trials must be undertaken, in spite of the logistical complexities. Indeed, two such trials had already been set up by the early 1990s, the Postmenopausal Estrogen/Progestin Interventions (PEPI) Trial and the Women's Health Initiative (WHI), to be discussed more fully in chapters 12 and 14. But it was not enough to sit back and wait for the results of these studies; PEPI, the smaller of the two trials, was not scheduled to conclude until the mid-1990s, and WHI, a much more ambitious venture, would not have results until after well after 2000. In the meantime, the FDA, physicians, and women needed guidance on the best course of action with regard to hormone replacement during the years after menopause.

A central theme in all of the Advisory Committee's deliberations revolved around the balance between the risks and benefits of hormone therapy. Some of the experts invited to testify before the committee calculated the relative advantages and disadvantages in terms of the potential number of lives saved versus the potential number of lives lost. At the June 1990 meeting a physician from the University of Southern California speaking on behalf of Wyeth-Ayerst Laboratories (the manufacturer of Premarin, which sought approval to include mention of the drug's cardio-protective benefits in its labeling), presented a series of actual and hypothetical statistics to make his case. He noted that cardiovascular disease killed 500,000 women annually, compared to cancer's toll of 200,000.[45] Then, he assumed, somewhat arbitrarily, a 50 percent reduction in death from cardiovascular disease, a 40 percent reduction in death from osteoporosis, a 10 percent increase in death from breast cancer, and a 300 percent increase in death from endometrial cancer among estrogen users to come up with the following numbers. For every 100,000 women on a daily dose of 0.625 milligrams of Premarin, he calculated that 5,250 lives would be saved from cardiovascular death and 563 from death related to osteoporosis. On the other side of this imaginary equation, 187 lives would be lost to breast cancer and 93 to endometrial cancer.[46] To drive his point home, he reminded his audience of the annual health care costs of allowing women to age without hormone replacement: $40 to $60 billion for cardiovascular disease and $7 to $10 billion for osteoporosis.[47] If the stark statistics of dead women didn't move the committee, perhaps the shock of the economic consequences would persuade them to approve the relabeling of Premarin.

Some observers doubted the motives of Wyeth-Ayerst in its application for expanded indications for Premarin. Cynthia Pearson, program director of the National Women's Health Network, addressed the Advisory Committee in the open public hearing at the same meeting in June 1990. After itemizing a laundry list of bad behavior by Wyeth-Ayerst, beginning with the company's support of Robert A. Wilson back in the 1960s, she pointed out that Wyeth-Ayerst's request for a cardio-protective indication for estrogen followed closely on the heels of its campaign to get the FDA to make the requirements for generic conjugated estrogens more strict (in 1991 the FDA deemed that generic conjugated estrogens were not bioequivalent to Premarin and ordered them taken off the market).[48] She saw both of these efforts as profit-driven, and not at all in the best interests of American women.

While the Advisory Committee members listened politely to Pearson's statement, they were much more interested in the scientific presentations. As medical researchers, they felt most comfortable dealing with data. The problem, however, was that the data refused to yield any neat conclusions. Unlike the correlation between estrogen and osteoporosis, which was mainly uncontested in the medical literature, the associations between hormone therapy, cardiovascular disease, and breast cancer were much more controversial. While a different group of advisors in 1984 had little trouble recommending a label change for Premarin to include its indication for postmenopausal osteoporosis, the 1990 committee did not feel the evidence was strong enough in the case of cardiovascular disease. Also, by 1990 the calculation of the risks and benefits of hormone replacement had become further complicated by the addition of another variable, progestin.

An editorial in the *New England Journal of Medicine* tried to make sense of the risks and benefits of aging with and without hormones.[49] The authors emphasized the difference between relative risk and absolute risk. Whereas relative risk compared the risk between two different groups (estrogen treated and untreated), absolute risk described an individual's risk of developing a disease (heart disease, breast cancer). As discussed above, meta-analyses of the research studies reported a 50 percent reduction in the relative risk of cardiovascular disease between estrogen users and nonusers. The cumulative absolute risk of dying from heart disease for white women aged 50 to 94 was 31 percent (or, put another way, 31 out of every 100 women would die from heart disease); thus, estrogen users would have an absolute risk of 16 percent (16 out of 100). The absolute risk of death from breast cancer was almost 3 percent (3 out of 100); thus, even a doubling of the risk with use of estrogen would translate to an absolute risk of 6 percent (6 out of 100), still much lower than the risk of dying from heart disease. The point of this mathematical exercise was to demonstrate that big increases in relative risk for less frequently occurring diseases (here, breast cancer) had to be weighed against moderate decreases in relative risk for common diseases (cardiovascular disease). In essence, this editorial made the same argument as did the physician who presented to the FDA lives saved versus lives lost, except in terms of an individual's personal chances of developing and dying from different diseases.

The danger in all of these risk-benefit calculations was the tendency to reify experimental results into accepted facts. As the variability among the dozens of research studies should have made clear, the estimates of relative risk for heart disease, and even more so for breast cancer, were by no means indisputable. Yet once these tentative figures were used to substantiate arguments for (or against) hormone replacement therapy, they tended to take on the aura of actuality. And once these numbers made their way out of closed FDA committee meetings and medical research journals and into the popular media, they had the power to influence millions of women's decisions.

Numbers aside, estrogen's diverse effects on so many bodily organs brought additional specialists into the debate. The potential benefit of a reduced risk of heart disease and the potential detriment of an increased risk of breast cancer interested cardiologists and oncologists, as well as the obstetrician-gynecologists and primary care physicians who regularly saw middle-aged and older women as patients. Estrogen's multiple consequences also created a sort of competition of fear: which disease was more frightening? An editorial in the *New England Journal of Medicine* captured the dilemma in its title: "Hormone-Replacement Therapy— Breast versus Heart versus Bone."[50]

It is worth reiterating that the controversy centered on long-term hormone replacement for the prevention of disease, not short-term therapy for the relief of menopausal symptoms. Few disputed the use of estrogen (and progestin for women who had a uterus) to cope with hot flashes and night sweats. In this case, the drugs were being used in the conventional sense: to alleviate physical discomfort. On the other hand, long-term hormone therapy to prevent cardiovascular disease or osteoporosis signified a new pharmaceutical paradigm of risk-reduction.

Both cultural conceptions and personal inclinations had as much influence as epidemiological statistics in shaping individuals' attitudes toward the chronic diseases of aging and the value of long-term hormone replacement. As the history of estrogen in the twentieth century has shown, the available scientific evidence has always been but one factor among several in the contemporary sociocultural context influencing attitudes toward and decisions about hormone therapy. The latest debates over the risks and benefits of long-term hormone use by postmenopausal women took place within a matrix of heightened political and popular consciousness about menopause, aging, and women's health care. In the early 1990s, menopause came out of the closet.

# CHAPTER TWELVE

# 1992: The Year of the Menopause

Nineteen ninety-two, according to the pundits, was the year of the woman in politics. Eleven women won primary contests as candidates for the United States Senate, four of whom prevailed in November's general election (Carol Moseley Braun of Illinois, Barbara Boxer of California, Diane Feinstein of California, and Patty Murray of Washington), bringing the total number of female senators to six (the two incumbents were Nancy Kassebaum of Kansas, elected in 1978, and Barbara Mikulski of Maryland, elected in 1986). While the tripling of their numbers was welcomed, the exceptionality of their success was shrugged off by Senator Mikulski, who told the media, "Calling 1992 the Year of the Woman makes it sound like the Year of the Caribou or the Year of the Asparagus. We're not a fad, a fancy, or a year."[1]

Mikulski had a point. Although their numbers were small at all levels of government, women's influence in the public sector was growing. Nowhere was this more evident than on the subject of women's health. In the decade from 1979 to 1988, an average of only five or six bills on topics pertaining to women's health were introduced each year in the House of Representatives and the Senate combined. In the 101st Congress, which met 1989–1990, that number leapt to thirty-eight new bills annually, and in the decade from 1991 to 2000, it increased tenfold from the previous decade, to an average of fifty-eight bills proposed every year.[2]

In September 1990 the efforts of Senator Mikulski and her colleagues in the Congressional Caucus for Women's Issues to increase the visibility of women and women's health concerns in the research agenda of the National Institutes of Health (NIH) paid off, with the formation of the Office of Research on Women's Health. Located within the Office of the Director of the NIH, this division of the nation's primary agency for medical research was charged with promoting, coordinating, and monitoring research on women's health, as well as ensuring the inclusion of women subjects in clinical trials. Although the proposed appropriations for this office were but a tiny fraction of the overall annual budget for the NIH

($20 million out of $7.5 billion, or less than three-tenths of 1 percent), the mere fact of its establishment held symbolic significance, namely, that women's health deserved special attention.[3] Six months later another symbolic milestone was reached in the realm of government science, when Bernadine Healy received confirmation as the director of the NIH, the first woman ever to hold that position.

This heightened awareness of and attention to women's health signaled the advent of a new women's health movement, one that differed in personnel, strategies, and priorities from the grassroots movement that took shape in the 1970s.[4] The earlier movement demanded that women be allowed to participate in medicine, as both consumers and providers, and some of its successes, such as the opening of medical schools to female students, helped to pave the way for the new movement. In the late 1980s and 1990s, women in positions of power—physicians, medical researchers, elected and appointed officials—worked to incorporate women's health into the established institutions of medicine and medical research. Whereas the earlier generation of feminists had challenged the hegemony of organized medicine (for example, by setting up self-help clinics with lay providers), women who became doctors in the 1970s and 1980s joined the ranks of medical professionals. Female physicians (and government officials) acted within the system to effect change, while feminist activists strived to change the system itself. Both groups shared the ultimate goal of improved health care for women, and they agreed on the immediate need for randomized controlled trials as the scientific basis for the evaluation of drugs, devices, and procedures used in the practice of medicine. But with the letters M.D. after their names or offices on Capitol Hill, the leaders of the 1990s advocacy efforts had greater authority and more clout to enact their agenda for women's health.

What is interesting about this agenda—indeed, about much of the discourse about women's health in the early 1990s—was the emphasis on middle-aged women. If 1992 was the year of the woman in politics, it could also be characterized as the year of the menopause. Starting the year before, menopause stepped into the limelight and engaged the interest of a wide audience, thanks to the attention it received from public officials and the popular press. That the subject of menopause appeared in the halls of Congress and on the covers of magazines in the early 1990s had a lot to do with the demographics of the American population, specifically the aging of the first baby boomers.

The baby boom generation consists of the seventy-six million Americans born between 1946 and 1964. In 1991, the oldest boomers turned 45, which, for many of the females in the group, also signaled the onset of menopause. Given that the median age of natural menopause is 51, then almost half of all women reach menopause in their forties (and a few even earlier, in their thirties), which means that this first cohort of baby boom women was beginning to experience menopause in the early 1990s, with millions to follow in the coming two decades. Because of their sheer force of numbers, baby boomers had been able to focus public attention on their shifting interests and concerns throughout the stages of their lives. This largest generation has made its influence felt in every decade: as teens in the 1960s, as young adults in the 1970s, as "yuppies" in the 1980s, and as they entered middle age in the 1990s. Just as birth control and abortion topped the women's health agenda in the 1970s, menopause became a priority two decades later. From Washington, D.C., to Walla Walla, Washington, it seemed as if everyone was talking about menopause and its pharmaceutical sidekick, estrogen.

The United States Senate's Subcommittee on Aging, of the Committee on Labor and Human Resources, turned its attention to menopause and estrogen with a one-day hearing in April 1991 on "examining the health of mid-life and older women, focusing on the health effects of menopause and its treatment (hormone replacement therapy)."[5] Senate subcommittees had investigated estrogen in previous decades: directly, in the wake of the association with endometrial cancer in 1976, and indirectly, as a treatment option, in the hearing on osteoporosis in 1985 (see chapter 7 and chapter 9). The latest hearing covered a wide range of topics and themes: the relationship between menopause and diseases of aging, the role of the federal government in supporting research on midlife and older women's health, and cultural conceptions of menopause and menopausal women in American society. As such, it offers a snapshot of what members of Congress felt to be pressing concerns for their female constituents in 1991.

The opening remarks of Senator Brock Adams, chairman of the Subcommittee on Aging, revealed a distinctly medicalized view of menopause. Adams conflated menopause and aging, by identifying menopause as the marker, if not the actual cause, of the onset of disease in older women. "Menopause and the loss of ovarian hormones plays [sic] an important role in the development of diseases and conditions affecting women," he announced.[6] He lamented the lack of research on menopause and the lack of consensus on hormone replacement therapy, and he urged the NIH to devote more time and resources to the study of midlife and older women's health concerns.

(Senator Adams apparently had his own personal interest in women. In March 1992, a month after he formally announced his bid for reelection, the *Seattle Times* reported that eight women had accused Adams of sexual harassment. Although he denied the charges of twenty years' worth of drugging, molesting, and even raping these women—all lobbyists and employees of the Democratic Party—Adams immediately dropped out of the election and ignominiously ended his political career.)

After Adams's introduction, Representative Patricia Schroeder, co-chair of the Congressional Caucus for Women's Issues, made a statement before joining the hearing as an honorary participant at the podium, sitting alongside the senator. A long-time feminist, Schroeder had served as an ardent advocate for women since she was first elected to Congress in 1972. She famously told a *New York Times* reporter in 1977, "I have a brain and a uterus, and I use them both." In her comments at the hearing, Schroeder called for more and better research on midlife and older women's health, but she cautioned against both the medicalization and the commercialization of menopause. She located the medical imperative to treat menopause pharmaceutically within the larger context of the nation's prescription drug culture, recalling the observation made by Betty Ford (who underwent treatment for drug and alcohol dependency in 1978) that "women got their drug habits not from the streets but some of the finest trained minds in America."[7] Schroeder also placed the problems faced by midlife and older women into a broader social context, blaming gender bias for the low status of aging women. Whether the problem was sexist ageism or ageist sexism, she noted that "there is still a notion in our society that men age but women rot."[8]

Schroeder's concern about the lack of consensus on menopause and hormone replacement therapy had compelled her, along with Representative Olympia Snowe, her co-chair on the Caucus for Women's Issues, to write a letter in October 1990 to the Office of Technology Assessment (OTA) requesting an investigation into the current state of knowledge on these topics. For twenty-three years, from 1972 until its funding was withdrawn in 1995, OTA researched and analyzed complex scientific and technical issues for members of Congress, translating specialist knowledge into lay language that the legislators could understand. OTA's report, *The Menopause, Hormone Therapy, and Women's Health*, was issued in May 1992. Since the research was already under way, a representative from OTA was invited to testify at the Senate hearing.

This representative, Dr. Kathi Hanna, gave a very rough estimate of the number of women on hormone replacement in America: between five and nine million. She described this as a "low rate of use," because it represented "only 15 percent . . . of those women who are potential candidates for therapy." Hanna acknowledged two possible reasons why only a minority of women chose to use HRT. First, "many women may not be suffering enough to seek treatment"; in other words, only those women experiencing serious discomfort from immediate menopausal symptoms sought medical help. This explanation seemed to characterize HRT as the short-term remedy of a half-century earlier. But research and practice of the past decade had expanded the clinical indications to include the long-term prevention of disease. Why, then, were 85 percent of women not taking advantage of this preventive therapy? The second reason was "related to real and perceived risks. Women are better consumers than they used to be." In spite of the work done in the 1980s, lingering fears about endometrial cancer and current concerns about breast cancer, along with prior negative experiences with the pill, DES, and the Dalkon shield, made these more savvy women suspicious of medical intervention in their reproductive health. Although there had been a good deal of research on menopause and hormones, Hanna noted that these studies "appear conflicting and contradictory, especially to the consumer." More and better research, she indicated, was critical to improving the information available to women choosing whether to take HRT.[9]

Other testifiers echoed the lament over the lack of consensus on hormone replacement. Anne Wentz, professor of obstetrics and gynecology at Northwestern University Medical School, announced, "Right now, a woman has to make a decision about hormone replacement therapy based on conflicting and confusing data."[10] To address what she saw as a pressing need, she had joined the team of Florence Haseltine, Marie Bass, and Joanne Howes in founding the Society for the Advancement of Women's Health Research in 1990.[11] The goals of this organization were threefold: to identify areas of women's health that need attention, to make sure women were included in clinical trials, and to increase the numbers of women as medical researchers and in positions of leadership in academic and government research facilities. To achieve these objectives, the Society coordinated a group of "leading medical, health, and scientific organizations," including the American College of Obstetricians and Gynecologists, the American Psychiatric Association, and the American Nurses Association, to work with the executive and legislative branches of the federal government. The composition and mission of this coalition was emblematic of the new women's health advocacy: an organization of medical experts and professionals working within the established system of federally funded medical research to tackle the medical problems of women's health.

Bernadine Healy, the newly appointed director of the NIH, also appeared before the Senate subcommittee as a representative of the 1990s iteration of women's health advocacy. She distanced the contemporary women's movement from its forebears, contrasting the earlier generation's emphasis "on the same-

ness of women with men" with the current recognition among medical professionals of the biological differences between the sexes.[12] In response to a question
from Representative Schroeder about the lack of research on older women, Healy
mused, "My own personal views are that, in general, women and women professionals in particular—and I guess I can fault myself in this regard—tried not to
emphasize their femaleness when they were functioning in a professional setting. And I think that if we look back over the past 20 years, I think women
professionals were focusing on political and social and economic issues, and
health was almost a luxury in that context."[13] Apparently, Healy gave little credence to the efforts of groups such as the National Women's Health Network and
the Boston Women's Health Book Collective, which had been active since the
1970s. From her perspective, physicians and medical scientists, not lay advocates,
were the proper personnel for attending to matters of women's health. Similarly,
the proper venue for research on women's health was in federally funded academic medical centers, as opposed to, say, feminist self-help clinics.

To this end, Healy reported what would be the biggest news story to come out of
the Senate hearings. She announced the launch of the NIH Healthy Women's Study
(later known as the Women's Health Initiative, or WHI), "the largest community-
based clinical, prevention and intervention trial ever conducted . . . to examine the
major causes of morbidity and mortality in women of all races and all socioeconomic
strata."[14] Expected to cost at least $500 million and to take at least ten years, the
design of this ambitious study consisted of three components: first, an observational
study to follow several hundred thousand women over time; second, a nationwide
community intervention and prevention study addressing differences among socioeconomic and ethnic groups; third, and perhaps most significant, double-blind
randomized controlled trials of preventive strategies, such as diet modification,
dietary supplements (calcium, vitamins), exercise, smoking cessation, and, of
course, hormone replacement therapy. It would be the outcome of the study of this
final variable that would eventually receive the most attention in the years to come,
dramatically reshaping attitudes toward and uses of estrogen in the twenty-first
century (the WHI is discussed more fully in chapter 14).

The following month, in May 1991, the House of Representatives' Subcommittee on Housing and Consumer Interests of the Select Committee on Aging held its
own hearing on women, menopause, and health care.[15] Whereas the Senate
inquiry dealt mainly with the state of medical research and knowledge about
menopause and hormone replacement therapy, the House investigation focused
more on improving the quality of consumer information and the equitable provision of health care services to midlife women. Representatives from both the

medical and lay communities testified on ways to address disparities in the dis-
tribution of medical information and services related to menopause and postmenopause. Cynthia Pearson, the program director of the National Women's Health
Network (NWHN), sat side by side on the same panel with Bernadine Healy.
Especially noteworthy were the points of congruence between their testimonies.
Pearson could hardly disagree with Healy's comment that "women have a right to
know how to prevent and reduce health problems throughout their entire life."[16]
And Healy must have been gratified to hear Pearson state that "there must be
much more research on estrogen. We need to know about the long-term risks and
benefits of estrogen use alone and estrogen in combination with progestin."[17]
Although the two women's health advocates had very different approaches—
Healy, of the NIH, worked within the system, while Pearson, of the NWHN,
critiqued from without—they concurred on the value of informing midlife women
about their health care options and the need for basing the medical use of hormone
replacement on rigorous scientific data. Both women, and the constituencies they
represented, held high hopes for the Women's Health Initiative.

Bernadine Healy kept women's health in the spotlight in 1991 with two editorials published in the same week in the nation's most prestigious and widely
read medical journals. In the July 24/31 issue of the *Journal of the American
Medical Association,* she presented "Woman's Health, Public Welfare" to accompany a report by the Council on Ethical and Judicial Affairs of the American
Medical Association on "Gender Disparities in Clinical Decision Making." She
described the multiple initiatives undertaken by the NIH to address these disparities at the levels of both the individual doctor-patient interaction and the
epidemiological study of diseases and treatments in populations. "Women's
health," she declared, "in terms of research, services, and access to care—has
come of age and become a priority medically, socially, and politically."[18]

Healy's editorial in the July 25 issue of the *New England Journal of Medicine*
received even more attention, owing perhaps to its intriguing title. "The Yentl
Syndrome" commented on two research studies published in that issue of the
*Journal,* both demonstrating sex bias in the medical management of coronary
heart disease.[19] It used Yentl, the nineteenth-century female protagonist of Isaac
Bashevis Singer's short story, who had to disguise herself as a man in order to study
Jewish scripture at school (and who was portrayed by Barbra Streisand in the 1983
film adaptation), as a metaphor for the unenviable dilemma faced by all women: to
act "just like a man" to achieve equality or to remain different from and therefore
second-class to men. The studies reported that women and men hospitalized with
chest pain and other less dramatic symptoms of coronary disease received dif-

ferent kinds of treatment: men were much more likely to be given aggressive therapy, such as cardiac catheterization or bypass surgery. Only when women had full-blown heart attacks—that is, when they displayed symptoms more typical of the male profile of coronary disease—did they receive similar treatment. Healy argued that medical researchers could no longer ignore women's "unique medical problems," but rather they had to study and treat "the diseases of women as different from the diseases of men but of equal importance."[20]

As director of the NIH, Healy used her position as a bully pulpit to demand the inclusion of women in clinical trials and to publicize the importance of the soon-to-be-launched Women's Health Initiative. She also pressed for a greater presence of women as leaders in biomedical research and encouraged the training of both male and female doctors to understand and acknowledge biological differences between the sexes.[21] For this female cardiologist, the attempt by earlier feminists to downplay these differences had done a disservice to women's health. Biology was not destiny, but it ought to be factored into medical practice and research.

Healy had opened her *Journal of the American Medical Association* editorial with a reference to a recent two-page article in the *Philadelphia Inquirer* on menopause.[22] She found the headline on the second page, "Menopause Comes of Age as Medical and Social Issue," to be a particularly apt description of the new emphasis on older women's health issues in government and medical research. The headline on the first page captured even better the spirit of the times: "Menopause Becoming 'Au Courant' as It Hits Women of Baby Boom."

Menopause was not uncharted territory for newspapers and magazines. As we have seen, magazines had periodically run articles on the topic since the 1940s, and newspapers regularly reported the results of major medical studies and government actions relating to hormone replacement therapy (e.g., the link with endometrial cancer and the FDA mandate for a patient package insert in the 1970s). In 1991 newspapers across the country informed their readers of the big estrogen stories. In January, news of the Leisure World retirement community study (which found that estrogen users lived longer than nonusers) made headlines in the *San Francisco Chronicle*, the *New York Times*, the *St. Louis Post-Dispatch*, and the *Washington Post*.[23] In April, the *Boston Globe* deemed the announcement of the Women's Health Initiative worthy of space on the front page; three months later, it gave the same priority to a story about the *New England Journal of Medicine* heart disease studies and Bernadine Healy's "Yentl Syndrome" editorial.[24] Estrogen made the front pages of the *New York Times*, the *Washington Post*, and the *St.*

*Petersburg (Florida) Times* in September upon the publication of the latest data from the Nurses' Health Study, which indicated that women who took postmenopausal estrogen almost halved their risk of heart attack.[25] Communication of these medical reports on estrogen was nothing special, and no more significant than any of the other myriad accounts of the latest medical findings that appeared in the nation's daily papers. News of health and medicine was popular with both print and television audiences, and journalists, broadcasters, and editors obliged with ample coverage of recent developments.

What was new and different was the attention paid by newspapers, magazines, and television to menopause as a cultural phenomenon, beginning in 1990 and accelerating in 1991 and 1992. Lengthy features in newsweeklies, women's monthlies, and daily papers reminded readers of the aging of the baby boom generation and portended great changes in how both individuals and society would come to view menopause. These articles often addressed the controversy over hormone replacement therapy, quoting the latest statistics and "expert" opinions on the relationships between estrogen and endometrial cancer, breast cancer, osteoporosis, and heart disease. Some even offered handy quizzes for women to assess for themselves the relative risks and benefits of HRT. Above all, the media's interest in menopause and HRT helped steer these topics onto the radar screen of public consciousness.

*Newsweek* led the pack with an article titled "Not Past Their Prime" in August 1990. In their attempts to explode the myths about menopausal women, the authors may have inadvertently perpetuated these stereotypes to a larger audience. The article began: "For women who came of age during the 1960s and later, the intricacies of their reproductive lives are as familiar and comfortable as the layout of their homes. But one room has remained shadowy and unexplored: menopause . . . It's the only phase of the female cycle that chatty women rarely chat about—even among themselves. And with good reason: what little they know of it sounds pretty grim. Hot flashes, the weeps, chin whiskers—who would *want* to discuss it?"[26] Those who read further learned that "biologically, menopause is a very big deal," and "many clinicians consider menopause a potential disease state."[27] This article presented HRT in a rosy light, downplaying any association with breast cancer and emphasizing its effectiveness in treating hot flashes, reducing the risk of heart disease, and "stav[ing] off the ravages of osteoporosis." With regard to aging baby boom women, it predicted, "They'll ask their doctors for the help they need—and then they'll get on with the rest of the prime of their lives."[28] This positive portrayal of confident, pro-active boomers also depicted menopause as a condition in need of medical and pharmaceutical management.

The debate about the medicalization of menopause took center stage in the *Philadelphia Inquirer* article the following year. The *Inquirer* cited opposing viewpoints—Diana Laskin Siegal, co-author of *Ourselves, Growing Older,* bemoaning medicalization and commercialization, and a New Jersey gynecologist who described estrogen loss as "a great health risk"—to highlight the confusion among both women and doctors about the use of HRT. It also recounted stereotypical images of the menopausal woman—"dried-up . . . over-the-hill non-person . . . with gray hair and flat breasts wearing orthopedic equipment"—in anticipating the imminent demise of such stigmas.[29]

It was just a matter of time, according to magazine writers, before baby boomers' openness about other reproductive issues extended to menopause. "In a world that's gotten so frank about everything from sex to childbirth," began an article in *McCall's*, "women have been surprisingly silent about menopause . . . Now all of a sudden everyone's talking about it."[30] The author of an article in *Good Housekeeping,* called "The Baby Boom Meets Menopause," agreed. "Their feisty spirit," she enthused, "—if not their sheer number—has already redefined femininity, work styles, and parenting, and it now promises to revolutionize society's views toward 'the change.' "[31]

However, other authors took a less sanguine approach to the ease with which baby boomers would navigate menopause. "These days," an article in *Ladies' Home Journal* began ominously, "the biggest medical battleground is the body of the middle-aged American woman, as baby boomers are faced with life-and-death decisions about hormone treatments."[32] This article provided a "take-charge guide," while *McCall's* developed "The estrogen test" and *Good Housekeeping* printed "The Women's Heart Test" as starting points for their readers in weighing the benefits and risks of HRT.[33] The *New York Times Magazine* asked, "Menopause: Is It a Medical Problem?," and lamented the dilemma faced by women in their late forties: "Women in this take-charge generation, who pride themselves on working out, getting enough calcium and grilling both swordfish and doctors, are torn between two conflicting views," namely, menopause as a "natural milestone" or, in the words of one female gynecologist, as "an endocrinopathy."[34]

The buzzwords "take charge" appeared often in media characterizations of baby boom women. In May 1992 the *New York Times* ran the first of a two-part series on menopause on the front page of the Sunday paper, predicting that "this group's size and take-charge attitude will transform how society thinks about menopause, how much time and money gets spent understanding it and how doctors behave."[35] Indeed, medicine and commerce were the focal points of this article and its sequel, which appeared two days later on the front page of the

Science Times section. The author of the first article quoted both Cynthia Pearson, program director of the NWHN, and Dr. Morris Notelovitz, author of *Stand Tall! Every Woman's Guide to Preventing Osteoporosis* (see chapter 9), as experts who feared that "this huge cohort could be a snake oil salesman's dream."[36] Quack nostrums aside, the real moneymaker in the menopause game was estrogen, which had become a $750-million-a-year business. The author of the second article, the *Times*'s longstanding Personal Health columnist Jane Brody, identified the "nagging question" for baby boomers: "to treat or not to treat menopause and the decades beyond with hormone replacements."[37] Brody pointed out that the debate over HRT focused not on its short-term use but rather on the long-term pharmaceutical management of postmenopause for the prevention of disease and the improvement of quality of life, a rather nebulous outcome, which included things like mood, memory, and sexual satisfaction. She also dramatized the terminus of the trend toward the medicalization of menopause and aging: "If manufacturers and enthusiastic doctors prevail, upwards of 90 percent of women will take replacement hormones for three to five decades."[38]

Of course, the likelihood of any drug attaining that degree of prevalence was remote; at the time, only 15 to 18 percent of postmenopausal women were using HRT. Brody did make the important observation that many women could not afford the expense of this long-term daily medication.[39] Indeed, all of the hubbub centered on the middle-class and affluent members of the baby boom generation, those who had both the time to think about and the money to spend on menopause and its management. Regardless of the actual size of the target group for hormone replacement and other therapies, the flurry of information in the media about menopause reached an enormous audience of women, and men, of all ages.

Just a week after its front-page article in the *New York Times*, menopause turned up on the cover of *Newsweek*.[40] This piece, like others, noted the relative paucity and ambiguity of medical information about menopause and HRT for consumers. To illustrate this point, it quoted the quick-witted Representative Patricia Schroeder: "If you get six menopausal women together, you'll find that their doctors are doing six different things. Our joke is that you might as well go to a veterinarian."[41] The authors anticipated that the well-educated consumers of the baby boom generation would demand greater access to better information about menopause and HRT, but conceded that that information was not forthcoming anytime soon, especially with the results of the WHI not due until after the turn of the century. In an attempt to provide some guidance to menopausal readers, *Newsweek* accompanied its cover story with a shorter article on the signs and symptoms of menopause and the practicalities of hormone replacement. The main article ended on an optimistic note: "As society's long obsession with youth slowly gives way with the aging population, that passage [menopause] will become even easier."[42] This forecast, however, was little more than wishful thinking, given that the American obsession with youth showed absolutely no signs of abating. Just two months earlier, *Newsweek* had run an article called "The Battle of the Bulges," which asked, "If there's no such thing as cellulite, why are aging boomers spending a fortune trying to get rid of it?"[43] And consumer surveys indicated that Americans between 30 and 44—the baby boomers—were the age group most receptive to anti-aging products, procedures, and ploys. "The cohort that created America's youth culture," noted an article in *Marketing Research*, "may work harder than previous generations to prolong its youth."[44]

In addition to being showcased in *Newsweek* and the *New York Times*, menopause made other national media appearances in the month of May 1992, including the CBS Evening News's "Eye on America" report.[45] Local papers also gave space to menopause; the *Houston Chronicle*, for example, ran a thirty-six-hundred-word feature on "The Change" in its Sunday Lifestyles section. All of these features presented what by then had become a familiar storyline: the eldest of the forty million baby-boom women were poised on the brink of menopause; these women had brought other private issues into the public arena, so menopause was next on the list; they wrestled with whether menopause was a passage or a disease and whether taking HRT was wise or unnecessary; doctors offered conflicting and confusing information; American culture sent ambiguous messages about sexual and social roles for aging women. Whether intentionally or not, the physiological and psychosocial aspects of menopause were commingled, as were statistical data and anecdotal evidence, to create a pastiche that defied any sort of meaningful interpretation. Perhaps the purpose of these stories was not so much to come to any sort of conclusion about the menopausal transition but rather merely to identify it as an emerging topic for public conversation. Or maybe they were written to capitalize on the hype being generated by a newly published book that would come to be one of the year's best sellers, *The Silent Passage: Menopause*, by Gail Sheehy.

Fifteen years earlier, Gail Sheehy had penned *Passages: Predictable Crises of Adult Life*, about the transitions in people's lives as they progress through their twenties, thirties, forties, and fifties. The book was wildly successful; it spent more than three years on the *New York Times Book Review* best sellers list and was translated into twenty-eight languages. A survey conducted for the Library of

Congress's Center for the Book named *Passages* as one of the top ten books that made a difference in readers' lives.

Sheehy became one of the nation's best-known journalists, writing for *Vanity Fair* and *New York* magazine. She wrote articles on influential leaders such as Newt Gingrich, Margaret Thatcher, Saddam Hussein, and Mikhail Gorbachev; the last portrait resulted in a book, *The Man Who Changed the World: The Lives of Mikhail S. Gorbachev*, published in 1990. In 1991, she turned her attention from public lives back to private lives with a lengthy article in the October issue of *Vanity Fair* on menopause, in part to rectify her own ignorance as she navigated her way through this transition. For second wave feminists, the personal was political. For Gail Sheehy, the personal was publishable. The article provided the foundation for her book *The Silent Passage* the following spring.[46]

The book was another unmitigated success for Sheehy. It debuted on the *New York Times Book Review* best sellers list on May 31 at number eight, rising to number one the following week. *The Silent Passage* remained on the list for 114 weeks straight, and was the ninth best selling nonfiction book of 1992. (It was also identified as one of the top ten most shoplifted books in America.) It was reviewed in magazines and newspapers across the country, setting off, as mentioned above, a torrent of talk about menopause and hormone replacement in the popular media. Although some reviewers chided Sheehy for being "slick" and "shrill"[47] and for her "weakness for buzzwords and pop sociology,"[48] almost all of them acknowledged the importance of the subject she tackled and gave her credit for exposing the lack of consensus on menopause and hormone replacement therapy within the medical community. However purple her prose, Sheehy had clearly struck a chord that resonated with a large segment of the American population.

To expand her source base beyond her own personal experience, Sheehy interviewed about a hundred women, most of whom were white, affluent, successful professionals like herself. While she did include some working-class women and women of color, she recognized, and glossed over, the fact that they faced a different set of issues at midlife. Sheehy also interviewed ninety "experts," mostly medical doctors and researchers, but also some anthropologists, alternative healers, and even primate researchers for an evolutionary perspective on menopause, aging, and gender. She used anecdotes from these conversations plus the findings of studies published in the medical record as evidence for what she interpreted as the main physiological, medical, psychological, and social challenges of this "passage."

Although she expressed indignation at the absence of medical consensus on the management of menopause, Sheehy came out decidedly in favor of hormone replacement. She drew on similar arguments to those made by Lila Nachtigall in her book on estrogen: in addition to protection from osteoporosis and heart disease, women who took HRT had more energy, better skin, and more and better sex. Sheehy's vivid descriptions of the appearances and experiences of women on and off estrogen were compelling. Early in the book, she told the story of a psychotherapist who gave up estrogen because it made her gain weight. "Now in her early fifties," Sheehy reported, "she does look parched and abruptly elderly."[49] Another woman, a "top literary agent," explained her rationale for taking estrogen. Even though she never had a hot flash or any other sign of menopause, she took a "look at three of my friends who *don't* take hormones and their skin looks like leather." That empirical evidence, combined with her doctor's assurance that estrogen "keeps your memory, and you don't wrinkle as badly," sold her on long-term HRT.[50]

Sheehy recognized the breast/bone/heart dilemma faced by women choosing whether or not to take HRT, but she echoed the rhetoric of estrogen advocates about the relative magnitude of heart disease and breast cancer as killers of older women. "So persuasive is the evidence of the multiple protective benefits of estrogen," Sheehy proclaimed, "many experts are now openly promoting estrogen replacement therapy."[51] She faithfully cited the findings of the Nurses' Health Study (estrogen users cut their risk of heart attack in half) and the Leisure World Study (estrogen users lived longer than nonusers).[52] Her dramatic imagery of the ravages of osteoporosis ("hipbones pulverized almost into powder"), disturbing memory loss, and "bone-dry" vaginal walls might have been enough to convince women to ask their doctors for estrogen prescriptions.[53]

Sheehy chided critics who pooh-poohed the significance of menopause as "polemicists [who] seriously misrepresent the fledgling movement to bring menopause out of the closet" and "women frozen in an outdated era of feminism."[54] To dismiss the experience of menopause as a purely cultural phenomenon, she argued, was wrong and did a disservice to the millions of women for whom physical and psychological symptoms were very real. For Sheehy, if medical science could ease the transition and keep women healthier for longer, then it would be foolish not to take advantage of the available technology, namely, HRT. That said, she did not offer unrestrained support to medical professionals, and she took them to task for not having paid enough attention to menopause. She also borrowed freely and heavily from the rhetoric of the women's movement, making much of the inner and outer strength and beauty of aging women. In the final analysis, *The Silent Passage* was less about criticism than about celebration, offering positive images of postmenopausal women who took care of themselves physically, emotionally, and spiritually.

With *The Silent Passage,* Sheehy claimed to "shatter the myths about menopause," this last taboo subject. But this claim seems dubious, considering that Rosetta Reitz set out to achieve a similar goal fifteen years earlier with her book *Menopause: A Positive Approach.* Both Barbara Seaman and Lila Nachtigall had also written about menopause and hormone replacement therapy, albeit from opposing standpoints. Why did Sheehy get to take credit for bringing menopause out of the closet? The answer is good timing. Just as the three most important things in real estate are location, location, and location, the three most important things in publishing must be timing, timing, and timing. While Reitz and Seaman were able to capitalize somewhat on the feminist critique of medicalization in the 1970s, Sheehy captured the attention of a generation on the edge of middle age. The women she interviewed were not baby boomers; they were part of that much smaller cohort born in the 1930s, many of whom became professionals later in life. But their stories served as object lessons for baby boomers, who were eager for any information about this upcoming stage in their lives.

As much as Sheehy claimed to be looking forward and to be breaking barriers in *The Silent Passage,* she also harked back to decades-old characterizations at several points in the book. She rehearsed the revitalized parallel between menopause and other endocrine disorders as justification for hormone replacement. She noted that "two recent American presidents have been on hormone replacement therapy" (adrenal for John F. Kennedy and thyroid for George Bush) and quoted a Los Angeles gynecologist's analogy between the estrogen deficiency of menopause and the insulin deficiency of diabetes.[55] Ten pages later, she cited a Columbia medical scientist who said, "We wear glasses when our eyes are bad, we use hearing aids, we get false teeth, and now we're putting in new knees and hips."[56] This comment was similar to the rationalization for HRT given twenty-five years earlier by the doctor who wrote in *Ladies' Home Journal* (see chapter 4), "It is perfectly natural for women to wish to slow up the aging process and to remain more attractive. They don't hesitate to use contact lenses for failing eyesight, color rinses for drab-looking hair or caps for their teeth."[57] To both of these commentators, estrogen was just another tool in the arsenal against aging. Toward the end of the book, Sheehy made a statement that was eerily reminiscent of William Masters's description of the "neutral gender" forty years before: "Once she is no longer confined to the culture's definition of a woman as a primarily sexual object and breeder, a full unity of her feminine and masculine sides is possible. As she moves beyond gender definition, she gains new license to speak her mind and initiate action."[58]

Of course, Sheehy's characterization of the postmenopausal woman was positive and life affirming, whereas Masters had depicted women over 50 as unproductive burdens on society. The lessons of the women's movement were not lost on Sheehy, and she fully supported and promoted the active participation of older women in public and private life. But each of these examples suggests that Sheehy was not above pouring old wine into new bottles to make her case.

According to Sheehy, with enough funding from the federal government, attention from the medical research community, and attitude adjustment from both women and men alike, Americans could "break the conspiracy of silence" and reverse the "scandalous politics" of menopause. Menopause, she insisted, was a health issue; in her willingness to embrace modern medicine to meet the challenges of menopause, Sheehy articulated the position of the new women's health advocacy of the 1990s. This position was directly challenged in another book that appeared in the United States in the fall of 1992, *The Change: Women, Aging and the Menopause*, by Germaine Greer.[59]

Germaine Greer was, unapologetically, an old-style feminist. Born in Australia in 1939, she moved to Britain at the age of 25, and earned her Ph.D. in English literature from Cambridge University. In 1970 she published *The Female Eunuch*, about the sexual repression of women, which became one of the seminal, albeit hotly contested, texts of the women's liberation movement. Twenty years later, having passed the age of 50, she drew upon the occasion of her own confrontation with menopause and aging for the inspiration and indignation that fueled the writing of *The Change*.

Although the topics were the same, *The Change* and *The Silent Passage* were very different books, in terms of style, tone, and message. *The Change* was a heftier volume, both physically and intellectually. Almost three times as long as Sheehy's book, Greer's was filled with historical and literary references to aging women, and Sheehy's breezy, sensationalist writing was no match for Greer's erudite prose. Where Sheehy encouraged women to keep up their appearances if it made them happy, Greer suggested that women cast off patriarchal notions of beauty to achieve "serenity and power." "Only when a woman ceases the fretful struggle to *be* beautiful," she argued, "can she turn her gaze outward, find the beautiful and feed upon it."[60] Not surprisingly, Greer rejected the hormone replacement therapy marketed by pharmaceutical companies and prescribed by mainstream physicians, because she saw no need for women to try to maintain smooth skin and firm breasts. To ease menopausal discomforts, Greer advocated alternative treatments, even those for which the evidence of their efficacy was purely anecdotal.

Greer also rejected the notion that older women must continue to have active sex lives. "Some women, the lucky ones," she wrote, "lose interest in sex after menopause."[61] She saw the loss of vaginal moisture and elasticity not as cause for alarm and for either prescription or over-the-counter lubricants, but rather as liberation from the burden of having to please men. "If we are to be well, we must care for ourselves."[62] Greer's own personal journey had taken her from sexual liberation at the age of 31 to liberation from sex at the age of 52.

*The Change* received a good deal of media attention; it was reviewed, quite favorably, in several major newspapers and magazines in the autumn of 1992.[63] Although it helped to keep alive the dialogue about menopause, aging, and HRT, it appears that more people may have talked about the book than actually read it. *The Change* spent just five weeks on the *New York Times Book Review* best sellers list, never rising higher than number thirteen. That fleeting popularity probably owed more to the notoriety of the author than to interest in the book itself. Compared to *The Silent Passage, The Change* was a difficult read. Greer skipped from Shakespeare to Collette, from Tunisia to the Tiwi of Melville Island, Northern Australia, from nineteenth-century psychiatry texts to twentieth-century medical journals in weaving a complex tapestry of menopause and aging from historical, literary, anthropological, and medical threads. Readers had to work hard to keep up with Greer, whereas Sheehy told simple, easy-to-follow stories about contemporary American women. Perhaps for these reasons, *The Change* was nowhere near as successful as *The Silent Passage* in terms of numbers of copies sold. Judging by the popularity of the two books, readers seemed more amenable to Sheehy's approach of working within the system than to Greer's total rejection of it; of course, they may have simply preferred Sheehy's facile writing style. Taken together, the two books made a significant contribution to the menopause buzz of 1992.

Queen Elizabeth referred to 1992 as her "annus horribilis," owing to the divorces of her two sons, Charles and Andrew, and the fire at Windsor Castle. In the United States, the year of the woman in politics was certainly no "annus mirabilis," but it did represent some sort of progress. Even then, however, the status of women was still being debated, more than twenty years after the start of the women's liberation movement. In 1992 the "backlash" (as described in Susan Faludi's 1991 book of that title) against 1970s-style feminism was in full force. Although much fun was made that year of Vice President Dan Quayle's condemnation of the fictional TV character Murphy Brown and her decision to become a single parent, his attack indicated deep-seated concerns among conservatives about professional women and "family values." As noted above, *The Silent Passage*, published in May, was the ninth best-selling nonfiction book of 1992, but Madonna's *Sex*, her controversial coffee table book of erotica and sexual fantasies, was released in late October and still managed to be the tenth most popular nonfiction book of that year.[64] Feminists argued over whether Madonna represented the best or the worst of sexual liberation.

In the midst of cultural ambivalence about women, sexuality, and aging, menopause and hormone replacement therapy became hot topics. In the 1970s, the women's health movement and the consumer movement, along with a growing health consciousness among middle-class Americans at that time, had contributed to an initial expansion in the quantity and variety of health information sources on menopause and estrogen. An even greater upsurge in public discourse on these topics came in the 1990s, as the first baby boomers reached menopause. This large demographic group of women participated in the building of a community of active, informed health care consumers who influenced both the provision of and access to medical information. What did all this talk mean for estrogen, the women who might have used it, and the doctors who might have prescribed it?

# CHAPTER THIRTEEN

# Meno-Boomers

## Another Generation Confronts Estrogen

The 1990s were boom years for estrogen. In 1992 the American College of Physicians (ACP) recommended that "all women, regardless of race, should consider preventive hormone therapy," as part of its "guidelines for counseling asymptomatic postmenopausal women about using hormone therapy to prevent disease and to prolong life."[1] In 1994, the *American Family Physician* published a "Clinical Opinion" that identified estrogen replacement therapy as one of eight underused prescriptions, along with aspirin (to prevent myocardial infarction), beta-blockers (after myocardial infarction to reduce mortality), nicotine substitutes (for smoking cessation), anti-inflammatory agents (to treat asthma), antidepressants, fiber, and regular exercise. The primary reason given for prescribing long-term estrogen was the prevention of heart disease, followed by the prevention of osteoporosis.[2] The ACP, the nation's largest society of medical specialists, published its guidelines in its journal, *Annals of Internal Medicine*, which went out to a membership of more than one hundred thousand doctors and medical students in internal medicine and related subspecialties. The *American Family Physician* was sent to the more than ninety thousand general practitioners who belonged to the American Academy of Family Physicians. Together, these two articles reached about a third of all doctors in the United States, specifically the two groups who, besides gynecologists, were most likely to see menopausal women in their offices. The message conveyed to these physicians was clear: long-term estrogen therapy made good sense.

And the news about estrogen just kept getting better, as medical researchers confirmed the alleged health effects associated with hormone use. In November 1994 the results of the first NIH-funded large-scale randomized double-blind placebo-controlled trial of combined hormone replacement therapy (estrogen and progestin) were announced at the annual meeting of the American Heart Association; the official report of this "gold standard" study was published in the *Journal of the American Medical Association* two months later.³ Known as the Postmenopausal Estrogen/Progestin Interventions (PEPI, for short) Trial, the study enlisted 875 women between the ages of 45 and 64 at seven clinical centers across the country to take one of three estrogen/progestin regimens, estrogen alone, or placebo. The objective of the trial was to determine the effect of hormone therapies on four known risk factors for heart disease: cholesterol (HDL and LDL levels), fibrinogen (as a marker for blood clotting), serum insulin (to measure carbohydrate metabolism), and blood pressure. It is important to point out that evidence of actual heart disease (such as angina or heart attack or death) was *not* measured. At the end of the three-year study period, the women who took estrogen or estrogen and progestin had better cholesterol profiles (more of the so-called good cholesterol, HDL, and less of the so-called bad cholesterol, LDL) and lower (that is, better) levels of fibrinogen than those who took the placebo. Insulin levels and blood pressure were unaffected. The researchers cautiously interpreted these results as evidence that the addition of progestin—necessary to prevent endometrial cancer in women with a uterus—did not vitiate the cardio-protective effect of estrogen. Journalists were much more enthusiastic; the *New York Times* gave the story a prominent spot on the front page, summarizing for its readers, "A mixture of the two hormones estrogen and progestin is better than estrogen alone, protecting women not only against the risk of uterine cancer but heart attacks as well."⁴

Medical researchers also discovered new benefits for women who used estrogen, and these findings were enthusiastically reported in the popular press. Those physicians who did not keep up with the medical literature would have come across the good tidings about estrogen while scanning the headlines of their local newspapers. In 1995 and 1996, hormone replacement therapy was associated with reduced risk of colon cancer, prevention of tooth loss, lower incidence of osteoarthritis, increase in bone mass, reduced risk of Alzheimer's disease, and lower rates of death from all causes.⁵ The finding about Alzheimer's was especially welcomed in the wake of increased publicity and heightened awareness about the ravages of the disease and the absence of effective treatment. In 1994 Americans had been saddened (but by that time not necessarily surprised) by President Ronald Reagan's disclosure that he had been diagnosed with Alzheimer's.

Even when the news about HRT was not so good, it was presented to the public in a relatively positive light. Thus, when the Nurses' Health Study demonstrated that users of combined hormone replacement had an increased risk of breast cancer (as did users of estrogen alone) as compared to nonusers, the front-page headline in the *New York Times* read, cryptically, "New Clues in Balancing the Risks of Hormones after Menopause." Readers were reminded that these latest findings had to be weighed against the cardio-protective and osteo-protective benefits of HRT.[6] One month later, when a study of women in King County, Washington, found no difference in breast cancer rates between HRT users and nonusers, the *Times* front-page headline was more straightforward: "Cancer Link Contradicted by New Hormone Study."[7]

This optimism about long-term hormone replacement therapy was also evident in medical textbooks' treatment of the subject. Although medical texts cannot keep fully up-to-date with the latest research findings, they do suggest the baseline knowledge of the average practicing physician.[8] They also reveal what established leaders in the field, as authors of these textbooks, considered imperative knowledge to impart to medical students at the start of their clinical training. Obviously, gynecology texts devoted more space to menopause than did primers for other specialties; 1990s editions usually included a separate chapter on the menopause or the climacteric. In his chapter on menopause in Copeland's *Textbook of Gynecology*, M. Yusoff Dawood made his stance crystal clear: "The menopause should be looked on as a hormone-deficient state. Therefore, the metabolic sequelae and changes accompanying such a hormone-deficient state can be reversed to a large extent by correcting the hormone deficiency."[9] He described four categories of "menopausal symptoms and problems": cardiovascular disease, psychoneuroendocrine effects, skin and urogenital effects, and bone effect. It is significant that cardiovascular disease was the first problem discussed, and the one to which the most space was devoted. Dawood listed calcium and exercise as management strategies for menopause but clearly favored HRT and took his cue from the ACP guidelines: "From a preventive and public health standpoint, all postmenopausal women who have no absolute contraindication to the use of estrogen should be candidates for estrogen replacement therapy because of the many therapeutic benefits that far outweigh the risks."[10]

Physicians practicing internal medicine and medical students training in that specialty encountered the subject of menopause in the context of disease in their textbooks. The twentieth edition of *Cecil's Textbook of Medicine* took up menopause in a chapter on "Endocrine and Reproductive Diseases."[11] *Harrison's Principles of Internal Medicine,* thirteenth edition, introduced a section on "Estrogen treatment of the menopause" (in a chapter called "Disorders of the Ovary and Female Reproductive Tract") with the assertion, "The rationale for use of estrogens in postmenopausal women is based on the belief that such therapy may relieve many of the disorders of the menopause, including osteoporosis, and indeed of aging itself."[12] A shorter handbook of medicine told those who consulted its chapter on endocrine disease, "The issue of hormonal replacement therapy should actively be addressed for *every* perimenopausal woman."[13] And specialists in cardiology read in the fifth edition of Braunwald's *Heart Disease: A Textbook of Cardiovascular Medicine* about the ACP's guidelines on preventive hormone therapy. The text noted that the prescription of long-term HRT for both primary and secondary prevention of heart disease (that is, in women without and with the disease) could "represent an enormous potential change in cardiovascular therapeutics and practice," but it cautioned that only the results of ongoing randomized trials could validate this practice for certain.[14] Thus textbooks for both specialists and nonspecialists in women's health endorsed the conventional wisdom of the relative benefits of long-term HRT.

As in years past, another source of information for physicians about prescription medications could be found in the advertisements liberally interspersed among the articles in medical journals. Although both journal articles and textbooks gave increasing currency to estrogen as a preventive therapy against cardiovascular disease, it did not have FDA approval for that indication. Since advertisements could not promote estrogen products as cardio-protective, they had to stick to the approved indications: vasomotor symptoms (hot flashes) and osteoporosis. Of course this did not preclude the prescription of estrogen for the prevention of heart disease. As with any other drug, doctors could use their own discretion in prescribing for conditions not specifically indicated in the product labeling.

Wyeth-Ayerst actively marketed the long-term use of Premarin for the prevention of osteoporosis. One popular ad campaign that ran frequently in medical journals in the 1990s showed a young-looking woman in a leotard, complete with headband and leg warmers, with the caption, "Calcium every day, Aerobics every week, Bone loss every year."[15] The text claimed that calcium and exercise alone could not prevent postmenopausal osteoporosis; estrogen (in the form of Premarin) was essential to keep this fit-looking woman healthy. This theme was repeated in other campaigns. One featured schematic drawings of a young woman standing with a ramrod-straight back (and measuring a respectable 5 feet 7 inches) and an older woman sitting with a bent spine in a wheelchair, the before and after of untreated osteoporosis. Another illustrated the devastating and rapid progress of bone loss with micrographs of normal bone and osteoporotic bone. In these ads, the short-term treatment of hot flashes took a back seat to the long-term prevention of osteoporosis.

Although Premarin dominated the field, it was by no means the only estrogen product available. Competitors touted their unique advantages over the market leader. Ciba promoted Estraderm, its estradiol transdermal system, as "the next generation of estrogen replacement therapy," illustrating this point by contrasting two photographs, one of a woman in 1940s garb with an old-fashioned bottle of pills and the other of her updated counterpart (sporting a trendy hairdo and power blazer) and the ultramodern skin patch.[16] Estraderm was initially approved only for the treatment of hot flashes, but when it won FDA approval to protect against postmenopausal osteoporosis, Ciba developed new advertisements prominently announcing this new indication.[17] Upjohn peddled Ogen as a "natural, plant-derived product," claiming that, among women who expressed a preference (and almost half of those contacted by telephone did not), most preferred their prescription drugs to be derived from plants as opposed to animals or chemicals.[18] Mead Johnson also described its Estrace, plant-derived estradiol, as "Like Nature's Own." Interestingly, Mead Johnson flouted the convention of not promoting off-label indications in a 1990 ad that listed "favorable effect on lipid proteins" as one of Estrace's benefits.[19] In a defensive move against these estrogen alternatives, Wyeth-Ayerst ran an ad with the caption "Nothing else is Premarin" that went on to describe "what makes Premarin different."[20]

Wherever they learned about the risks and benefits of estrogen—whether in journal articles or advertisements or textbooks or continuing education seminars or at the water cooler in the group practice—physicians who saw menopausal women as patients were, by and large, enthusiastic prescribers of HRT. A survey of gynecologists who were members of the North American Menopause Society (NAMS) in 2000 reported that 92 percent "routinely offered HRT to their menopausal patients." Moreover, 86.4 percent prescribed hormone replacement for durations of at least ten years, which certainly could be characterized as long-term use.[21]

Other studies tried to identify which doctors would be most likely to prescribe HRT. A national survey of 1,383 cardiologists, internists, family doctors, and general practitioners reported that most (82%) prescribed HRT for menopausal and postmenopausal patients; of this group, two-thirds (66%) prescribed it for prevention of coronary heart disease.[22] A survey conducted in the state of Washington found that gynecologists prescribed HRT more frequently than did family practitioners, and older doctors were more apt than younger doctors to offer their patients HRT.[23] It correlated HRT prescribing practices with a belief in the cardioprotective benefit of the drug. And a study done in Boston a few years earlier found that prevention of cardiovascular disease was the third most common reason for prescribing hormone replacement, accounting for 32 percent of prescriptions (56 percent were for menopausal symptoms and 38 percent were for prevention of osteoporosis).[24] In spite of the FDA's refusal to approve estrogen for the prevention of heart disease, many doctors went ahead and prescribed it for that reason anyway.

Perhaps the most interesting discovery in both the Boston and Washington studies was that women physicians prescribed HRT more often than their male counterparts did. The Boston researchers reported that "controlling for potential confounding variables, patients seen by female physicians were more than five times as likely to begin ERT."[25] They were unwilling to speculate as to whether the gender disparity in prescribing estrogen reflected differences in the way male and female doctors interacted with patients or differences in their knowledge of or attitudes about hormone replacement. Nor could the Washington group account for their finding of a similar gender difference in HRT prescribing practices on the other side of the country.

Women physicians also used HRT themselves at a much higher rate than did the general public. Three physician-researchers at Emory University asked almost fifteen hundred postmenopausal female doctors between the ages of 40 and 70 whether they used hormone replacement.[26] Overall, 47.4 percent said yes; this number was significantly larger than the proportion of nonphysicians using HRT. The authors hypothesized that "higher rates of HRT use by women physicians . . . may presage greater use of HRT by U.S. women in the future" for two reasons.[27] First, they noted that changes in physicians' health-related behaviors tended to precede similar changes among the rest of the population; physicians, for example, had led the decline in smoking since the Surgeon General's Report thirty years earlier. Second, they conjectured that "physicians who use HRT may be more likely to recommend HRT to their patients."[28] Although this potential correlation was not formally investigated in their study, it seemed worthy of mention. Whether or not physicians counseled patients based on their own personal experiences, the very large proportion of postmenopausal female doctors on HRT denoted a high level of acceptance of the medical-pharmaceutical model of menopause and aging among this group.

Thanks in part to the general enthusiasm for HRT among American physicians, Premarin became the best selling drug in the nation in 1992, a position it held for the next several years. Part of this triumph was due to Wyeth-Ayerst's success in defeating its generic competitors. Since the 1970s, several drug companies had manufactured and sold generic versions of conjugated estrogens. By that time, the patent for Premarin had long since expired. The advantage to a generic drug, of course, is that its price is lower than that of the brand-name medication. Wyeth-Ayerst maintained that these generic products did not exactly match the mixture of compounds found in Premarin (recall that Premarin, derived from horse urine, consisted of about ten different conjugated estrogens) and that, therefore, they were not "bioequivalent" to Premarin, meaning that the different compositions of the drugs led to different actions or uptakes in the human body. In 1991, the FDA agreed and ordered the generics off the market because of concerns about their efficacy and safety. In 1997, the FDA upheld the ban, maintaining that the murkiness of the mechanism by which Premarin worked—it was unknown how many of the estrogen components were active ingredients—precluded successful copies of its formula.

In November 1995 Wyeth-Ayerst achieved another victory with the FDA approval of Prempro, which combined estrogen and progestin in a single tablet. Within a few months, the company had launched a splashy ad campaign in the medical journals, touting the convenience of its new product for both women and physicians. "Simplicity of a single tablet," the ad copy promised, "One prescription for you, one tablet for your patients."[29] In its first year on the market, Prempro sold more than 6.5 million prescriptions, vaulting to number fifty-four on the list of best-selling drugs. The editors of *American Druggist*'s annual review of the top two hundred prescription drugs called it "the year's most dramatic success."[30] That same year, 39.6 million prescriptions were written for Premarin, which dropped to number two on the list, probably because so many women switched to Prempro. Overall, however, patient visits resulting in prescriptions for HRT increased by 13 percent from 1995 to 1996, perhaps reflecting the positive results of the PEPI Trial, which seemed to reinforce the cardiovascular benefits of hormone replacement.[31]

Together, Premarin and Prempro accounted for almost 70 percent of the total number of HRT prescriptions in the United States. Other oral estrogens (e.g., Estrace) and the skin patch (e.g., Estraderm) made up the bulk of the remaining prescriptions. The total number of HRT prescriptions rose meteorically in the 1990s, from 36.5 million in 1992, to 58.3 million in 1995, to 75.8 million in 1997, and to 89.6 million in 1999.[32] Given that a prescription was typically written for two months' worth of pills, this meant that the number of women using estrogen or estrogen and progestin multiplied from about 6 million in 1992 to approximately 15 million just seven years later.

Some of the success of Premarin and Prempro can be attributed to direct-to-consumer (DTC) advertising. In 1985 the FDA approved the publication of DTC ads, so long as they supplied sufficient information about the product's risks as well as its benefits. Since the brevity of television spots precluded the announcement of risk information (even in an announcer's rapid-fire delivery), television ads were restricted to "see-your-doctor" ads, in which consumers were told to ask their doctors about treatments for the specified disease symptoms, and ads for specific products in which no mention was made of their therapeutic purposes. Then, in 1997, the FDA decided to allow product-specific television ads, provided they relayed the risk information by simply referring to a concurrently running print advertisement, naming a toll-free number for consumers to call or a website to access, or advising consumers to talk to their doctors or pharmacists.[33] Pharmaceutical manufacturers leapt at the chance to promote their products directly to potential consumers. The amount of money spent on DTC advertising by the drug industry skyrocketed from $12 million in 1989, to $156 million in 1992, to $595 million in 1996, to $1.58 billion in 1999, a 130-fold increase in just ten years.[34]

Wyeth-Ayerst stepped up its DTC advertising for its Premarin family of products in the late 1990s. In 1998, Premarin ranked eleventh among prescription drugs with the greatest amount of DTC advertising spending with $37.1 million, most of which was used to purchase print ads in women's magazines.[35] The following year, Wyeth-Ayerst spent $34.7 million to promote Prempro directly to consumers, making that product the eighteenth most heavily advertised drug.[36] Initially, the company concentrated its DTC advertising dollars on women's magazines. A typical ad was a two-page spread that ran in *Better Homes and Gardens.*[37] It featured on the left side a large photograph of an attractive brunette (with just a hint of laugh lines around her eyes), and on the right, amid several paragraphs of text, a smaller inset of the same woman jogging in athletic clothes. The banner headline that stretched across both pages read, "Every day they're discovering more about estrogen loss. That's why I'm glad I take my Premarin." Because Premarin was approved only for osteoporosis and the specific menopausal symptoms of hot flashes, night sweats, and vaginal dryness, these were the only indications mentioned in the ad.

Wyeth-Ayerst got around this regulatory obstacle by running informational advertisements sponsored by its Women's Health Research Institute, created in 1993. Paraded as a vaguely philanthropic organization, this institute bore some resemblance to the Wilson Research Foundation, also funded in part by Ayerst, thirty years earlier. Whereas the Wilson Research Foundation developed informational pamphlets to be distributed to women in doctors' waiting rooms, Wyeth-Ayerst's Women's Health Research Institute ran "messages" in popular women's magazines. One produced in 1998 showed two women (one who looked very much like the woman in the Premarin ad and another equally fit-looking gray-haired woman) in sneakers, spandex tights, and polar fleece jackets walking and talking together. They were surrounded by eight blurbs about the effects of meno- pause on brain, eyes, teeth, heart, "uncomfortable symptoms," colon, sexuality, and bone. Lines ran from each blurb to the corresponding body part in one or the other woman ("uncomfortable symptoms" were identified as being located some- where in the vicinity of the gray-haired woman's armpit). The heading read, "When considering menopause and the consequences of its associated estrogen loss, consider the entire body of evidence."[38] What was interesting about this full- page "message" was the association made between estrogen lost after menopause and cognitive functioning, memory, Alzheimer's disease, tooth loss, colon cancer, and cardiovascular disease, correlations that had been explored in medical re- search studies but were not approved indications for Premarin. Presumably, Wyeth-Ayerst hoped that women would follow the recommendation to "Talk to your doctor," who would then prescribe the appropriate medication, namely, Premarin or Prempro.

Although the number of women receiving prescriptions for HRT increased greatly during the 1990s, some in the medical profession expressed two con- cerns: first, that the proportion of eligible postmenopausal women who received HRT prescriptions was still too low, and second, that women who did receive prescriptions were not having them filled at the pharmacy and thus were not actually taking the pills as directed. A significant literature arose to address the question of "compliance" among middle-aged and older women. Some strove to identify the characteristics of women who chose not to follow their doctors' or- ders; others suggested ways to reduce "noncompliance." Researchers agreed that while some 20 percent of postmenopausal women began hormone replacement therapy each year, only 40 percent of them stuck with it for more than one year.[39] For advocates of long-term therapy to prevent osteoporosis and heart disease, these figures were disappointing. One study found, perhaps not surprisingly, that older women (65 and over) were more likely than were younger women (aged 50– 55) to stop using prescribed hormones. The women in their fifties usually started HRT to get some relief from hot flashes and related menopausal symptoms; the older women, by contrast, began HRT in hopes of treating or preventing os- teoporosis. The majority of women in both age groups cited unwelcome side effects (such as the return of periodic vaginal bleeding) as the main reason for discontinuation of therapy.[40]

In the 1990s researchers also began to look at demographic, socioeconomic, and educational data to determine who was and was not taking HRT. Recall that patterns of estrogen use differed according to geographic region, as discussed in chapter 10, with the highest level of use in the West and the lowest level in the East. Overall, the rate of use in 1995 was estimated to be 33 percent of women between the ages of 50 and 74 (and 27 percent of all women over 50).[41] A national study of 4,365 Medicare beneficiaries (women over 65) that same year found that 13 percent of elderly women used estrogen; this group tended to be white, with high income and private health insurance.[42] Other studies supported the conclusion that women of color, women with lower incomes, and women with less education were less likely to use HRT, while women who had hysterectomies were more likely to do so. While socioeconomic factors and insurance type correlated with access to care and thus HRT prescriptions for postmenopausal women, race turned out to be an independent predictor of HRT use. Black women were only about half as likely as white women to take estrogen.[43] It was unclear whether this difference was due to different attitudes toward menopause (a few small studies had reported more positive attitudes among black women), to prescribing biases among physicians (owing perhaps to the dearth of data on the risks and benefits of long-term HRT for non-Caucasian women or to the data indicating that African American women were less prone to osteoporosis), or to other as yet undetermined factors.[44]

Some investigators were convinced that education was the answer to the problem of low rates of estrogen use and compliance. A group in Wisconsin proudly reported in their state medical journal that they had increased the rate of HRT use among a sample of postmenopausal patients at their medical center in La Crosse from 31 percent to 64 percent over a five-year period, by targeting educational programs at both physicians and patients.[45] Another group surveyed women in Georgia to assess their awareness of hormone replacement therapy, particularly for the prevention of heart disease and osteoporosis, and found that race, education, and menopausal status correlated with knowledge about HRT. The authors of this report urged their colleagues to step up public health efforts to inform African American women and women with less education about the preventive health benefits of estrogen.[46]

The absence of minority women in research studies on HRT was documented —belatedly—in a 1999 report. An analysis of thirty studies of cardiovascular disease and HRT and eleven studies of osteoporosis and HRT conducted from 1966 to 1998 found that African American women made up only 0.1 percent of subjects in the former and 0.4 percent of subjects in the latter.[47] The investigators noted also the relative absence of Asian and Latina women in these studies, although they were unclear as to what influence, if any, these factors may have

had on physicians' prescribing practices. In the spirit of enthusiasm about estrogen that permeated much of the medical profession in the 1990s, minority
women were seen as an underserved population that ought to be encouraged
to reap the alleged benefits provided by long-term postmenopausal hormone
replacement.

What did women in the 1990s think about HRT? Several studies tried to assess
women's opinions of hormone replacement therapy; others asked women about
their attitudes toward menopause. Most of these studies obtained similar results:
they found that the main reasons women gave for starting HRT were to relieve
menopausal symptoms, to prevent osteoporosis, and to follow their physician's
advice.[48] This last explanation surprised investigators, especially in surveys of
well-educated women (one of 337 Stanford University alumnae, the other of 91
residents of the St. Louis area, two-thirds of whom were college graduates), who
perhaps were expected to think more independently. Very few women cited prevention of heart disease as a major incentive for hormone replacement, a puz-
zling contradiction to the studies of physicians' prescribing practices, which
found cardio-protection to be a significant indication for HRT. These findings
prompted one set of authors to recommend "a more far-reaching educational
message about CVD risk prevention [to] foster greater awareness among women
of the cardioprotective benefits of HRT."[49]

Researchers from a variety of specialties (epidemiology, obstetrics and gynecology, cardiology) acknowledged that different age cohorts would regard HRT in
different ways; obviously, symptom relief was much more important to midlife
women than to older women. A survey that queried 7,667 "nonblack" women
over 65 in Baltimore, Minneapolis, Portland, Oregon, and western Pennsylvania,
in an attempt "to understand the low prevalence of estrogen use among older
women," seemed to report the obvious: women never started or discontinued
HRT because they said "they felt they did not need it."[50] But there seemed to be an
almost universal conviction among these investigators that women of all ages—
those approaching the age of menopause, those going through menopause, and
those well past the milestone of menopause—should be informed of the preventive health benefits of long-term hormone replacement therapy. Furthermore,
there was an implicit assumption that better education would encourage more
women to take up and stick with HRT. The authors of a study of women in the
Chicago area expressed dismay that "more than one-half (54%) of these menopausal respondents seem to be unaware that menopause is an ongoing condition
. . . continuing for the rest of a woman's life, and associated with defined medical

risks." They called for direct education of women about menopause "and its continuing medical risks."[51]

Part of this educational effort would have to challenge women's fears of breast cancer, which clearly outweighed their concerns about heart disease. The authors of the St. Louis study concluded, "Perhaps the most important goal, then, is to educate menopausal women to the facts as we now know them—that the risk of cancer appears to be quite small relative to the benefits of HRT in preventing CVD, the number one cause of death of women in America."[52] Another problem —perhaps less easily overcome by education—was that women objected to the side effects brought about by HRT, especially the return of menstrual periods. Contrary to the outlook of the Chicago physician-investigators, few women perceived of menopause as "a time of disease risk." Only 4 percent of participants in the Seattle Midlife Women's Health Study defined menopause in this way; the vast majority (79%) identified menopause simply as the "cessation of periods."[53] And for most women, the end of menstrual bleeding was cause for celebration. The very tangible and immediate hassle of resuming the monthly purchase and use of pads and tampons eclipsed the intangible benefit of warding off the possible development of disease in the distant future.

The medical literature of the 1990s advocated the use of educational materials to encourage more women to use hormone replacement therapy; although authors recommended that women be advised of the risks as well as the benefits of HRT, they implied that an informed patient was one who agreed that the benefits of estrogen outweighed the risks. Who supplied this information was a source of contention. "Unless physicians and other women's health providers seek to meet these needs [for information about the risks and benefits of HRT]," warned one article, "this important educational task may be left to less-informed sources."[54] Physicians worried that nonmedical informants would turn women away from the medical model of menopause and aging. And these concerns were legitimate: the profusion of books, magazine articles, television reports, and internet sites on menopause, aging, hormone replacement, and alternative therapies offered women what appeared to be a wide variety of opinions and advice.

One of the ways doctors and other health care professionals asserted control over the dissemination of information was through menopause seminars and clinics offered by hospitals or private practices or women's health centers. Indeed, as Sandra Morgen observed in her history of the women's health movement, the rise of hospital-based women's health centers in the 1980s and 1990s "demonstrates both the dramatic effects of the women's health movement on mainstream medicine and the resilience and power of mainstream medicine to co-opt the ideology and practices of the movement."[55] These often for-profit centers incorporated some of the features of feminist self-help clinics of the 1970s, such as cozy décor and feminist-inflected rhetoric, but ultimately they were driven by the competitive market economy of health care. Women's health centers—of which there were some thirty-six hundred in the United States by the mid-1990s—focused on reproductive health care, but as their clientele aged, menopausal and postmenopausal health became a part of their mission.[56] Significantly, the target audience for these centers consisted of women with health insurance, as opposed to the previous generation of feminist clinics that sought to provide health care to the underserved. Thus women who had the time and money to seek out medical care at menopause—women described by Bernadine Healy as "active and effective consumers of health care"[57]—had a growing array of providers and resources from which to choose.

Although menopause clinics garnered some attention from the popular media, and even more from physicians specializing in menopause, they seem not to have been noticed by many women. A nationwide survey of midlife women conducted by the North American Menopause Society (NAMS) found that only a handful of the 750 women interviewed mentioned menopause clinics as a source of information. The most popular educational source was the popular media; 59 percent of respondents cited books, magazines, newspapers, newsletters, television, and radio as their main resources, while 43 percent asked their gynecologists or primary care practitioners about menopause-related issues.[58] The results of this poll confirmed the findings of two previous studies, which demonstrated that, since the 1970s, the authoritative voice of the physician had to compete with other opinions, advice, and information readily accessible to women. In 1993 NAMS conducted its first survey, phoning 833 women (ages 45–60) who reported their reliance on print and broadcast media to learn about menopause; 40 percent consulted magazines, newspapers, books, and television, while 36 percent looked to their physicians for information.[59] Two years earlier, in 1991, *McCall's* had surveyed its readers about menopause and found that 83 percent gleaned their information from magazines, newspapers, books, and television; only half that number (44%) included doctors on their list. Forty percent identified the media as their best educational resource; doctors were credited by just 18 percent.[60] Although doctors fared better in the NAMS surveys, they still took a back seat to popular sources.

Women who sought to educate themselves could find plenty of choices at the local bookstore or newsstand. A spate of mass-market books appeared in the wake of Gail Sheehy's *Menopause: The Silent Passage* and Germaine Greer's *The Change*. From 1992 to 1995, more than a hundred books were published by both medical and lay authors on the subject on menopause.[61] Newspaper and magazine editors sensed the sea change in interest about menopause and began to publish more articles on menopause, hormone replacement therapy, and osteoporosis. In all of the popular magazines indexed in the *Reader's Guide to Periodical Literature,* there was never more than a total of a dozen articles on menopause or related topics in any year prior to 1985. The numbers began to rise in the late 1980s, and by the 1990s there was an average of more than thirty-two articles a year. Articles with titles such as "The Baby Boom Meets Menopause" and "Menopause and the Working Boomer" recognized that "an entire generation is approaching a big change."[62]

Articles in the 1990s often addressed the controversy over hormone replacement therapy, quoting the latest statistics and "expert" opinions on the relationships between estrogen and endometrial cancer, breast cancer, osteoporosis, and heart disease. In 1991 the National Women's Health Network surveyed a sample of twenty-five articles from popular magazines and found that most (83%) gave a balanced assessment, mentioning both risks and benefits of hormone replacement therapy, and more than half (52%) suggested alternative therapies, such as herbal remedies, dietary options, and exercise regimens.[63] *Better Homes and Gardens* exhorted its readers: "Educate Yourself. In the absence of complete medical data, the best advice is to learn all you can about hormone replacement and make a careful assessment of your own risks . . . the biggest health risk of all—not being an educated consumer."[64] The language of this recommendation was significant in its addressing women as active consumers in charge of their health care, not as passive patients at the mercy of their doctors.

Television and radio jumped on the menopause bandwagon, too. Whereas the subject of menopause was rarely heard on the airwaves prior to 1990, it rapidly became a popular topic on news broadcasts and talk shows as the decade progressed.[65] Paula Zahn, co-host of "CBS This Morning," introduced her 1990 interview with the authors of a book called *Managing Your Menopause* by saying, "Women have traditionally dreaded approaching menopause. Now medical science and a woman's own awareness can eliminate much of that dread and many of the problems."[66] A 1992 CBS Evening News "Eye on America" segment explored how baby boomers on the verge of menopause were "demanding explanations from doctors and talking openly among themselves."[67] Even situation comedies broached the subject, with menopause featured in episodes of "The Golden

Girls" and "The Cosby Show."[68] With restrictions lifted on direct-to-consumer advertising of prescription drugs on television in 1997, Wyeth-Ayerst began to run commercials for Premarin during prime time. A few years later, ex-supermodel Lauren Hutton starred in one television spot for Premarin, in which she proclaimed, "Knowledge is power. Information is how you get it."[69] She advised women to ask their doctors for that information and offered her own experience as a cautionary tale: having lost one inch in height in the year after menopause, she decided to take her doctor's advice and started a regimen of HRT. Wyeth-Ayerst also hired the singer Patti LaBelle to endorse Premarin on television, perhaps in an effort to draw African American women into the market for HRT.

This expanded coverage meant that more women were exposed more often to talk—in news, in entertainment, in advertising—about menopause and hormone replacement therapy. Learning now included a sort of passive reception: while watching the evening news or listening to a daytime talk show, a woman might hear something about menopause or HRT—a brief news report, an in-depth story, or perhaps an interview with an author of the latest self-help guide. The point here is that the recipient of this information was not actively seeking knowledge; as menopause became popularized, women had to work less hard to learn about it.

Of course, for the woman who wanted to conduct her own research, the available information had multiplied dramatically since the previous decade. Not only were there more books and articles available in print, from both mainstream and alternative presses, but also the new medium of the Internet provided a whole new educational resource and information clearinghouse. For example, the web site called "Power Surge" came on line in 1994, offering women the opportunity to chat via email with doctors, authors, and other women going through menopause. The heading to one of the links provided by this site proclaims, "Knowledge is Power. Educate your body by educating yourself."[70]

While many observers welcomed this influx of information, the feminist magazine *Ms.* scoffed at the increased attention to menopause. A 1993 article began, "1992 was the year of the . . . full-scale launch of the meno-boom, inundating women with 'information' [the quotation marks signifying irony] that detailed all the loss, misery, humiliation, and despair suddenly in store for us."[71] And a feminist columnist for *The Nation,* a liberal weekly publication, satirized the plethora of magazine and newspaper articles as follows: "Invariably beginning with a tribute to upper-income baby-boom women—you know, educated medical consumers used to speaking out and getting what they want—the article moves in swiftly for the kill: hot flashes, osteoporosis, irritability, heart disease, insomnia, migraines, declining libido, wrinkles and job loss."[72] Why did the latest batch of menopause articles merit such derision?

Journalists writing in the 1980s and 1990s had successfully incorporated the rhetoric of the women's health movement, encouraging their readers to become educated health consumers. However, many articles still perpetuated the idea of medicalized menopause and portrayed the image of menopausal women "in decline," if not in the main text itself, then in the accompanying illustrations. *Newsweek*'s cover story on menopause in 1992 graced its cover with the image of a woman's face represented as a tree losing its leaves; three years later, *Time*'s cover story on "The Estrogen Dilemma" included a similar drawing.[73] The autumnal imagery of deciduous tree branches (accompanied by a troubled look on the woman's face) presented menopause and the years beyond as cause for concern. This tendency to biologize menopause and to suggest personal health remedies (pharmaceutical, nutritional, or otherwise) rather than to evaluate the social status of older women irritated feminists who had hoped for broader cultural change.

Furthermore, information about menopause and HRT was bound up with prescriptive literature on *how* women should age. The implicit and explicit message in the 1990s was that women ought to be pro-active in the management of their menopausal and postmenopausal years. The concept of aging gracefully was replaced by aging dynamically; that is, women were expected to do everything they could to fight the physical and mental decline associated with growing older. At the same time, there was a concurrent literature that talked about a new world for aging Americans, one to be shaped by the giant demographic cohort of baby boomers. These two lines of thinking appeared sometimes together and sometimes separately, sometimes reinforcing and sometimes contradicting each other. In the 1990s readers encountered a myriad of contested meanings and conflicting messages about the health and social aspects of growing older as a woman in America.

The feminist Simone de Beauvoir had theorized aging back in 1970 in *The Coming of Age*, but at that time, as discussed in chapter 6, aging issues had not made it onto the main radar screen of Americans, appearing only at the margins of popular and political consciousness. When Betty Friedan, doyenne of the women's movement of that era, tackled the subject a generation later in her 1993 book *The Fountain of Age*, she had a potentially much larger audience of readers—the very women she had so intoxicated with *The Feminine Mystique* thirty years earlier. *The Fountain of Age* earned good reviews from several major newspapers and magazines and spent seven weeks on the *New York Times Book Review* best sellers list. For those unwilling to labor through the more than six-hundred-page tome, excerpts were published in *Time* and *Good Housekeeping*, thus securing an even larger audience for Friedan's paean to positive aging.

The 1990s saw a number of books on aging written by women who had become public figures in the 1960s and 1970s. In 1994, two of the icons of the sexual revolution—Helen Gurley Brown, long-standing editor of *Cosmopolitan* magazine and author of *Sex and the Single Girl* (1962), and Erica Jong, poet and author of the erotic novel *Fear of Flying* (1973)—imparted their thoughts on life after 50, which seemed to be the agreed-upon dividing line between the young and the old. In *Fear of Fifty: A Midlife Memoir*, Jong offered her own experience as an example of how the avant-garde of the baby boom generation were facing the later decades of their lives. "I figure that if I'm confused, you are too," she wrote in her introduction. "I make the assumption that I am not so different from you or you. I want to write a book about my generation. And to write about my generation and be fiercely honest, I can only start with myself."[74] Brown, ever in denial of her own chronological age, waited until she was 71 before conceding to give advice to her fellow seniors in *The Late Show: A Semiwild but Practical Survival Plan for Women over 50*. Brown's plan for aging was simple: don't. To avoid the appearance and feeling of physical and psychological deterioration, she advocated cosmetic surgery, sexual activity (marital or extramarital), and estrogen.[75]

Feminist scholars also turned their attention to the subject of growing older. Lois Banner's *In Full Flower: Aging Women, Power, and Sexuality*, Margaret Lock's *Encounters with Aging*, and Margaret Morganroth Gullette's *Declining to Decline: Cultural Combat and the Politics of the Midlife* won critical acclaim within the academy for their gendered analyses of aging. No longer relegated to journals of gerontology and geriatric medicine, age joined race, class, and gender as a category worthy of social and historical investigation.

But the average reader probably did not engage the critical study of age in scholarly texts. Instead, she would have come across the topic in articles in newspapers and popular magazines, as aging became an increasingly popular subject for journalistic scrutiny. Contemporary observers made much of the fact that the baby boom generation was approaching middle age, in articles with titles such as "Kicking and Screaming, Baby Boomers Begin to Talk about Aging."[76] Some looked to the future, trying to predict the local and global effects of this impending demographic shift toward an older population. Others offered personal reflections on the consequences, and unexpected joys, of growing older; one columnist described her sixties as "a gift decade."[77] Much ink was spilt on the subject of wrinkles as the most obvious marker of age, with opinions along the full spectrum from acceptance to rejection; some writers claimed to celebrate their acquisition of a lined face as evidence of a life well lived, while others defended their decisions to use cosmetics or surgical procedures to smooth out their skin.

The discourse on wrinkles was representative of the media's mixed messages on aging. Often these split signals came within the very same article. A front-page story in the Style section of the *New York Times* in 1994 began by listing recent advertising campaigns—by Clinique, Revlon, and Nike—which featured older female models. "Could it be?" the article posited. "Is it finally permissible to grow older in this youth-sodden culture?" In spite of the seemingly positive trend in that direction, most of the women interviewed for the article answered in the negative. They did not see themselves in advertising or movies or any other outlets of popular culture. One scholar, who identified herself as "in her thirties," described a double standard vividly reminiscent of the one depicted by Susan Sontag twenty-two years earlier: "a man of 45 looks distinguished, but a woman of the same age is over the hill." In spite of some marketers' efforts to capture the over-50 market, the article's author conceded, "The extent to which the culture is ready to embrace aging women remains painfully limited."[78]

Indeed, some feeble attempts to encourage women to embrace their grayer, heavier, more wrinkled menopausal selves and to cast off the yoke of keeping up their looks for the sake of pleasing men notwithstanding, the tendency was to advise women how to maintain a youthful appearance.[79] Older women were asked to walk a fine line between welcoming their maturity and hiding its external manifestations, although rarely were the contradictions in this balancing act addressed. In this regard, the articles and advertisements in women's magazines differed little from those of previous decades. Take, for example, the April 1994 issue of *Ladies' Home Journal*. A reader flipping through this popular monthly periodical would have come across ads for L'Oreal's Plenitude Advanced Overnight Replenisher ("Reduces the signs of aging"), Revlon's Results Rest and Renewal Night Cream Concentrate ("Now your skin doesn't have to act its age. [Or look it!]"), and Elizabeth Arden's Alpha-Ceramide Intensive Skin Treatment ("Ten years younger? You be the judge."). Although these anti-aging products did not contain estrogen, their ads read as identical to those for estrogen skin creams from a half-century before. Two additional ads—one for the Estraderm patch, the other for Tums calcium supplement—played on women's fears of menopause and osteoporosis. And an article in the "Beauty and Fashion Journal" section of the magazine offered "Four Steps to Ageless Lips." The message here was that older women could, and should, still turn heads. The way to achieve this goal was to make use of all the means at their disposal, readily available for purchase at the local pharmacy.

Two new publications tried to capitalize on the growing market of women over 40. In 1988, Frances Lear, recently divorced from the Hollywood producer Norman Lear, used $25 million of her $112 million divorce settlement to start a magazine, *Lear's*, "for the woman who wasn't born yesterday."[80] The inaugural editorial proclaimed that "age is integral to beauty," but the tacit assumption in the articles and advertisements that followed was that money was necessary to facilitate this merger. Articles encouraging women to drink champagne and hire personal shoppers spoke only to affluent consumers; such luxuries were beyond the means of many, if not most, older women. As one skeptical reviewer noted acerbically, "The median income for older black women is less than the price of LEAR's fashionable spring jumpsuit."[81] Furthermore, the implicit standard for beauty continued to be youth; skin creams, were, after all, intended to minimize, not embellish, the appearance of wrinkles. Although the first issue claimed, "It is the right new magazine appearing at precisely the right time," in fact it may have debuted prematurely, before there was a sufficient audience. *Lear's* limped along for seven years before folding in 1995.

Three years later, another magazine, *More*, popped up to take its place. By this time, the population of over-40 baby boomers had reached critical mass. The (mixed) message remained the same: the secret to successful aging was in looking, feeling, and acting young and sexy. For example, an article in 1998 called "The Second Look Test" sent a 42-year-old model out onto the streets of Manhattan in three different outfits to see if she could get men to look at her. Neither her "casual garb" nor her "professional uniform" caused any male heads to turn, but with "the time-tested tricks: high-heeled sandals, short black dress, ripe red lips" she successfully completed her mission. "Invisible no more," the article crowed. Thirty-five years after Robert Wilson wrote about "desexed women found on our streets today" who "pass unnoticed," this magazine posited that invisibility to the male gaze was still something of a concern for older women. Feminist efforts to the contrary, aging baby boomers appeared to be receptive to this kind of thinking; within seven years, *More* magazine more than tripled its readership from an initial circulation of 320,000 in 1998 to 1.1 million subscribers (and a total of 4.4 million readers) in 2005. *More*'s success can also be measured by the growth of its advertising revenue. As the number of advertising pages consistently increased year after year, the publication earned a spot on *Advertising Age*'s "A List" of top ten magazines in 2003.

These advertisements displayed a myriad of products competing for the dollars of well-to-do older women. In the category of health and beauty, scores of creams and lotions and hair dyes and vitamins vied in the anti-aging marketplace. As mentioned above, hormone replacement therapies were advertised directly to the public in women's magazines. Both Wyeth-Ayerst's Women's Health Research Institute and Lilly's Center for Women's Health also placed prominent full-page ads in *More* about menopause and estrogen; the goal of these ads was to get women to ask their doctors about pharmaceutical treatments "to stay strong and vital in the years after menopause."[82] However, in the 1990s, prescription drugs had to reckon with a serious contender in the arena of postmenopausal women's health: the so-called alternative therapies.

In 1993 a survey published in the *New England Journal of Medicine* reported that one in three Americans had made use of alternative medicine in 1990. Moreover, visits to alternative healers exceeded visits to regular primary care physicians, and as much money was spent on alternative therapies as on hospitalizations.[83] The researchers had asked respondents about sixteen interventions neither taught in medical schools nor available in hospitals, such as herbal medicine, massage, megavitamins, self-help groups, folk remedies, energy healing, and homeopathy. When they repeated the survey in 1997, they found that alternative healing had grown even more popular: 42 percent of Americans were using some form of alternative medicine, the annual number of visits to alternative medicine practitioners had increased by 50 percent, and expenditures on alternative health services and therapies had almost doubled.[84]

Although these revelations took the regular medical profession by surprise, alternative medicine was not a new phenomenon; it had a long and rich history in the United States.[85] So-called irregular practice dated back to the early 1800s; throughout the nineteenth century, Thomsonians, homeopaths, hydropaths, eclectics, osteopaths, and chiropractors competed with regular (or allopathic) physicians in the medical marketplace. While orthodox and scientific medicine eclipsed most other medical systems in the first half of the twentieth century (with the exception of osteopathy and chiropractic, which succeeded in coexisting), Americans continued to consult alternative healers and to subscribe to alternative practices, even in the so-called golden age of medicine at midcentury.[86]

Alternative medicine underwent a major revival in the 1970s, as many Americans rejected the reductionism of biomedicine in favor of a more holistic approach to health. Fueled by the counterculture and more organized movements such as consumer advocacy and feminism, the use of alternative therapies became increasingly widespread. According to historian James Whorton, sales of homeopathic remedies increased by 1,000 percent in the 1980s.[87] Although physicians initially dismissed this trend as quackery, they soon realized that its popularity could not be ignored. In 1991 the United States Congress passed legislation allocating funding to establish an Office of Alternative Medicine within the National Institutes of Health, and by 1999 alternative medicine had moved out of the Office of the Director into its very own independent center, the National Center for Complementary and Alternative Medicine. Indeed, the addition of the term *complementary* indicates the growing visibility and viability of alternative healing within mainstream medicine.

Most people used unconventional therapies for chronic conditions; not surprisingly, menopause sent many women to alternative health sources in search of relief. Both Rosetta Reitz and Barbara Seaman had introduced alternative remedies to readers of their 1977 books on menopause. As hormone replacement therapy rose in popularity in the 1980s and 1990s, a parallel interest in alternative therapies grew as well. Some women did try estrogen, but abandoned it because of unwelcome side effects, such as monthly bleeding, bloating, and weight gain. Some were advised by their doctors not to take estrogen, because of breast tumors or other conditions. Others were only looking for short-term relief, as opposed to the long-term prevention of disease allegedly offered by HRT. These women who suffered from the hot flashes, night sweats, and vaginal dryness of menopause turned to alternative remedies.

Women who did not subscribe to the principle of the medical model of menopause also pursued nonpharmaceutical ways to deal with their symptoms. Since menopause was not a disease, they reasoned, there was no need to take drugs for it. The women who explicitly rejected medicalized menopause found support in the books of two female physicians who promoted this antiestablishment position. Dr. Susan Love, a breast surgeon who had educated millions about breast cancer with *Dr. Susan Love's Breast Book* (first edition, 1990), turned her attention to menopause and postmenopause in 1997 with *Dr. Susan Love's Hormone Book: Making Informed Choices about Menopause.* Love objected vehemently to the notion of menopause as disease. She stated at the outset that she was "not against hormones," but she did ask her readers, "Do you really want to take a hormone pill every day for the rest of your life?" Her book gave equal attention and credence to lifestyle changes, alternative treatments, and pharmaceuticals for dealing with menopausal and postmenopausal symptoms.[88] Dr. Christiane Northrup, trained as an obstetrician/gynecologist, based her mini-empire of alternative health care in Maine. She incorporated the language of New Age spiritualism into her enormously popular books, *Women's Bodies, Women's Wisdom* (1994) and *The Wisdom of Menopause* (2001). Northrup encouraged women to experiment with a variety of healing agents: "The wisdom in nature is user-friendly, and you have a lot of it within you already. To tap in to it, just pick the herb, the formula, or the foods that seem to jump out at you and say 'Try me.'"[89] Although these two physicians swam against the tide of medical opinion (the *New Yorker*'s review of Love's *Hormone Book* was titled "The Estrogen Question: How Wrong Is Dr. Susan Love?"), they touched a nerve with women who felt alienated by mainstream medical approaches to their health.[90]

The variety of herbs, formulas, and foods promoted as menopausal remedies was staggering. Herbal remedies included black cohosh, chaste tree berry, dong quai, ginseng, evening primrose oil, motherwort, red clover, and licorice. Vitamin E pills, topical wild yam creams (which contain progesterone), and soy products (which contain phytoestrogens) were also touted as effective against hot flashes. Acupuncturists and behavioral therapists offered services, not consumables, to alleviate menopausal symptoms. While some women were lucky enough to rely solely on diet and exercise as health maintenance strategies, many more turned to pharmacy shelves, health food stores, and alternative healers in pursuit of that goal.

It is difficult to ascertain how many older women made use of alternative therapies for menopausal symptoms. The North American Menopause Society telephone survey in 1997 found that 10 percent of respondents reported use of plant estrogens (soy foods or supplements) in addition to, or instead of, HRT, 9 percent used herbal therapies, 3 percent used acupuncture, and 70 percent took vitamins.[91] Soy and herbal supplements really took off in the late 1990s, when manufacturers began to spend big money to advertise their products to the public. Thanks to the passage of the Dietary Supplement Health and Education Act in 1994, herbs were reclassified as dietary supplements (sometimes referred to as "neutraceuticals") and not as drugs, which reduced the regulatory power of the FDA over these products. In 1998 the Australian company Novogen launched a $5 million promotional campaign for Promensil, made from red clover. Enzymatic Therapy, a Wisconsin distributor, spent almost a million dollars in just the first three months of that year to advertise Remifemin, made from black cohosh.[92] These efforts paid off. In 1999, Americans bought $34 million worth of black cohosh products, more than three times the $11 million dollars spent the year before.[93] Sales of soy isoflavones also tripled from 1998 to 1999, as did sales for "all-in-one" menopausal formulas, which the *New York Times* described as "contain[ing] low doses of every supplement and herb that has ever been used in the same sentence with the word 'woman.' "[94]

Consumers chose among alternative remedies on the basis of anecdotal evidence. If black cohosh worked for one's neighbor, or a relative swore by vitamin E, a woman might be willing to try one of those therapies. There is some irony in the way that these relatively new herbs and dietary supplements were readily accepted by word of mouth, even as estrogen, commercially available and employed in medical practice for some sixty years, was criticized for not being sufficiently tested. This paradox raises the question: what counts as evidence or validation that a treatment "works"?

Although abundant anecdotal evidence existed that estrogen alleviated hot flashes, the short-term use of the hormone was never really in dispute. Rather, the debate centered on the long-term use of HRT to prevent disease. Alternative therapies did not profess to forestall the development of osteoporosis or heart attacks or Alzheimer's disease. Menopause researchers cautioned that untested herbal formulations could have dangerous side effects, but these warning seemed to fall on deaf ears. The market for alternative therapies showed no sign of abating.

The growth of alternative therapies for menopause by no means detracted from the continuing popularity of hormone replacement therapies. In the late 1990s, there was ample room for both in the menopause marketplace. Indeed, both the medical model and the alternative model shared in the commercialization of menopause and aging. Whether women chose to use pharmaceuticals or nutraceuticals, their purchases contributed to the coffers of either the drug industry or the supplement industry.

At the end of the twentieth century, women had access to much more health-related information than any of their predecessors had, but these resources did not seem to clear the path through menopause and postmenopause.[95] First, the presentation and interpretation of advice differed according to the author's (and reader's) position on the medicalization of menopause and postmenopause. Drug manufacturers, not surprisingly, encouraged the readers of their educational brochures, advertisements, and other promotional materials to consult their doctors about hormone replacement therapy. Feminist health writers, on the other hand, argued that aging was a natural process that did not require medical or pharmaceutical management. Popular magazine articles often combined these two approaches, offering information about both medical and nonmedical regimens, leaving the reader to make her own decisions.[96]

Second, the personal and cultural implications of the medicalization of aging and the dilemma of growing old in a youth-centered society presented issues that remained unsettled and unsettling. Many of the available educational resources on menopause and aging gave equal weight to issues of health and appearance. Women in postmenopause were threatened not only with osteoporosis and heart disease but also with wrinkled skin and sagging breasts. The dominant cultural valuation of youth as the standard for beauty encouraged women to try to impede the aging process, not only to prevent disease but also to forestall the physical signs of growing old.[97] Ambivalence about the status of older women in American society further confounded the messages implicit within allegedly neutral health information sources.

The feminist mandate that more information about the topics of menopause and hormones be made freely available in the public domain had come to pass, but the substance of those reports, at least in the mainstream press, tended to privilege the medical-pharmaceutical management of menopause and postmenopause with hormone replacement. Popular media coverage reflected the confidence among medical researchers and physicians that the Women's Health Initiative, the ongoing clinical trial of hormone replacement therapy scheduled to be completed in 2005, would, like the PEPI trial that concluded in 1994, confirm the cardio-protective benefits of HRT and justify its widespread use among older women. Once again, the conventional wisdom about estrogen could not have been more wrong.

# CHAPTER FOURTEEN

# The "Gold Standard"

## Estrogen and the Randomized Controlled Trials

Estrogen underwent yet another reversal of fortune in July 2002. That month, the Women's Health Initiative—the multiyear, multisite, multimillion-dollar, federally funded clinical trial of hormone replacement therapy—was stopped three years before its scheduled endpoint, because the study's Data and Safety Monitoring Board concluded that the risks of adverse effects outweighed the benefits of continuing the study. The risk of breast cancer, in particular, exceeded the predetermined limit of safety; additionally, women taking the estrogen-progestin pills were at increased risk for heart attacks, strokes, and blood clots. Letters were sent to the sixteen thousand study participants with instructions to discontinue use of the drugs. The story made the front pages of newspapers across the country and was reported on national television news broadcasts, because it affected the lives of the millions of American women who were using or thinking about using HRT.[1] This development flew in the face of accepted medical practice in the United States and threw the consumers of monthly prescriptions of this received wisdom into a state of panic.

The shock waves from this report traveled quickly. In rapid response to the news, the price of Wyeth's stock fell by more than 24 percent.[2] A survey of physicians taken ten days after the release of the report found that they had told about half of their patients to stop taking Prempro or to switch to an alternative treatment.[3] Within a year, prescriptions for Prempro dropped by 33 percent, Premarin's plummeted by 66 percent, and the total number for all brands of HRT declined by more than a third.[4] The fallout from the Women's Health Initiative (WHI) was both immediate and long lasting, but it was also confusing and contested.

The WHI was not, however, the first indication of bad news about estrogen. Four years earlier, in August 1998, a different randomized controlled trial, the

Heart and Estrogen/Progestin Replacement Study (HERS), found that hormone replacement therapy did not reduce the rate of heart attacks in women who already had coronary disease. Funded by Wyeth-Ayerst, the maker of Premarin and Prempro, this four-year study of more than twenty-seven hundred women cast the initial shadow on the alleged cardio-protective benefits of estrogen.

The genesis of the HERS trial was intertwined, although unintentionally, with that of the WHI, back in 1992. A group of researchers at the University of California, San Francisco, decided that it was time to put the hypothesis that estrogen reduced the risk of heart disease to the "gold standard" test of a randomized, double-blind, placebo-controlled clinical trial. (The PEPI trial had been a randomized controlled trial, but it had only measured intermediate outcomes, such as blood lipid levels; a more conclusive study would actually evaluate disease outcomes.) These investigators, proponents of the movement toward evidence-based medicine (the application of the scientific method to medical practice), wanted to design a study that would either confirm or refute the evidence from observational studies that correlated postmenopausal estrogen use with lower rates of heart disease. Some of them had been involved in writing the guidelines on hormone replacement therapy for the American College of Physicians, and they had felt uneasy about basing their recommendation that all women consider preventive use of HRT on observational, as opposed to experimental, evidence.[5]

The group wrote a proposal for a large-scale trial to be conducted at medical centers around the country, recruiting a cadre of well-known, experienced investigators to participate in carrying out the work. As a "secondary prevention trial," the study would test the effect of hormone therapy in women who already had coronary heart disease. If estrogen could prevent heart attacks in these high-risk women, then it could be concluded that the drug would have a similar benefit in the larger population of postmenopausal women, as a "primary preventive" of heart disease. Some people believed that such a trial should *not* be done, because it would be unethical to randomize women to placebo, thereby denying them the presumed benefits of hormone therapy. Indeed, the investigators had high hopes that the randomized trial would substantiate what they accepted as persuasive evidence: the strength and consistency of the observational data—dozens of studies had found a reduced risk of heart disease—were supported by a biologically plausible mechanism that had been demonstrated in laboratory and animal studies. They reasoned that the National Institutes of Health (NIH) would be more than willing to support a trial that could provide conclusive evidence to justify a major public health initiative to prevent heart disease in American women.

Thus it came as a great surprise when the director of the institute to which they had applied for funding, the National Heart, Lung, and Blood Institute (NHLBI), rejected the proposal. Furthermore, he announced that, contrary to standard NIH practice, the group would not be permitted to resubmit their application at a later date. The reason he gave was that NHLBI planned to put out a request for proposals for primary prevention studies under the auspices of the WHI; since HERS was a secondary prevention trial, it would not be considered. The San Francisco researchers were very disappointed; they had poured an enormous amount of time and energy into their massive investigator-initiated proposal and despaired at the thought that all their hard work would be for naught.

Soon thereafter, the HERS group was contacted by representatives from Wyeth-Ayerst. The company was interested in testing its new product Prempro (which combined estrogen and progestin in a single pill) for cardio-protection. Wyeth-Ayerst offered to fund the HERS study, with one important qualification: the company was interested only in testing the estrogen-progestin combination pill against placebo. The investigators had wanted to include a second arm of estrogen-only (Premarin) against placebo, but the manufacturer reasoned that if estrogen-progestin could be shown to reduce the risk of heart disease, that conclusion could then be generalized to estrogen alone. The company would save a lot of money by not funding a study of what much of the medical profession already accepted, namely, that estrogen reduced the risk of heart disease. Confirmation from a randomized controlled trial for estrogen *plus* progestin would enable the company to expand the market for its estrogen products to almost the entire postmenopausal population.

After negotiating the details of the arrangement, the investigators agreed to proceed with the HERS study with the financial support of Wyeth-Ayerst. The company gave the scientists complete jurisdiction over the conduct of the study, the analysis of the data, and the writing of the report. Having ordained the design of the study, the manufacturer was content to relinquish control. After all, any hint of industry involvement (beyond check writing) in the experiment would taint the results and invalidate the conclusions in the eyes of many observers. In January 1993 the study began to enroll participants at twenty clinical centers across the nation. By September of the next year, 2,763 women had been registered. Half the group would receive hormone therapy; the other half would receive the placebo. The women ranged in age from 44 to 79 years; their average age was 67. The stated objective: "to determine if estrogen plus progestin therapy alters the risk for CHD [coronary heart disease] events in postmenopausal women with established coronary disease."[6]

As is common with large-scale clinical trials, the study included a Data and Safety Monitoring Board to evaluate the data every three to six months. About two years into the study, the board noticed an excess of blood clots and an excess of cardiovascular events (e.g., heart attacks) in one group as compared with the other. The board requested that the groups be identified (up to this point, they were labeled simply as Group A and Group B). The investigators assumed that these adverse effects were occurring in the placebo group, but much to their astonishment, it turned out to be the hormone-treated group who experienced more blood clots and heart attacks. So baffled were they by this unexpected turn of events that the investigators asked Wyeth-Ayerst to check the chemical composition of the Prempro and placebo pills to make sure they hadn't mixed up the two formulations. They had not. The board decided to allow the study to continue, provided that the investigators informed the participants (by sending each one a letter) and the medical community (by publishing a notice in a scientific journal). HERS was completed in the summer of 1998; by that time, the participants had been followed for an average of four years.

When the data were compiled and analyzed, the final results were as shocking as the interim findings: hormone therapy did not protect women with heart disease against heart attacks or heart disease-related deaths. Further, the use of HRT almost tripled participants' risk of having a blood clot.[7] After checking and rechecking the data and analysis to make sure they hadn't erred, the HERS Research Group published the results of the study in the *Journal of the American Medical Association* in August.

The article was accompanied by an editorial subtitled "Experiment Trumps Observation." Although the author applauded the design and execution of HERS, she was reluctant to make any clinical recommendations based on the findings: "The beneficial effects of ERT and HRT on bone and menopausal symptoms have been established clearly in randomized trials. Discussions of HERS findings between women and their physicians do not need to occur today, this week, or even this month. HERS identifies no new risks, and there is no emergency."[8]

Other commentators were equally circumspect in their interpretation of the significance and utility of the HERS findings in clinical practice. "It is disappointing that $40 million and a lot of effort by a lot of people did not yield definitive results," one wrote in an article in *Maturitas*. "All of the available evidence must be included in this decision making, not just the results of the HERS trial, and this evidence must be balanced with individual patient needs."[9]

Some of the Monday morning quarterbacking focused on the timing of the heart attacks and deaths that occurred during the study. Since the number of these events was greatest in the first year, some speculated that this early increase in risk masked a later increase in benefit; continue the study for longer, they argued, and an overall reduction in the risk of recurrent heart disease would become apparent.[10] The HERS Research Group did not reject this hypothesis; rather, they decided to test it by extending the study for another three years. More than twenty-three hundred women (93% of the surviving participants) agreed to continue taking their prescribed pills (hormones or placebos) as part of the follow-up (the results of HERS II, as the study was called, will be discussed below).

For many researchers in the fields of menopause and cardiovascular disease, it was simply unfathomable that a single study could vitiate all of the data from observational studies, animal experiments, and clinical trials with intermediate endpoints (such as PEPI). Although a few estrogen advocates dismissed the results of HERS as erroneous, most observers took a cautious "wait-and-see" approach, looking to the results of primary prevention trials, such as the WHI, before changing their opinions on the cardio-protective benefits of hormone replacement therapy.[11] Just a few months after the HERS report, Bernadine Healy, the former director of NIH, recommended estrogen therapy to prevent heart attacks during a televised interview on the CBS Evening News.[12] Two years later, a group of Belgian researchers decided to assess the lasting impact of HERS on medical opinion about estrogen. They surveyed researchers and physicians who had published articles in medical journals on HRT and cardiovascular disease (half from Europe, half from the United States) and found that, in spite of HERS, a third of these experts still believed HRT to be cardio-protective in women with heart disease. An even larger proportion of the group remained convinced of the benefits of HRT in preventing heart disease in women without the disease.[13]

The media reflected this skepticism within the medical research community in their coverage of the release of the HERS results. The *New York Times* buried its story on page 20.[14] Nor did any other major newspapers—for example, *Wall Street Journal, Washington Post, Chicago Sun-Times*—consider the story to be front-page news.[15] ABC, CBS, and CNN did broadcast short reports of the study on their evening newscasts, but there were no follow-up reports or in-depth analyses.[16] Even the usually cautious *Consumer Reports* weighed the results of the HERS trial against the observational studies and was persuaded by the latter, recommending that women without heart disease consider taking hormones to prevent its development.[17]

Practicing physicians also tended to disregard the results of HERS. Since the number of hormone therapy prescriptions continued to rise throughout 1998 and 1999, women, too, apparently remained convinced of the merits of both short-term and long-term hormone replacement therapy.[18] In spite of the rigorous methodology of HERS and its adherence to the "gold standard" requirements of a randomized controlled trial, the evidence generated by this one study was not sufficient to overcome the solid base of support for estrogen that had developed over the past two decades. It would take the evidence produced from the massive Women's Health Initiative, a study an order of magnitude larger than HERS, to shake the foundations of the estrogen empire.

Although Bernadine Healy announced the Women's Health Initiative in the spring of 1991, enrollment of participants did not begin until 1993. It took two years to plan, coordinate, and fund proposals to carry out this study and another five years to register the full complement of participants. By 1998 the WHI had enlisted 161,809 postmenopausal women (aged 50–79) across the country into four clinical trials and one observational study. One-tenth of these women (16,608) were randomized into the estrogen-progestin versus placebo component of the clinical trial (to test whether HRT prevented heart disease). About 100,000 joined the observational study, in which no medications were taken nor any lifestyle changes made. The rest were randomized into the estrogen-only trial (for women without uteruses, also to test the hypothesis that estrogen prevented heart disease), the dietary modification trial (to test whether a low-fat diet would prevent breast and colorectal cancer), or the calcium and vitamin D trial (to test whether these supplements would reduce osteoporotic fractures). The estrogen-progestin trial was the one watched most closely and whose results were most eagerly anticipated by the investigators, the NIH sponsors, physicians, and midlife and older women.

The WHI marked the arrival of "big science" in medical research, much like the Manhattan Project in physics and the Human Genome Project in biology. It was the largest randomized controlled trial in women ever undertaken, administered by the NHLBI in cooperation with the National Institute of Arthritis and Musculoskeletal and Skin Diseases (NIAMS), the National Cancer Institute (NCI), and the National Institute on Aging (NIA). It involved more than two hundred investigators, and hundreds more research assistants and administrators and nurses, at forty clinical centers and seven coordinating centers in twenty-seven states. The initial cost of the project was projected to be $625 million over fourteen years; the final tab came closer to a billion dollars.

In 1993 the U.S. House of Representatives' Appropriations Committee expressed concerns about the study's size, complexity, and cost. It asked NIH to contract with the Institute of Medicine, a division of the National Academy of Sciences, to review the WHI. The Institute put together a review committee, which published a report later that year that found much to fault in the WHI. First, it took issue with the cost estimates, predicting that the study would run over budget. Second, it felt that the WHI was not sufficiently subjected to peer review, from either within or without NIH. The report noted, somewhat caustically, "Nagging doubt exists, however, that political concerns influenced the design, timetable, and budget of the study."[19] Third, it speculated that the need for the study might be made moot, once results of the other randomized clinical trials (PEPI and HERS) became available.[20] In other words, the committee, like so many in the medical research community, believed that evidence was forthcoming to confirm the cardio-protective benefit of HRT. The report also picked on some specific points in the study's design, citing the proposed informed consent measures as inadequate and the consideration of initiating HRT in the oldest cohort of women (ages 70–79) as insufficiently addressed with respect to risks and benefits. The committee was skeptical that any of its concerns could be dealt with, because the study was already well under way, when "public funds had already been committed to institutions and investigators, and public expectations had been raised."[21] In the final analysis, it offered begrudging, qualified support for the project.

The Institute of Medicine report received virtually no publicity, and the WHI proceeded according to plan. The study hit stumbling blocks in autumn 1999, spring 2000, and spring 2001, when the Data and Safety Monitoring Board discovered an elevated number of adverse cardiovascular events. While none of these developments was deemed serious enough to halt the study, the monitoring board did direct the study's steering committee to notify participants by letter in 2000 and again in 2001. The letters were vague, reporting a small increase in the risk of heart attacks, strokes, and blood clots. They cautioned that these findings were preliminary and encouraged the women to continue to stick with the study. Participants appeared not to be too concerned by this turn of events; very few dropped out in the wake of the notification letters. Only the 2000 letter caught the media's attention. Both the *Washington Post* and the *Los Angeles Times* broke the story on 4 April; the *New York Times* gave it a spot on the front page the following day.[22] Although these preliminary results may have raised some eyebrows, they had little effect on the prescription and use of HRT beyond the WHI, as the annual number of prescriptions in the United States hit its zenith in 2001, with ninety-one million dispensed to some fifteen million women.[23]

The stumbling blocks became a massive roadblock on 31 May 2002, when the board met for the tenth time since the WHI began. Upon review of the data, it found that not only did adverse cardiovascular events continue to occur, but also the number of breast cancer diagnoses among participants exceeded the predetermined level of acceptability. The estrogen-progestin arm of the study had to be stopped.

The *Journal of the American Medical Association* "fast-tracked" the article describing the results of the clinical trial and the reasons for its premature conclusion, so as to get the clinical information to physicians as quickly as possible. The paper was scheduled to come out in the 17 July issue, but the *Journal* also published an advance version on its Web site on 9 July, to coordinate with the press release that day from the NHLBI and the receipt of letters by study participants (which told them to stop taking their pills, both hormones and placebos). While most physicians did not read the full article until they received their paper copies of *JAMA* in the mail the week of 17 July, they could not have missed the screaming headlines in newspapers across the country on 9 July—"Study Says Halt Hormone Therapy," "Study Is Halted over Rise Seen in Cancer Risk," "Health Risk to Women Halts Hormone Study"—or the reports on television later that night, when all three major networks, plus CNN, covered the story in their evening broadcasts.[24] Nor did the news escape the millions of women currently taking HRT, creating alarm and anxiety among them.

What, exactly, was contained within the medical and popular reports? The risks of HRT, based on an average of 5.2 years of use among study participants, were expressed in both relative and absolute terms in the *JAMA* article.[25] Women who took the estrogen-progestin pills, as compared with those in the control group who took placebo pills, increased their risk of breast cancer by 26 percent (relative risk of 1.26), coronary heart disease by 29 percent (1.29), stroke by 41 percent (1.41), and pulmonary embolism (blood clot) by 213 percent (2.13). The hormone users experienced a *decreased* risk of colorectal cancer (0.63), endometrial cancer (0.83), and hip fracture (0.66). Expressed in absolute terms, these figures translated into eight more breast cancers, seven more coronary heart disease events, eight more strokes, eight more pulmonary embolisms, six fewer colorectal cancers, and five fewer hip fractures per ten thousand women each year. The investigators concluded, "Results from WHI indicate that the combined postmenopausal hormones . . . should not be initiated or continued for the primary prevention of CHD [coronary heart disease]."[26]

Newspaper articles reported these statistics in various forms. Some opted to give the relative risks, some listed the absolute risks, and some gave no figures at all, focusing instead on a qualitative interpretation of the study's results. The presentation of the statistical data affected its reception: a more than doubled risk of getting a blood clot, for example, seemed more ominous than the risk of being one of eight women out of ten thousand who would have a clot as a result of HRT use. The *Houston Chronicle* translated the statistics into an even scarier prediction: "If 6 million women take the drug . . . that works out to 11,400 additional strokes or cases of breast cancer." This article offered the opposing viewpoints of two local female physicians, one who described the risk to benefit ratio for Prempro as "not favorable," and another who "reacted angrily to the study," saying, "I think we're causing an irrational fear that I'm not sure is warranted."[27] Similar reports, often with a local angle included (for example, the *Atlanta Journal-Constitution* mentioned that 286 women from metropolitan Atlanta were involved in the study at Emory University), appeared in papers across the country.[28]

The release of the WHI results brought about a collision of worlds: the alleged objectivity and expert vocabulary of scientific experimentation, the public nature and lay language of consumer information, and the art and empiricism of medical practice. Researchers debated the design of the study, the significance of its findings, and the implications of its premature termination. Physicians and women wondered how to apply these population-based epidemiological statistics to clinical recommendations for individual women. Some physicians resented the intrusion of biostatisticians and epidemiologists into their relationships with patients. Women read the news reports and, wanting more information, turned to their physicians, many of whom had little or no further information to give. Individuals in each of these three groups—medical researchers, menopausal and postmenopausal women, and practicing physicians—responded to the WHI news in ways that reflected their stakes in the use, prescription, and understanding of HRT.

Many researchers found the results to be persuasive, because the design of the study adhered to the rigorous standards of a randomized controlled trial. For those who put their faith in the scientific method, the WHI answered the question of whether combined estrogen-progestin therapy could prevent heart disease with a resounding "no." The conclusions of the WHI received additional corroboration from the recently released results of the extended HERS trial. Published in *JAMA* just one week before the release of the WHI bombshell, HERS II confirmed the 1998 findings of HERS I: postmenopausal hormone therapy did not decrease the risk of heart attacks in women with heart disease. Although HERS received considerably less fanfare than its larger and better-known peer, the complementary findings of HERS and WHI seemed to work synergistically to convince observers of their validity and to overturn the medical consensus on HRT.

Furthermore, this camp believed that the findings of the WHI could and should be translated into guidelines for clinical practice, according to the tenets of "evidence-based" medicine. Although neither Prempro nor Premarin nor any other hormone replacement drug had ever received FDA approved for prevention of heart disease, this indication had become part of common medical practice in the 1990s. The editorial that accompanied the *JAMA* article noted that "long-term use [of hormone therapy] has been in vogue to prevent a range of chronic conditions, especially heart disease."[29] After reviewing the results of the WHI, the authors of the editorial concluded with a positive evaluation of the study and a definitive recommendation for readers: "The WHI provides an important health answer for generations of healthy postmenopausal women to come—do not use estrogen/progestin to prevent chronic disease."[30]

While few nonphysician women would have read either the medical journal article or editorial, many would have heard about the early termination of the WHI in the blanket print and television coverage of this news from the world of clinical research. Physicians reported that their offices were inundated with phone calls from patients asking whether they, too, should discontinue their use of hormone therapy. Newspapers reported this reaction, which in turn fueled even more anxiety among women. The *New York Times* described it as "a day of reckoning" and then quoted several women who were loath to give up their hormones.[31] The responses of these women were reminiscent of those who wrote to the FDA three decades earlier to protest the mandated patient package insert for estrogen products; in 2002, as in 1976, individual women perceived that the current benefits they received from hormone replacement therapy far outweighed the risk of contracting cancer in the future. However, many more women took seriously the conclusion that HRT did more harm than good and tossed their remaining pills in the trash.

The National Women's Health Network hailed the WHI results as a victory. For years, this group had battled against the claims of the estrogen empire. As Executive Director Cynthia Pearson put it, the promotion of HRT by its advocates "was sexist and ageist," with its message that women should "Stay healthy. Stay sexually vital. Be less of a pain to your husband."[32] These attitudes, however, proved hard to challenge. Pearson and her colleagues had more success in confronting Wyeth-Ayerst on the specific issue of the approval of Premarin to prevent heart disease. Pearson had testified at the FDA hearings in the early 1990s against this petition, arguing that a randomized controlled trial had to be conducted before such a decision could be made. Indeed, that estrogen, one of the older and most widely prescribed drugs on the American market, had never been subjected to a rigorous large-scale trial was one of the main spurs for the WHI. While Pearson's prediction that a clinical trial would disprove many of the claims made for estrogen was borne out in 2002, the larger goals of the women's health movement for postmenopausal health and well-being, articulated by feminists in the 1970s in terms of greater social and cultural status for aging women, were left unmet. The authority of medical science validated feminists' thirty-year battle against the long-term use of estrogen as a "treatment" for postmenopause, but it did not address social concerns about the roles and positions of aging women in American society. In the dismantling of the estrogen worldview, science trumped social change.

Physicians who saw menopausal and postmenopausal women in their clinical practices struggled to provide their patients with advice in the wake of the WHI announcement. Many ran group information sessions, as a way of dealing with the flood of inquiries. Lila Nachtigall sent a two-page letter to her patients, which began: "I see hundreds of menopausal patients in my New York practice, many taking some type of hormone replacement therapy (HRT). I think I got a call from every one of them each week!" She went on to describe her response to the study as "disappointed (but not shocked)" and concluded, "There is no one-size-fits-all response."[33]

Physicians' groups scrambled to publish recommendations for their constituents, based on the HERS and WHI findings. The U.S. Preventive Services Task Force, an independent panel of preventive and primary care experts, revised its 1996 recommendations on HRT; the updated version recommended against the routine use of estrogen and progestin for the prevention of chronic conditions in menopausal women. This update was published in the *Annals of Internal Medicine* and the *American Family Physician* and on the Web site of the task force's sponsor, the Agency for Healthcare Research and Quality of the Department of Health and Human Services. The largest group of menopause researchers and physicians, the North American Menopause Society (NAMS), convened an advisory panel to develop a set of guidelines, which was presented to the membership at the society's annual meeting in Chicago in October 2002. While the panel agreed on certain issues (most notably, that estrogen-progestin should not be used for prevention of heart disease), there were several others on which the group was unable to reach consensus (such as how to make a distinction between short-term and long-term therapy and whether there remained any indications for long-term use of hormone replacement).[34]

At the same meeting, NAMS presented the results of a Gallup Poll it had commissioned to survey the effect of HRT on quality of life. Using a scale developed by the society's founder, Wulf Utian, it asked six hundred postmenopausal women between the ages of 50 and 64 questions about their physical, emotional, sexual, and occupational lives and found that HRT improved the quality of life for women who experienced menopausal symptoms, so much so that their scores exceeded those of nonmedicated asymptomatic women. After about half of the surveys had been completed (by telephone), the WHI announcement was made, so the pollsters added a question to find out if subsequent interviewees were aware of the study and its findings. Not only had most women (77%) heard the news, but also those queried after the announcement reported lower quality of life scores, especially those women who had been using hormones for more than two years. Utian used this finding to highlight what he saw as the critical—and not necessarily favorable—influence of the media in shaping popular opinion on medical matters.[35]

Many women continued to insist that they felt better on hormones, expressing this conviction both in person to their doctors and in letters written to the editors of their newspapers, in which they offered personal anecdotes to support the pre-WHI Gallup Poll finding of improved quality of life. The *Pittsburgh Post-Gazette* published a full page of readers' responses a couple of weeks after the announcement in July. One woman said of Prempro, "It has made me a happy, wonderful person." Another described her menopause as "much worse than I could ever imagine." "Will I continue with HRT?" she asked. "Absolutely."[36] Others hailed the long-term benefits, such as the prevention of osteoporotic fractures (which the WHI confirmed). A hormone user wrote to the *New York Times*, "Do I prefer a shorter but more active life span, or longevity, with possibly more disabilities? No contest! For me, quality of life wins, even though it may not be quite so long."[37]

Nor was everyone in the medical community content to immediately reverse decades of menopause and postmenopause therapeutics in the wake of the WHI. The British authors of an editorial in the *British Medical Journal* were circumspect in their assessment of the meaning of the study for clinical practice. They made much of the fact that the WHI results applied only to one of the many estrogen-progestin formulations on the market and speculated that different results might be obtained with different estrogens and progestins, different doses, or different modes of delivery (e.g., patch versus pill). They also noted that the continuing estrogen-only arm of the WHI might reveal progestin to be the culprit. Their recommendation did not call for a radical change in medical practice: "At present, long term hormone replacement therapy should be given only on an individual basis, depending on the needs and risk factors of the patient. Long term therapy could still be considered for prevention of osteoporosis, used as part of the management of women with particular cardiovascular risk factors, and used for better quality of life."[38]

Other observers took issue with the design of the WHI, contesting the decision to test the effect of HRT in older postmenopausal women (the average age of the WHI participants was 63) rather than in newly menopausal women (late forties and fifties). They charged that the WHI did not study the population most likely to take HRT, namely, women experiencing adverse menopausal symptoms. They hypothesized that the older women may have had "sub-clinical" (and hence, undetectable) cardiovascular disease that eventually appeared (in the form of a heart attack or other event), whether promoted or not by the estrogen and pro-gestin. They also postulated that perhaps estrogen worked by a different mecha-nism in younger women (whose own supplies of estrogen had just recently been reduced) as compared to older women (whose bodies may have adjusted to a decade or more with very low estrogen).

The counterargument to these criticisms pointed to the infeasibility of doing a similar trial in younger women. Since the occurrence of heart disease was rela-tively rare in women in their fifties, tens of thousands more women would have to be enrolled in order for the study to achieve sufficient statistical power. Moreover, effective agents, both drugs and behavioral modifications, already existed to re-duce the risk of heart disease, such as aspirin, beta-blockers, statins, low-fat diets, exercise, and the cessation of smoking. Given the other negative outcomes associ-ated with HRT use (increased risk of breast cancer, stroke, and blood clots), this line of reasoning saw no logic in continuing to pursue research on the potential cardio-protective effects of HRT. And, finally, it was unlikely that the federal government, having already spent close to a billion dollars on the WHI, would allocate a similarly large sum to redo the study in a slightly younger population.

The federal government had its own responses to the WHI. First, officials at NIH and FDA announced a formal name change for the hormone drugs taken at and after menopause. "Hormone replacement therapy" would henceforth be known as "menopausal hormone therapy," to reflect the position that "the hor-mone treatment never was a replacement and never did restore the physiology of youth."[39] Not only did the new name counteract the notion of replacement, it also implied that hormones were to be taken at menopause, not during the decades of postmenopause. Second, in early January 2003, the FDA mandated that all es-trogen and estrogen-progestin products for menopausal use had to include a boxed warning on their labels about the increased risk of heart attacks, strokes, blood clots, and breast cancer. Furthermore, the indications for the drugs were modified, so that they were recommended for vaginal dryness *only* in moderate to severe conditions and for osteoporosis *only* in very high-risk women. Otherwise, physicians should counsel their patients to use topical creams for the former and other drugs (such as bisphosphonates) for the latter. The indication for treating the temporary hot flashes and night sweats of menopause remained unchanged.[40]

As 2002 turned into 2003, the news for Prempro went from bad to worse. More studies were published based on the data from the WHI. In May, the lead article in the *New England Journal of Medicine* announced that the estrogen plus progestin regimen did not improve women's quality of life, in terms of general health, vitality, mental health, depressive symptoms, and sexual satisfaction.[41] Three weeks later, three papers published in the same issue of *JAMA* hammered more nails into the coffin of long-term hormone replacement therapy. The first described how hormone users had a twofold increased risk of developing dementia (usually, Alzheimer's disease).[42] The second found that the group taking estrogen plus progestin did not demonstrate improved cognitive functioning as compared to the group taking placebos.[43] The third presented a more detailed analysis of the risk of stroke, which was 31 percent higher in the treated group than in the control group.[44] The following month, *JAMA* published two papers in the same issue on hormone therapy and breast cancer. The news from the WHI was even more threatening than the initial report a year earlier: even relatively short-term use of HRT increased the risk of breast cancer. Furthermore, cancers in hormone users were diagnosed at a more advanced stage than those in the placebo users.[45] These data were reinforced by a case-control study, reported in the second paper, of some two thousand women in three counties in the state of Washington. The investigators found that the longer women used combined hormone therapy, the greater their risk of developing breast cancer.[46]

Each of these groups of scientific articles was accompanied by an editorial, which duly noted the adverse effects of combined estrogen-progestin therapy. Two of the editorials reiterated the recommendation to prescribe HRT only for the short-term relief of severe menopausal symptoms; the one accompanying the breast cancer articles offered the most strongly worded judgment: "The WHI trial of estrogen plus progestin therapy is as close to definitive as can be expected . . . [it] provide[s] further compelling evidence against the use of combination estrogen plus progestin therapy."[47] Each report was also covered by the national media. The *New England Journal* released the results of the quality of life study to the popular press a full seven weeks before its actual publication, because "they were too important to hold."[48]

Whereas the medical journal articles stuck to the prescribed format of presenting their data, analysis, and conclusions in the dispassionate vocabulary of science and statistics, newspaper articles tackled the emotional and subjective implications of the findings. The *New York Times* report of the quality of life study contrasted the findings with the experiences of physicians and patients, asking "whether women and doctors will believe the findings." But the next day, the *Times*'s editorial staff threw its weight behind the scientific data, discounting individual women's own perceptions. "Bit by bit," the editorial began, "the evidence is accumulating that most women are foolish if they keep taking hormone pills for years at a time. . . . [The results] should also shake the confidence of everyone who has believed, on the basis of anecdotal reports and less rigorous scientific studies, that hormone treatments made women feel better."[49]

As Prempro lost case after case in the court of scientific opinion, the jury remained out on the matter of Premarin, since the estrogen-only arm of the WHI had not been stopped in 2002 along with the estrogen-progestin arm. Although the Data and Safety Monitoring Board had in 2000 and 2001 required the steering committee to inform participants in the estrogen-only trial, along with those in the estrogen-progestin trial, of increases in heart attacks, strokes, and blood clots, it had allowed the estrogen-only arm to continue, because no increase in breast cancer rates had been detected. In February 2004, NIH handed down its decision on Premarin, and it was almost as damning as the pronouncements on Prempro. NIH decided to end the study a year ahead of its scheduled termination date, because estrogen use appeared to have no effect on the risk of heart disease. Given the increased risk of stroke and the likelihood that one more year of follow-up would not yield convincing evidence of a decrease in risk for heart disease or breast cancer, it was deemed unacceptable to subject healthy women in a prevention trial to this risk.[50]

Upon closer inspection, the data on estrogen were not so bad. The relative risk for estrogen users as compared to placebo users was statistically insignificant for coronary heart disease (0.91) and colorectal cancer (1.08). The risk was increased for stroke (1.39; 39% increase), and pulmonary embolism (1.34; 34% increase) and decreased for breast cancer (0.77; 23% reduction) and hip fracture (0.61; 39% reduction). Spun another way, these results might have been presented in a positive fashion, with the decreased risks for breast cancer and hip fracture offsetting the increased risks for stroke and blood clots. In fact the global index of health risks and benefits (which represented the first event for each participant of any adverse event or death) was 1.01: virtually indistinguishable between the two groups.[51] But the tide of opinion had already turned against estrogen, with or without progestin, as a long-term therapy for postmenopausal women. With no clear overall benefit to estrogen therapy, the investigators echoed the FDA recommendation that women use conjugated estrogens "only for menopausal symptoms at the smallest effective dose for the shortest possible time."[52]

The documentation of null to slightly negative effects of long-term estrogen in women without uteruses continued to pile up. In June, the Women's Health Initiative Memory Study (WHIMS), a subset of the larger WHI clinical trial that looked at women over the age of 65, reported that estrogen, like estrogen plus progestin, increased the risk of dementia.[53] Similarly, both estrogen and estrogen-progestin had adverse effects on cognitive functioning.[54] A year later, any remaining hopes for estrogen were dashed with the publication of results of the health-related quality of life evaluation for the estrogen-only arm of the trial. After one year, estrogen users had slightly improved sleep (fewer disturbances), slightly lower social functioning, and neither improvement nor deterioration in general health, physical functioning, pain, vitality, role functioning, mental health, depressive symptoms, cognitive function, or sexual satisfaction. After three years, no significant benefits were detected. Even women aged 50 to 54 years with moderate to severe vasomotor symptoms—the group most likely to take estrogen to gain relief from menopausal symptoms—did not experience improvement along any of the axes after one year of treatment.[55] While 72 percent of women in the treatment group who started with moderate or severe hot flashes no longer reported them after one year, the same was true for a majority of women in the control group (56%).[56] This finding raised an interesting question about the possibility of a placebo effect in the mitigation of menopausal discomfort. Some researchers continued to voice their objections about the addition of these "secondary" endpoints to the WHI. The study had been planned to test the primary outcome of cardiovascular disease, they argued, and it was not well suited to test other variables such as quality of life, especially with the study population skewed heavily to the older end of the age range.[57] But these estrogen advocates were no longer part of the mainstream of opinion; now, their voices were drowned out by the chorus of rejection.

Women's responses to the WHI are hard to gauge. On the one hand, studies of prescription databases and large-scale surveys of postmenopausal women indicated a rapid and precipitous decline in the use of hormone therapy. From the second quarter of 2002 (April through June), just before the release of the first batch of WHI results, to the last quarter of 2003 (October through December), prescriptions for all forms of hormone therapy decreased by 43 percent. The decline for Prempro was even more dramatic, dropping by 80 percent in the same time period.[58] A national telephone survey of women over 50 found that the number of women reporting use of hormone therapy decreased 57 percent from the first half of 2002 (when 28 percent reported use) to the first half of 2004 (when 12 percent reported use).[59] A local study of tens of thousands of women, ages 50 to 74, participating in the San Francisco Mammography Registry, documented that the rate of decline in hormone use was 18 percent per quarter; by March 2003, just nine months after the termination of the estrogen-progestin arm of the WHI, 45 percent of these Bay Area women had stopped taking their hormone pills.[60] In other words, these studies indicate that between four and six out of every ten women who had been taking estrogen, with or without progestin, stopped their medication in the wake of the WHI.

On the other hand, anecdotal evidence suggested that many women were reluctant or unable to give up hormone therapy. In August 2003 the *Washington Post* illustrated an article called "Hormone Therapy Proves Tough to Quit" with the stories of two women who refilled their prescriptions after several uncomfortable months of trying to go without.[61] After an unsuccessful attempt to live hormone-free, one of the women said to her husband, "Honest to goodness, I'd rather live 10 years less than live the rest of my life like this."[62] Four months later, the *Pittsburgh Post-Gazette* article "HRT Update: Many Women Go Back On" quoted Wulf Utian, executive director of the North American Menopause Society, who said, "More than a third of my patients have gone back on it. Some physicians are seeing 60 percent."[63] On the one-year anniversary of the WHI, *More* magazine profiled ten women, all of whom had been taking some form of hormone therapy, and the decisions they made about whether to stop or to continue. Five of the women chose to resume or remain with the treatment.[64]

The anecdotal evidence did not contradict the statistical data: while many women quit hormone replacement therapy, just as many stuck with it. In 2004, more than twenty million prescriptions for Premarin were dispensed, along with four and a half million prescriptions for Prempro.[65] Although these figures were drastically reduced from the pre-WHI days (by almost 50 percent for Premarin and by almost 75 percent for Prempro), they represented a significant number of menopausal and postmenopausal women—more than four million—using these two brands of HRT. Factor in the other brands on the market (about twenty different products), and some eight million women were still using estrogen (and progestin) in menopause and postmenopause. Although the data do not reveal why these women were on HRT, the reason was more likely to be the relief of menopausal symptoms than the prevention of chronic disease. With other drugs on the market to prevent osteoporosis, such as the bisphosphonates, and other drugs to prevent heart disease, such as statins and beta-blockers, physicians could offer women pharmaceutical alternatives to estrogen for long-term preventive care. What they did not have was a good substitute for estrogen in treating the hot flashes of menopause.

Wyeth (American Home Products, the parent company of Wyeth-Ayerst, took on the name Wyeth in 2002) made the most of this lone remaining indication for estrogen. The company did its best to attract women to Premarin and Prempro, after lying low for several months after the WHI came out. In the spring of 2003, it won approval from the FDA for one lower-dose formulation of Premarin and two lower-dose formulations of Prempro. One year later, it launched a major national advertising campaign for low-dose Prempro, with a sixty-second television commercial and a direct-to-consumer print advertisement that appeared in popular magazines. Although promotional spending for regular-dose Prempro had declined steeply (from $19.2 million in the quarter before the WHI to $3.5 million in the last quarter of 2003, a decrease of 82 percent), Wyeth gave low-dose Prempro a healthy advertising budget of $13.5 million in the fourth quarter of 2003 and even more in 2004 to finance the new ad campaign.[66] The ads emphasized short-term use for menopausal symptoms, appealing to the four thousand women reaching menopause each day in the United States.

The almost 1.5 million American women who entered menopause annually in the first years of the twenty-first century were healthier, better educated, and more gainfully employed than previous cohorts. These women were the middle, not the vanguard, of the baby boom generation. Their older sisters had been the trail blazers; these younger boomer women expected to participate fully in all aspects of their lives, private and public. The author of the *More* magazine article on women's HRT choices declared, "We are the first menopausal women to see ourselves as equal partners with the medical people who help us sort through the science . . . We do not relinquish authority over our own bodies to anyone."[67] According to this view, these women had imbibed the messages of both women's health movements, the activist-driven movement of the 1970s and the professional-driven one of the 1990s. The former had instructed them to take charge of their bodies and their health care; the latter had taught them to work *with,* not *against,* their doctors and health care providers.

The women featured in the article were more than just menopausal and post-menopausal bodies. Each had an active career outside the home: attorney, therapist, archaeologist, investment banker. Surely, these individuals matched the target demographic for the magazine, but they also represented a new approach to aging, one that did not include fading gracefully into the background. *Time* magazine picked up on this theme with an article in May 2005 about the midlife crises of modern-day women. However facile its characterizations might have been, *Time*'s reflections of contemporary society usually provided a fairly accurate representation, without too much distortion. In this instance, the magazine seemed to capture a new and liberating approach to aging among women. "With that endearing sense of discovery that baby boomers bring to the most enduring experiences . . . women are confronting the obstacles of middle age and figuring out how to turn them into opportunities."[68] From this perspective, menopause was a minor distraction, compared with the other challenges women faced at midlife. "However disruptive menopause may be for some women, the changes that matter most are often more psychic and spiritual than physical."[69] Since estrogen could no longer provide the key to successful aging, physical concerns took a back seat to emotional issues, which women were empowered to manage on their own, without the help of physicians and pharmaceuticals.

A sidebar to this feature article specifically addressed the topic of menopause, which was, according to the author, "in the midst of a makeover."[70] If so, then this makeover was retro, a return to the world of the 1930s, '40s, and '50s, when menopause was also considered a natural and temporary transition. Her matter-of-fact recommendation to readers—"If you need relief, hormone replacement is worth considering. It's best to start with as low a dose as is effective. But many women find they do just fine without it"—echoed the advice given five decades earlier. Estrogen, toppled from its pedestal as an anti-aging wonder drug, found itself back where it had started more than sixty years ago, as a short-term treatment for the small minority of women who experienced severe menopausal symptoms.

Although the life cycle of estrogen appears to have come full circle, its trajectory brought it to a place in the twenty-first century very different from its origins in the twentieth. The information age that began in the 1970s, and particularly the expansion of the Internet in the 1990s, brought about an enormous increase in the quantity and quality of information on hormone therapy, menopause, and countless related topics. Anyone with access to a computer could log on to read the latest updates on the Web sites of government agencies such as NIH and FDA and news digests on consumer health Web sites such as WebMD. Women could join a listserve or chat room to learn about others' experiences and to share their own. Independent researchers could download original scientific articles in medical journals from the vast database called PubMed. With enough time and determination, a woman could become almost as well informed as her doctor on the topic of HRT.

This leveling of the information playing field helped to transform the doctor-patient relationship into a doctor-patient partnership. By the turn of the twenty-first century, it had become customary for physicians to lay out the various treatment options for their menopausal patients and to let the patients make the final decision on what course of action, if any, to take. In this situation, patients were not passive recipients of doctors' orders but rather active consumers of health care goods and services in the medical marketplace. Women had come a long way from the days when Rosetta Reitz couldn't get a straight answer from her doctor or adequate information from published sources. Now the challenge was sifting through the massive volume of available data to become well informed enough to team up with the doctor to make smart choices.

Gone, too, were the days when doctors treated their female patients, and particularly their older female patients, like children. Women of all ages garnered more respect in American society in the twenty-first century. Also, conceptions of aging changed, as the baby boom generation refused to let go of their youthful ways. On the one hand, their behavior encouraged the celebration of youth that had characterized American society since the 1920s. On the other hand, it expanded the possibilities for older people to engage more actively in public life. The baby boomers redefined what it meant to be 50, and as they arrive at 60, the notion of senior citizenship will almost certainly be made over.

The women of this cohort rejected the equation of menopause with aging. Menopause was just one of many transitions that women encountered at midlife; it did not symbolize the beginning of the end. Of course this attitude was not new; it had been present in both popular and medical thought for much of the twentieth century. What was different was its pervasiveness and its precedence over the medical model of menopause that had reigned since the 1950s.

Perhaps nowhere was this rejection of the medicalization of menopause made more clear than in the March 2005 NIH State-of-the-Science Conference Statement on the Management of Menopause-Related Symptoms. The NIH Consensus Development Program (which had produced the statements on estrogen use in 1979 and osteoporosis in 1984) convened a panel of twelve experts, who wrote their report based on a review of the existing literature, presentations on more than two dozen topics from researchers and physicians, and both open and closed discussion sessions. The sea change in the professional conceptualization of menopause was evident from the first sentence, "Menopause is a natural process that occurs in women's lives as part of normal aging," through to the final paragraph, "Menopause is 'medicalized' in contemporary U.S. society. There is great need to develop and disseminate information that emphasizes menopause as a normal, healthy phase of women's lives and promotes its demedicalization. Medical care and future clinical trials are best focused on women with the most severe and prolonged symptoms. Barriers to professional care for these women should be removed."[71] The discussion of estrogen was minimal, meriting just five paragraphs in the eight-page report. It was considered along with ten other prescription, over-the-counter, behavioral, and alternative interventions for symptom relief. Thanks to the WHI, not only was estrogen no longer recommended for the prevention of chronic disease, it had also lost its place as the automatic therapy-of-choice for women suffering from menopausal symptoms.

It was science that brought about the downfall of hormone replacement therapy. By the twenty-first century, the randomized controlled trial had become the primary means of adjudicating the success or failure of a clinical intervention. Paradoxically, as Americans lost faith in the products of science (e.g., pharmaceuticals), they continued to respect the processes of science. Thus physicians and patients turned away from estrogen and progestin drugs because they were convinced by the conclusions of the WHI and HERS experiments.

Some people maintained their objection to the clinical application of the WHI and HERS conclusions, believing that hormone replacement still had merit as a therapy when started at or soon after menopause. In 2004 the Kronos Longevity Research Institute of Phoenix, Arizona, announced that it would sponsor a five-year, $12-million study of conjugated estrogens (Premarin) and the estradiol patch (both administered cyclically with progesterone) in some seven hundred women between the ages of 42 and 58 who were within three years of their last menstrual period, to test whether HRT could prevent cardiovascular disease if begun early enough.[72] The Kronos Early Estrogen Prevention Study (KEEPS, for short) aimed to answer the questions that some observers felt the WHI had left hanging.

Other investigators took their research in a different direction. Since estrogen had been on the market as the number one treatment for the vasomotor symptoms of menopause for so long, surprisingly little work had been done to understand the physiology of hot flashes and to develop alternative therapies. SERMs (selective estrogen receptor modulators), botanical products, and behavior modifications all received attention as potential replacements for hormone replacement therapy. None of these interventions would gain acceptance, however, unless they passed the obligatory test of the randomized controlled trial.

Women continued to pick and choose menopausal remedies from the myriad available in pharmacies and health food stores. Some turned to "bio-identical" hormones, individually compounded by pharmacists and touted as "natural" because they were composed of estrogens derived from plant sources, as opposed to the pharmaceutical estrogens derived from horse urine (such as Premarin, which, sixty years earlier, had been promoted as more "natural" than synthetic estrogens such as DES). Mainstream medicine expressed skepticism about these hormones, because there was no scientific proof of their efficacy or safety; the American College of Obstetricians and Gynecologists warned that they were not FDA-approved, and the North American Menopause Society recommended against their use. But bio-identical hormone replacement got a big boost in publicity in 2004 from the actress Suzanne Somers, who wrote an advice book called *The Sexy Years: Discover the Hormone Connection.*[73] The book hit the *New York Times Book Review* best sellers list, and Somers hit the talk show circuit, spreading the gospel of bio-identical hormone replacement therapy on *The Today Show* and *Larry King Live.*[74] She was interviewed in *Good Housekeeping;* the cover headline read, "Suzanne Somers: How to Stay Feminine after 50."[75] Robert Wilson would have approved, if not of the particular product itself, then certainly of Somers's promise that "bio-identical hormone therapy can help you lose weight, reinvigorate your sex life, and fight aging."[76] Indeed, the promotion of bio-identical hormone replacement as an anti-aging remedy in 2004 harkened back to the *Feminine Forever* rhetoric of the 1960s and the ensuing decades of false hopes for HRT as a preventive against the diseases of aging. Advocates for bio-identical hormone therapy would do well to review the history of pharmaceutical hormone therapy, because if this history can teach us anything, it is that there is no such thing as an estrogen elixir.

## Introduction

1. Robert A. Wilson, *Feminine Forever* (New York: M. Evans, 1966), front cover; Barbara Seaman, *The Greatest Experiment Ever Performed on Women: Exploding the Estrogen Myth* (New York: Hyperion, 2003).

2. For other studies of drugs and high expectations, see Elizabeth Siegel Watkins, *On the Pill: A Social History of Oral Contraceptives, 1950–1970* (Baltimore: Johns Hopkins University Press, 1998); Sheila M. Rothman and David J. Rothman, *The Pursuit of Perfection: The Promise and Perils of Medical Enhancement* (New York: Pantheon Books, 2003); Meika Loe, *The Rise of Viagra: How the Little Blue Pill Changed Sex in America* (New York: New York University Press, 2004); John Hoberman, *Testosterone Dreams: Rejuvenation, Aphrodisia, Doping* (Berkeley: University of California Press, 2005).

3. Adele E. Clarke et al., "Biomedicalization: Technoscientific Transformations of Health, Illness, and US Biomedicine," *American Sociological Review* 68 (April 2003), 161.

4. On the authority of medical science, see Paul Starr, *The Social Transformation of American Medicine* (New York: Basic Books, 1982), and the articles in the special issue of the *Journal of Health Politics, Policy and Law* called "Transforming American Medicine: A Twenty-Year Retrospective on *The Social Transformation of American Medicine*," vol. 29 (August–October 2004). On clinical studies, see Harry M. Marks, *The Progress of Experiment: Science and Therapeutic Reform in the United States, 1900–1990* (Cambridge: Cambridge University Press, 1997).

5. On medicalization, see Clarke et al., "Biomedicalization," 161–94; Robert Crawford, "Healthism and the Medicalization of Everyday Life," *International Journal of Health Services* 10 (1980), 365–89; Carroll L. Estes and Elizabeth A. Binney, "The Biomedicalization of Aging: Dangers and Dilemmas," *The Gerontologist* 29 (1989), 587–96; Renee C. Fox, "The Medicalization and Demedicalization of American Society," *Daedalus* 106 (1977), 9–22; Catherine Kohler Riesmann, "Women and Medicalization: A New Perspective," *Social Policy* (Summer 1983), 3–18; Irving Kenneth Zola, "Medicine as an Institution of Social Control," *Sociological Review* 20 (November 1972), 487–504. On women's bodies, see Thomas Laqueur, *Making Sex: Body and Gender from the Greeks to Freud* (Cambridge, MA: Harvard University Press, 1990); Wendy Mitchinson, *The Nature of Their Bodies: Women and Their*

*Doctors in Victorian Canada* (Toronto: University of Toronto Press, 1991); Ornella Moscucci, *The Science of Woman: Gynaecology and Gender in England, 1800–1929* (Cambridge: Cambridge University Press, 1990). On ovarian research, see Chandak Sengoopta, *The Most Secret Quintessence of Life: Sex, Glands, and Hormones, 1850–1950* (Chicago: University of Chicago Press, 2006).

6. Lois Banner, *In Full Flower: Aging Women, Power, and Sexuality* (New York: Knopf, 1992); Margaret Morganroth Gullette, *Declining to Decline: Cultural Combat and the Politics of the Midlife* (Charlottesville: University Press of Virginia, 1997); Margaret Morganroth Gullette, *Aged by Culture* (Chicago: University of Chicago Press, 2004).

7. Janet Golden addresses the role of the media in publicizing health issues in *Message in a Bottle: The Making of Fetal Alcohol Syndrome* (Cambridge, MA: Harvard University Press, 2005).

8. Barbara Seaman, *The Greatest Experiment Ever Performed on Women: Exploding the Estrogen Myth* (New York: Hyperion, 2003).

9. The literature on hysterectomy is extensive. A starting point is U.S. Senate Subcommittee on Aging, *Unnecessary Hysterectomies: The Second Most Common Major Surgery in the United States: A Hearing Before the Subcommittee on Aging of the Committee on Labor and Human Resources,* United States Senate, 103rd Congress, 1st Session (5 May 1993). See also Steven Bernstein et al., *Hysterectomy: A Review of the Literature on Indications, Effectiveness, and Risks* (Santa Monica, CA: Rand, 1997).

10. My hope is that this book will lay the groundwork for transnational comparative histories, to complement the many fine cross-cultural studies undertaken by anthropologists. The most extensive of these is Margaret Lock, *Encounters with Aging: Mythologies of Menopause in Japan and North America* (Berkeley: University of California Press, 1993). See also J. M. A. Richters, "Menopause in Different Cultures," *Journal of Psychosomatic Obstetrics and Gynecology* 18 (June 1997), 73–80.

11. Marcia Angell, *The Truth about Drug Companies: How They Deceive Us and What to Do about It* (New York: Random House, 2004), 3; Greg Critser, *Generation RX: How Prescription Drugs Are Altering American Lives, Minds, and Bodies* (New York: Houghton Mifflin, 2005), 2.

## One • Beginnings

1. George W. Corner, "The Early History of the Oestrogenic Hormones," *Journal of Endocrinology* 31 (January 1965), xiv.

2. Ruth Schwartz Cowan, "Edgar Allen," *Dictionary of Scientific Biography* (New York: Scribner's, 1970), 1:123.

3. "Edward A. Doisy—Biography," *Nobel Lectures, Physiology or Medicine 1942–1962* (Amsterdam: Elsevier Publishing), www.nobel.se/medicine/laureates/1943/doisy-bio .html.

4. Corner, "Early History," xiii.

5. Edward A. Doisy, "Isolation of a Crystalline Estrogen from Urine and the Follicular Hormone from Ovaries," *American Journal of Obstetrics and Gynecology* 114 (1 November 1972), 701.

6. Ibid.

7. Corner, "Early History," xiii–xiv. Corner quotes from a letter he received from Doisy explaining the genesis of the partnership with Allen. See also Doisy, "Isolation of a Crystalline Estrogen," 701.

8. Edgar Allen, "Ovarian Hormone and Female Genital Cancer," *Journal of the American Medical Association* 114 (25 May 1940), 2108.

9. George W. Corner, *The Hormones in Human Reproduction* (Princeton: Princeton University Press, 1942), 83.

10. Doisy, "Isolation of a Crystalline Estrogen," 701.

11. Ibid., 701–2.

12. For a detailed history of scientific studies of ovarian function from the 1840s to the 1920s, see Chandak Sengoopta, "The Modern Ovary: Constructions, Meanings, Uses," *History of Science* 38 (2000), 425–88.

13. A. A. Berthold, "Transplantation der Hoden," *Archiv fur Anatomie, Physiologie und Wissenschaftliche Medizin* (1849), 42–46, as translated by D. P. Quiring in *Bulletin of the History of Medicine* 16 (November 1944), 399–401.

14. Georges Canguilhem, *Etudes d'histoire et de philosophie des sciences* (Paris: Vrin, 1968), as translated by Arthur Goldhammer in Francois Delaporte, ed., *A Vital Rationalist: Selected Writings from Georges Canguilhem* (New York: Zone Books, 1994), 115–28.

15. Ibid., 120–21.

16. Corner, "Early History," vii.

17. Quoted in ibid.

18. Ernest Henry Starling, *The Croonian Lectures on the Chemical Correlations of the Body* (London: Women's Printing Society, 1905), 6; Merriley Borrell, "Origins of the Hormone Concept: Internal Secretions and Physiological Research, 1889–1905" (Ph.D. dissertation, Yale University, 1976), 187.

19. Paul Starr, *The Social Transformation of American Medicine* (New York: Basic Books, 1982), 94–95.

20. Merriley Borrell, "Brown-Sequard's Organotherapy and Its Appearance in America at the End of the Nineteenth Century," *Bulletin of the History of Medicine* 50 (Fall 1976), 311.

21. See, for example, Ronald G. Walters, *Primers for Prudery: Sexual Advice to Victorian America* (Baltimore: Johns Hopkins University Press, 2000), 32–48.

22. Borrell, "Brown-Sequard's Organotherapy," 312.

23. Ibid., 309–10.

24. Ibid., 313. The book was edited by Newall Dunbar and published by J. G. Cupples Company in 1889.

25. Ibid., 316.

26. David Hamilton, *The Monkey Gland Affair* (London: Chatto and Windus, 1986). See also Julia Ellen Rechter, "'The Glands of Destiny': A History of Popular, Medical and Scientific Views of the Sex Hormones in 1920s America" (Ph.D. dissertation, University of California at Berkeley, 1997), 173–211.

27. Hamilton, *The Monkey Gland Affair*, 32.

28. "Bone Grafting in Army," *New York Times* (4 December 1914), 1. See also Hamilton, *The Monkey Gland Affair*, 9.

29. Hamilton, *The Monkey Gland Affair*, 20–21.

30. Ibid., 72.

31. Ibid., 66.

32. Ibid., 26, 68.

33. W. G. Clugston, "Goat Glands Figure in Kansas Campaign," *New York Times* (2 November 1930), E6; "J. R. Brinkley Dies; Goat Gland Doctor," *New York Times* (27 May 1942), 24. See also Hamilton, *The Monkey Gland Affair,* 96.

34. Rechter, "The Glands of Destiny," 203.

35. V. C. Medvei, *The History of Clinical Endocrinology* (New York: Parthenon, 1993), 232–33.

36. Rechter, "The Glands of Destiny," 206.

37. Ibid., 207.

38. Corner, "Early History," vi.

39. Lawrence D. Longo, "The Rise and Fall of Battey's Operation: A Fashion in Surgery," *Bulletin of the History of Medicine* 53 (Summer 1979), 244.

40. See, for example, Walter E. Dixon, "The Ovary as an Organ of Internal Secretion," *The Practitioner* 66 (1901), 525–29.

41. John H. Hannan, *The Flushings of the Menopause* (London: Bailliere, Tindall and Cox, 1927), 34.

42. Ibid., 36.

43. A. J. Carlson, "Glandular Therapy: Physiology of the Mammalian Ovaries," *Journal of the American Medical Association* 83 (13 December 1924), 1923.

44. Diverse sources reveal a number of American and European pharmaceutical companies engaged in the manufacture and sale of ovarian extracts in the 1920s, including: Parke, Davis and Co., Reed and Carnrick, Schering-Kahlbaum A. G., E. R. Squibb and Sons, and Organon. See Hannan, *The Flushings of the Menopause,* 39; Gary L. Nelson, ed., *Pharmaceutical Company Histories* (Bismarck, ND: Woodbine Publishing, 1983), 127; U.S. Trademark #254, 304; U.S. Trademark #318, 536; Nelly Oudshoorn, *Beyond the Natural Body: An Archeology of Sex Hormones* (London: Routledge, 1994), 92–93.

45. Oudshoorn, *Beyond the Natural Body,* 80.

46. Parke, Davis and Company, *Annual Report for the year ending December 31, 1930* (Detroit, 1931), 9.

47. John Parascondola, "Industrial Research Comes of Age: The American Pharmaceutical Industry, 1920–1940," *Pharmacy in History* 27 (1985), 13.

48. Ibid., 14.

49. Examples are Merck in 1933 and Lilly in 1934. Ibid., 15–16.

50. Marcel C. LaFollette, *Making Science Our Own: Public Images of Science, 1910–1955* (Chicago: University of Chicago Press, 1990), 172.

51. Ibid., 172.

52. "The Endocrine Glands," *Fortune* 8 (November 1933), 76–94.

53. Ibid., 76.

54. Ibid., 92–93.

55. Ibid., 86.

56. Ibid., 89.

57. Ibid.

58. Ibid., 94.

59. United States Patent No. 1,967, 350 and No. 1, 967, 351 (24 July 1934).

60. Parke, Davis and Company, *Annual Report for the year ending December 31, 1931*, 11.

61. United States Trademark No. 318,536; Elmer L. Sevringhaus and Joseph S. Evans, "Clinical Observations of the Use of an Ovarian Hormone: Amniotin," *American Journal of the Medical Sciences* 178 (November 1929), 639.

62. Nelly Oudshoorn, *Beyond the Natural Body*, 91–92; Christopher Kobrak, *National Cultures and International Competition: The Experience of Schering AG, 1851–1950* (Cambridge: Cambridge University Press, 2002), 119–20.

63. Alison Li, "Marketing Menopause: Science and the Public Relations of Premarin," in Georgina Feldberg, Molly Ladd-Taylor, Alison Li, and Kathryn McPherson, eds., *Women, Health, and Nation: Canada and the United States since 1945* (Montreal: McGill-Queen's University Press, 2003), 103. For more on Collip, see Alison Li, *J. B. Collip and the Development of Medical Research in Canada* (Montreal: McGill-Queen's University Press, 2003).

64. *New and Nonofficial Remedies* (Chicago: American Medical Association, 1940), 369–71.

65. Elmer L. Sevringhaus, "The Use of Folliculin in Involutional States," *American Journal of Obstetrics and Gynecology* 25 (March 1933), 361.

66. Ibid., 362–63.

67. For an excellent analysis of physicians' attitudes toward menopause in the 1930s, see Judith A. Houck, "Common Experiences and Changing Meanings: Women, Medicine, and Menopause in the United States, 1897–1980" (Ph.D. dissertation, University of Wisconsin-Madison, 1998), 51–52. See also her new book *Hot and Bothered: Women, Medicine, and Menopause in Modern America* (Cambridge, MA: Harvard University Press, 2006).

68. Elmer L. Sevringhaus, "The Relief of Menopausal Symptoms by Estrogenic Preparations," *Journal of the American Medical Association* 104 (23 February 1935), 624–25.

69. Kobrak, *National Cultures and International Competition*, 370.

70. Sevringhaus, "The Relief of Menopausal Symptoms by Estrogenic Preparations," 626.

71. Bernhard Zondek, "Mass Excretion of Oestrogenic Hormone in the Urine of the Stallion," *Nature* (10 February 1934), 209.

72. Oudshoorn, *Beyond the Natural Body*, 74–75.

73. "Synthesized Sex Hormones," *Newsweek* 13 (24 April 1939), 28; "Theelin: Research Step Three: Hormone Made Chemically," *Newsweek* 8 (22 August 1936), 24–25; "Synthetic Theelin" *Time* 28 (24 August 1936), 39.

74. *Female Sex Hormone Therapy: A Clinical Guide* (Bloomfield, NJ: Schering Corporation, 1941). The oral form was called Progynon-DH, for α-estradiol dipropionate; the injectable form was Progynon-B, for α-estradiol benzoate.

75. E. C. Dodds, L. Goldberg, W. Lawson, and R. Robinson, "Oestrogenic Activity of Certain Synthetic Compounds," *Nature* 141 (5 February 1938), 247–48.

76. For a discussion of the relationship between DES and the medicalization of menopause, see Susan E. Bell, "Changing Ideas: The Medicalization of Menopause," *Social Science and Medicine* 24 (1987), 535–42. For an alternative interpretation, see Houck, "Common Experiences and Changing Meanings," 120–23.

77. Emil Novak, "The Management of the Menopause," *American Journal of Obstetrics*

*and Gynecology* 40 (October 1940), 594; S. Charles Freed, "Present Status of Commercial Endocrine Preparations," *Journal of the American Medical Association* 117 (4 October 1941), 1178; AMA Council on Pharmacy and Chemistry, "Diethylstilbestrol," *Journal of the American Medical Association* 119 (20 June 1942), 634.

78. Novak, "The Management of the Menopause," 592.

79. Houck, "Common Experiences and Changing Meanings," 155–57.

80. Nelson, ed., *Pharmaceutical Company Histories*, 7. See also Alison Li, "Marketing Menopause," 103.

81. United States Patent No. 2,429,398 (21 October 1947).

82. S. C. Freed, W. M. Eisin, and J. P. Greenhill, "The Oral Effectiveness of Estrone Sulfate (Conjugated Estrogens-Equine) in Women," *Journal of Clinical Endocrinology* 3 (February 1943), 89–91; Laman A. Gray, "Clinical Study of a New Type of Estrogenic Preparation for Oral Use," *Journal of Clinical Endocrinology* 3 (February 1943), 92–94; S. J. Glass and Gordon Rosenblum, "Therapy of Menopause: Superiority of Conjugated Estrogens-Equine over Diethylstilbestrol," *Journal of Clinical Endocrinology* 3 (February 1943), 95–97; Elmer L. Sevringhaus and Ruth St. John, "Oral Use of Conjugated Estrogens-Equine," *Journal of Clinical Endocrinology* 3 (February 1943), 98–100.

83. Glass and Rosenblum, "Therapy of Menopause: Superiority of Conjugated Estrogens-Equine over Diethylstilbestrol," 97.

84. J. R. Goodall, "Premarin in Some Post-menopausal Complications," *Journal of Obstetrics and Gynaecology of the British Empire* 49 (December 1942), 660.

85. Susan E. Bell, "The Synthetic Compound Diethylstilbestrol (DES), 1938–1941: The Social Construction of a Medical Treatment" (Ph.D. dissertation, Brandeis University, 1980), 1. For a thorough study of the DES story, see Roberta J. Apfel and Susan M. Fisher, *To Do No Harm: DES and the Dilemmas of Modern Medicine* (New Haven: Yale University Press, 1984).

86. Peter Temin, *Taking Your Medicine: Drug Regulation in the United States* (Cambridge, MA: Harvard University Press, 1980), 33–34.

87. Ibid., 38.

88. For a thorough analysis of the 1938 law, see ibid., 43–48.

89. In 1951 the Durham-Humphrey Amendment formally established the two classes of drugs, prescription and over-the-counter, and gave FDA the authority to categorize drugs as prescription-only.

90. "Notice to Manufacturers of Preparations of Ovary" TC-13 (1 December 1939) in Vincent A. Kleinfeld and Charles Wesley Dunn, *Federal Food, Drug, and Cosmetic Act Judicial and Administrative Record, 1938–1949* (New York: Commerce Clearing House, 1949), 574.

91. In 2001 the Freedom of Information Office at the Department of Health and Human Services denied my request to see the new drug application for Premarin because the product is still active.

92. Susan E. Bell, "Gendered Medical Science: Producing a Drug for Women," *Feminist Studies* 21 (Fall 1995), 475.

93. Susan E. Bell, "A New Model of Medical Technology Development: A Case Study of DES," *Research in the Sociology of Health Care* 4 (1986), 13.

94. Temin, *Taking Your Medicine*, 125.

95. Bell, "Gendered Medical Science," 489.

96. Quoted in ibid., 491–92.

97. Ibid., 490.

98. According to Bell, "A New Model," 28, the twelve companies were Abbott Laboratories; Armour Laboratories; Ayerst, McKenna and Harrison; George A. Breon.; Charles E. Frosst; Eli Lilly; Merck; Sharp and Dohme; E. R. Squibb and Sons; Upjohn; Winthrop Chemical; John Wyeth and Brother.

99. Cowan, "Edgar Allen," 123.

100. "Edward A. Doisy—Biography," *Nobel Lectures, Physiology or Medicine 1942–1962*.

101. A. S. Parkes, "The Rise of Reproductive Endocrinology, 1926–1940," *Journal of Endocrinology* 34 (March 1966), xx–xxxii. Parkes attributes the comment about the heroic age of reproductive endocrinology to Guy Marrian, on p. xx.

102. *Physicians' Desk Reference to Pharmaceutical Specialties and Biologicals* (Rutherford, NJ: Medical Economics Inc., 1946), iv.

## Two • From the "Neutral Gender" to "Feminine Forever"

1. John Galbraith Simmons, *Doctors and Discoveries: Lives That Created Today's Medicine* (Boston: Houghton Mifflin, 2002), 341.

2. Richard Severo, "William H. Masters, a Pioneer in Studying and Demystifying Sex, Dies at 85," *New York Times* (19 February 2001), B7.

3. Ibid.

4. Simmons, *Doctors and Discoveries*, 341.

5. William H. Masters, "Endocrine Therapy in the Aging Individual," *Obstetrics and Gynecology* 8 (July 1956), 62.

6. Judith A. Houck, "Common Experiences and Changing Meanings: Women, Medicine, and Menopause in the United States, 1897–1980" (Ph.D. dissertation, University of Wisconsin-Madison, 1998), 136.

7. W. O. Johnson, "Menopause from the Viewpoint of the Gynecologist," *Kentucky Medical Journal* 45 (June 1947), 211.

8. Robert A. Ross, "Definitive Sex Endocrine Therapy in the Female," *North Carolina Medical Journal* 8 (January 1947), 14.

9. Houck, "Common Experiences and Changing Meanings," 153.

10. Hugh C. McLaren, "The Present Status of Hormone Therapy at the Menopause," *The Practitioner* 171 (November 1953), 505.

11. Houck, "Common Experiences and Changing Meanings," 151–52.

12. "Estrogen Therapy—A Warning," *Journal of the American Medical Association* 113 (23 December 1939), 2324.

13. Rita S. Finkler, "The Use and Misuse of Estrogens in Menopause," *Medical Women's Journal* 52 (February 1945), 28–30; W. F. T. Haultain, "Oestrogenic Therapy: Its Uses and Abuses," *The Medical Press* 217 (14 May 1947), 405–7; J. M. Habel Jr., "Indiscriminate Use of Estrogens in the Menopause," *Virginia Medical Monthly* 75 (October 1948), 517–20.

14. Habel, "Indiscriminate Use," 517.

15. Robert A. Kimbrough and S. Leon Israel, "The Use and Abuse of Estrogen," *Journal of the American Medical Association* 138 (25 December 1948), 1220.

16. Edgar Allen, "Ovarian Hormone and Female Genital Cancer," *Journal of the American Medical Association* 114 (25 May 1940), 2113–14.

17. Edward A. Doisy, "Glandular Physiology and Therapy: The Estrogenic Substances," *Journal of the American Medical Association* 116 (8 February 1941), 505.

18. Samuel H. Geist, Robert I. Walter, and Udall J. Salmon, "Are Estrogens Carcinogenic in the Human Female?" *American Journal of Obstetrics and Gynecology* 42 (August 1941), 247.

19. Houck, "Common Experiences and Changing Meanings," 165.

20. Historical Note, William B. Kountz Papers, Bernard Becker Medical Library, Washington University School of Medicine, http://becker.wustl.edu/ARB/find/kountz.

21. Paul G. Anderson, Associate Professor and Archivist, Bernard Becker Medical Library, Washington University School of Medicine, personal communication, 20 January 2004.

22. William H. Masters and Willard M. Allen, "Female Sex Hormone Replacement in the Aged Woman," *Journal of Gerontology* 3 (July 1948), 183.

23. Ibid.

24. William H. Masters, "Sex Steroid Replacement in the Aging Individual," in *Hormones and the Aging Process: Proceedings of a Conference Held at Arden House, Harriman, New York, 1955,* ed. Earl T. Engle and Gregory Pincus (New York: Academic Press, 1956), 249.

25. See Harry M. Marks, *The Progress of Experiment: Science and Therapeutic Reform in the United States, 1900–1990* (Cambridge: Cambridge University Press, 1997), esp. 129–63.

26. Bettye McDonald Caldwell and Robert I. Watson, "An Evaluation of Psychologic Effects of Sex Hormone Administration in Aged Women: I. Results of Therapy after Six Months," *Journal of Gerontology* 7 (April 1952), 229.

27. Ibid., 242.

28. William H. Masters, "The Rationale and Technique of Sex Hormone Replacement in the Aged Female: A Preliminary Result Report," *South Dakota Journal of Medicine and Pharmacy* 4 (November 1951), 297.

29. Ibid., 298.

30. Ibid.

31. William H. Masters and Marvin H. Grody, "Estrogen-Androgen Substitution Therapy in the Aged Female: II. Clinical Response," *Obstetrics and Gynecology* 2 (August 1953), 146.

32. Caldwell and Watson, "An Evaluation of Psychologic Effects of Sex Hormone Administration in Aged Women," 239.

33. Ibid., 237.

34. Ibid., 241–42.

35. Ibid., 243.

36. Masters, "The Rationale and Technique of Sex Hormone Replacement in the Aged Female," 296.

37. Ibid.

38. Ibid.

39. Ibid.

40. William H. Masters, "Sex Steroid Replacement in the Aging Individual," 251.

41. William H. Masters and John W. Ballew, "The Third Sex," *Geriatrics* 10 (January 1955), 1.

42. Ibid., 2–3.

43. William H. Masters, "Rationale of Sex Steroid Replacement in the 'Neutral Gender,'" *Journal of the American Geriatrics Society* 3 (June 1955), 394.

44. Masters and Ballew, "The Third Sex," 3.

45. William H. Masters, "Endocrine Therapy in the Aging Individual," *Obstetrics and Gynecology* 8 (July 1956), 61.

46. Masters and Ballew, "The Third Sex," 1.

47. Dr. C. Lee Buxton, a prominent New Haven obstetrician-gynecologist, made this comment in the discussion following Masters's presentation at the 80th annual meeting of the American Gynecological Society in Hot Springs, VA, in May 1957. The discussion was included in William H. Masters, "Sex Steroid Influence on the Aging Process," *American Journal of Obstetrics and Gynecology* 74 (October 1957), 744. The "panacea" quote is from Masters, "Endocrine Therapy in the Aging Individual," 66.

48. Masters and Ballew, "The Third Sex," 1.

49. Fuller Albright, Esther Bloomberg, and Patricia H. Smith, "Post-Menopausal Osteoporosis," *Transactions of the Association of American Physicians* 55 (1940), 305.

50. See, for example, Fuller Albright, Patricia H. Smith, and Anne M. Richardson, "Postmenopausal Osteoporosis," *Journal of the American Medical Association* 116 (31 May 1941), 2465–74; Ian A. Anderson, "Postmenopausal Osteoporosis, Clinical Manifestations, and the Treatment with Oestrogens," *Quarterly Journal of Medicine* 19 (January 1950), 67–96; Philip H. Henneman and Stanley Wallach, "A Review of the Prolonged Use of Estrogens and Androgens in Postmenopausal and Senile Osteoporosis," *AMA Archives of Internal Medicine* 100 (November 1957), 715–23.

51. Ibid., 722. It is interesting to note that Henneman and Wallach listed Ayerst Laboratories, the makers of Premarin, as one of the financial sponsors of their research.

52. Masters, "Sex Steroid Influence on the Aging Process," 736.

53. John H. Wuest Jr., Thomas J. Dry, and Jesse E. Edwards, "The Degree of Coronary Atherosclerosis in Bilaterally Oophorectomized Women," *Circulation* 7 (June 1953), 801–8.

54. George C. Griffith, "Oophorectomy and Cardiovascular Tissue," *Obstetrics and Gynecology* 7 (May 1956), 479–82; Clyde L. Randall, "Ovarian Function and Women after the Menopause," *American Journal of Obstetrics and Gynecology* 73 (May 1957), 1000–1010.

55. Roger W. Robinson, Norio Higano, William D. Cohen, Ronald C. Sniffen, and Joseph W. Sherer, "Effects of Estrogen Therapy on Hormone Functions and Serum Lipids in Men with Coronary Atherosclerosis," *Circulation* 14 (September 1956), 365–72.

56. Roger W. Robinson, William D. Cohen, and Norio Higano, "Estrogen Replacement Therapy in Women with Coronary Atherosclerosis," *Annals of Internal Medicine* 48 (January 1958), 95–101.

57. Geist, Walter, and Salmon, "Are Estrogens Carcinogenic in the Human Female?" 242.

58. Henneman and Wallach, "A Review of the Prolonged Use," 722.

59. Randall, "Ovarian Function and Women after the Menopause," 1003.

60. Masters, "Sex Steroid Replacement in the Aging Individual," 249.

61. The phrase "puberty to grave" comes from Robert A. Wilson and Thelma A. Wilson, "The Fate of the Nontreated Postmenopausal Woman: A Plea for the Maintenance of Adequate Estrogen from Puberty to Grave," *Journal of the American Geriatrics Society* 11 (April 1963), 347–62.

62. Herbert S. Kupperman, Ben B. Wetchler, and Meyer H. G. Blatt, "Contemporary Therapy of the Menopausal Syndrome," *Journal of the American Medical Association* 171 (21 November 1959), 1632.

63. Stanley Wallach and Philip H. Henneman, "Prolonged Estrogen Therapy in Postmenopausal Women," *Journal of the American Medical Association* 171 (21 November 1959), 1642.

64. E. Kost Shelton, "The Use of Estrogen after the Menopause," *Journal of the American Geriatrics Society* 2 (October 1954), 628.

65. Ibid., 629.

66. Ibid.

67. E. Kost Shelton, "The Pros and Cons of Estrogen Administration after the Menopause," *Journal of the American Geriatrics Society* 4 (April 1956), 348.

68. Shelton, "The Use of Estrogen after the Menopause," 632.

69. Ibid., 630–31.

70. Ibid., 632.

71. Shelton, "The Pros and Cons of Estrogen Administration after the Menopause," 350.

72. "Hormones Rejuvenate Women Past 65, St. Louis Doctor Reports at Conference," *New York Times* (13 July 1950), 27.

73. Waldemar Kaempffert, "Aging Processes Are Arrested by the Injection of Male and Female Sex Hormones," *New York Times* (30 August 1953), E9.

74. Allan C. Barnes, "Is Menopause a Disease?" *Consultant* 2 (1962), 22.

75. Ibid., 23.

76. Houck, "Common Experiences and Changing Meanings," 272; Robert A. Wilson, "The Roles of Estrogen and Progesterone in Breast and Genital Cancer," *Journal of the American Medical Association* 182 (27 October 1962), 327–31.

77. Wilson and Wilson, "The Fate of the Nontreated Postmenopausal Woman," 347–62.

78. Ibid., 355.

79. Ibid., 347.

80. Ibid., 356.

81. Robert A. Wilson, "The Obsolete Menopause," *Connecticut Medicine* 27 (December 1963), 735; Robert A. Wilson, "The Obsolete Menopause," *Delaware Medical Journal* 36 (January 1964), 20.

82. Robert A. Wilson, Raimondo E. Brevetti, and Thelma A. Wilson, "Specific Procedures for the Elimination of the Menopause," *Western Journal of Surgery, Obstetrics and Gynecology* 71 (May–June 1963), 110–21.

83. Wilson and Wilson, "The Fate of the Nontreated Postmenopausal Woman," 356, 355.

84. Houck, "Common Experiences and Changing Meanings," 273. The sales figure comes from "Feminine Forever," *Newsweek* 69 (3 April 1967), 55.

85. See, for example, "How to Live Young at Any Age," *Vogue* (August 1965), 61–64; "Change of Life," *Good Housekeeping* (September 1967), 19–20; "No More Menopause," *Newsweek* (13 January 1964), 53; Helen D. Borel, "The Book that Ends Menopause," *Science Digest* 59 (June 1966), 28.

86. The annual number of estrogen prescriptions increased from 15.5 million in 1966 to 28 million in 1975. Dianne L. Kennedy et al., "Noncontraceptive Estrogens and Progestins: Use Patterns over Time," *Obstetrics and Gynecology* 65 (March 1985), 442.

87. Noel S. Weiss et al., "Increasing Incidence of Endometrial Cancer in the United States," *New England Journal of Medicine* 294 (3 June 1976), 1261.

88. B. V. Stadel and N. Weiss, "Characteristics of Menopausal Women: A Survey of King and Pierce Counties in Washington, 1973–1974," *American Journal of Epidemiology* 102 (1975), 209–16.

89. Peter A. van Keep and Christian Lauritzen, eds., *Ageing and Estrogens* (Basel: S. Karger, 1973); Peter A. van Keep and Christian Lauritzen, eds., *Estrogens in the Post-Menopause* (Basel: S. Karger, 1975); P. A. van Keep, R. B. Greenblatt, and M. Albeaux-Fernet, eds., *Consensus on Menopause Research* (Lancaster, England: MTP Press, 1976).

90. Wilson, Brevetti, and Wilson, "Specific Procedures for the Elimination of the Menopause," 111.

91. For a more detailed discussion of birth control pills, see Elizabeth Siegel Watkins, *On the Pill: A Social History of Oral Contraceptives, 1950–1970* (Baltimore: Johns Hopkins University Press, 1998).

92. Although both therapies consisted of sex hormones, they differed in terms of the types and dosages of estrogen and progestin. In birth control pills, the main constituent was the progestin, which inhibited ovulation; the estrogen component reduced the likelihood of breakthrough bleeding. In hormone replacement therapy, estrogen was the primary agent; progestin, when prescribed, was intended to encourage sloughing of the uterine wall (the addition of progestin to the regimen is discussed in chapter 8).The first birth control pill, Enovid, made by G. D. Searle and Company, consisted of 10 milligrams synthetic progestin (norethinodrel) and 0.15 milligrams synthetic estrogen (mestranol). By the late 1960s, there were several brands and formulations on the market, most of which consisted of 2.5 or 5 milligrams synthetic progestin (norethinodrel, norethindrone, levonorgestrel) and 0.05 milligrams synthetic estrogen (mestranol or ethinyl estradiol). This first generation of oral contraceptives contained much higher doses of hormones than products on the market today (the so-called tri-phasic pills, which attempt to mimic more closely a woman's hormonal cycle, contain no more than 1 milligram progestin and 0.035 milligrams estrogen). Premarin pills consisted of 1.25 milligrams conjugated equine estrogens, which were derived from horse urine, not synthesized in the laboratory. When researchers demonstrated in the late 1960s that birth control pills caused increased rates of potentially fatal blood clotting, they believed that the synthetic estrogen component was to blame. By contrast, conjugated equine estrogens were thought to be less potent, and therefore "harmless." (See C. C. Edwards and V. A. Drill, "Estrogens: Why Harmless as Menopausal Therapy but Hazardous in the 'Pill'?" *Journal of the American Medical Association* 215

[18 January 1971], 492–93.) This stance would change after 1975 (see chapter 5), but in the 1960s, prior to the discovery of the link between the pill and blood clotting, both hormone therapies were hailed as modern medical successes.

93. Houck, "Common Experiences and Changing Meanings," 174.

94. U.S. Department of Health, Education, and Welfare—Public Health Service, *Health, United States, 1975* (DHEW Publication No. HRA 76-1232), 51.

95. "Ayerst Premarin Goes to Sudler & Hennessey," *New York Times* (6 January 1970), 64.

96. Patricia A. Kaufert and Sonja M. McKinlay, "Estrogen-Replacement Therapy: The Production of Medical Knowledge and the Emergence of Policy," in Ellen Lewin and Virginia Olesen, eds., *Women, Health, and Healing: Toward a New Perspective* (New York: Tavistock Publications, 1985), 114.

## Three • Selling Estrogen to Doctors

1. *Journal of the American Medical Association* 197 (15 August 1966), 43.

2. *Obstetrics and Gynecology* 31 (February 1968).

3. Gary L. Nelson, ed., *Pharmaceutical Company Histories* (Bismarck, ND: Woodbine, 1983), 10, 14.

4. Russell R. Miller, "Prescribing Habits of Physicians: A Review of Studies on Prescribing of Drugs," *Drug Intelligence and Clinical Pharmacy* 8 (February 1974), 82.

5. Ibid., 83.

6. American Medical Association, *Opinions of AMA Members—1973* (Chicago: American Medical Association, 1973), cited in Colman M. Herman and Christopher A. Rodowskas Jr., "Communicating Drug Information to Physicians," *Journal of Medical Education* 51 (March 1976), 191.

7. Robert A. Kimbrough and S. Leon Israel, "The Use and Abuse of Estrogen," *Journal of the American Medical Association* 138 (25 December 1948), 1216.

8. Miller, "Prescribing Habits of Physicians," 85.

9. M. Gershenson, "Pharmaceutical Detailman Survey," *The Internist* (February 1971), 4–5, cited in Russell R. Miller, "Prescribing Habits of Physicians," 84.

10. Franklin T. Branch, "A Comparison of Direct-Mail and Magazine Cost in Pharmaceutical Advertising," *Journal of Business of the University of Chicago* 18 (April 1945), 94.

11. Robert Ferber and Hugh G. Wales, *The Effectiveness of Pharmaceutical Promotion* (Urbana: University of Illinois, 1958), 24–26, 57–58.

12. Colman M. Herman and Christopher A. Rodowskas Jr., "Communicating Drug Information to Physicians," *Journal of Medical Education* 51 (March 1976), 190; Jeremy A. Greene, "Attention to 'Details': Etiquette and the Pharmaceutical Salesman in Postwar USA," *Social Studies of Science* 34 (2004), 2.

13. Abbott, Armour, Ayerst, Breon, Chicago Pharmacal, Ciba, Endo Products, G. W. Carnrick, Lakeside, Lederle, Merrell, Ortho, Parke Davis, Reed and Carnrick, Roche-Organon, Schering, Schieffelin, Searle, Smith-Dorsey, Squibb, Upjohn, Wallace, White, Winthrop, Wyeth.

14. Dianne L. Kennedy et al., "Noncontraceptive Estrogens and Progestins: Use Patterns over Time," *Obstetrics and Gynecology* 65 (March 1985), 442.

15. The *New York Times* reported in late 1975 that Premarin accounted for 75–80 percent of the estrogen market. See Robert Metz, "Market Place," *New York Times* (12 November 1975), 64. Two years later, *Chemical Week* reported that Premarin held 63 percent. See "Estrogens hurting, corticoids healthy," *Chemical Week* (23 November 1977), 23. A 1985 study from the Drug Use Analysis Branch of the Food and Drug Administration estimated that Premarin had accounted for 70 percent of estrogen use over time. See Kennedy et al., "Noncontraceptive Estrogens and Progestins: Use Patterns over Time," 444. In testimony at a 1976 Senate hearing investigating the association between estrogen and endometrial cancer, an Ayerst executive claimed that Premarin held less than 40 percent of the estrogen market share. Chester J. Cavallito, executive vice-president for scientific affairs, Ayerst Laboratories, quoted in U.S. Senate, 94th Congress, 2nd Session, *Oral Contraceptives and Estrogens for Postmenopausal Use,* Joint Hearing before the Subcommittee on Health of the Committee on Labor and Public Welfare and the Subcommittee on Administrative Practice and Procedure of the Committee on the Judiciary (21 January 1976), 213.

16. *American Journal of Obstetrics and Gynecology* 27 (January 1934), 11.

17. Ibid., 33 (April 1937), A-4.

18. Ibid., 33 (June 1937), 23.

19. Ibid., 46 (July 1943), A-2.

20. Ibid., 41 (January 1941), A-8.

21. *Journal of the American Medical Association* 124 (5 February 1944), 33.

22. *American Journal of Obstetrics and Gynecology* 46 (September 1943), inside cover.

23. Mary M. Schweitzer, "World War II and Female Labor Force Participation Rates," *Journal of Economic History* 40 (March 1980), 92.

24. *American Journal of Obstetrics and Gynecology* 46 (July 1943), A-8.

25. *Journal of the American Medical Association* 124 (5 February 1944), 9.

26. Ibid., 124 (1 January 1944), 23.

27. *American Journal of Obstetrics and Gynecology* 50 (July 1945), A-2.

28. Ibid., 54 (December 1947), 17.

29. *Journal of the American Medical Association* 137 (5 June 1948), 25.

30. Ibid., 137 (1 May 1948), 37.

31. Ibid., 143 (15 July 1950), 21.

32. Ibid., 130 (6 April 1946), 27.

33. Ibid., 167 (10 May 1958), 57.

34. *American Journal of Obstetrics and Gynecology* 50 (September 1945), A-6.

35. *Journal of the American Medical Association* 148 (22 March 1952), 21; *Journal of the American Medical Association* 148 (26 April 1952), 32; *Obstetrics and Gynecology* 1 (May 1953), xvii; *Journal of the American Medical Association* 173 (4 June 1960), 292.

36. *American Journal of Obstetrics and Gynecology* 50 (September 1945), 12.

37. Ibid., 50 (August 1945), 17.

38. *Journal of the American Medical Association* 130 (16 March 1946), 84.

39. *American Journal of Obstetrics and Gynecology* 62 (October 1951), A-2.

40. *Journal of the American Medical Association* 137 (8 May 1948), 15.

41. Ibid., 167 (17 May 1958), 41.

42. *American Journal of Obstetrics and Gynecology* 53 (April 1947), 15.

43. Allan C. Barnes, "The Menopause," *Clinical Obstetrics and Gynecology* 1 (1958), 210.

44. *American Journal of Obstetrics and Gynecology* 62 (October 1951), 15; *American Journal of Obstetrics and Gynecology* 66 (September 1953), 33.

45. *Journal of the American Medical Association* 148 (26 April 1952), 22–23.

46. *American Journal of Obstetrics and Gynecology* 66 (July 1953), 27.

47. Mickey C. Smith, *A Social History of the Minor Tranquilizers: The Quest for Small Comfort in the Age of Anxiety* (New York: Pharmaceutical Products Press, 1991), 35. See also David Healy, *The Antidepressant Era* (Cambridge, MA: Harvard University Press, 1997).

48. Susan L. Speaker, "From 'Happiness Pills' to 'National Nightmare': Changing Cultural Assessment of Minor Tranquilizers in America, 1955–1980," *Journal of the History of Medicine* 52 (July 1997), 34.

49. *American Journal of Obstetrics and Gynecology* 74 (October 1957), 45. This ad ran frequently through the 1960s.

50. *Obstetrics and Gynecology* 20 (July 1962), 85.

51. Ibid., 39 (March 1972), 28–29.

52. *Journal of the American Medical Association* 189 (7 September 1964), 218.

53. *Obstetrics and Gynecology* 33 (January 1969).

54. Ibid., 35 (January 1970).

55. Ibid., 37 (January 1971), 14–15.

56. *Journal of the American Medical Association* 208 (16 June 1969), 2206–9.

57. *Obstetrics and Gynecology* 40 (December 1972).

58. Ibid., 39 (April 1972), 92–95.

59. Ibid., 46 (October 1975), A38–A39.

60. "Ayerst's First Profiling Program Rated a Success," *Scoreboard* 33 (December 1970), 16.

61. "New Premarin Advertising Campaign in Evidence," *Scoreboard* 34 (November 1971), 2

62. Ibid., 5.

63. Ibid., 3.

64. "MD's Wife Purrs for PREMARIN," *Scoreboard* 34 (September 1971), 21.

65. "Doctor's Secretary Vouches for PREMARIN," *Scoreboard* 34 (December 1971), 15.

66. "One PREMARIN Script-1, 000!" *Scoreboard* 34 (May 1971), 22.

67. "Ayerst's First Profiling Program Rated a Success," 16.

68. *Obstetrics and Gynecology* 45 (February 1975), A14–A16.

69. Ibid., 45 (February 1975).

### Four • Selling Estrogen to Women

1. Robert A. Wilson, *Feminine Forever* (New York: M. Evans and Company, 1966), 196.

2. Ibid., 194–95.

3. Ibid., 195–96.

4. Ibid., 203.

5. Judith A. Houck, "Common Experiences and Changing Meanings: Women, Medi-

cine, and Menopause in the United States, 1897–1980" (Ph.D. dissertation, University of Wisconsin-Madison, 1998), 272–75.

6. Carl G. Hartman, "Sex Education for the Woman at Menopause," *Hygeia* 19 (September 1941), 699, 747.

7. Helen Haberman, "Help for Women over 40," *Hygeia* 19 (November 1941), 898–99.

8. Bernadine Bailey, "Fair, Fit and Forty," *Hygeia* 25 (December 1947), 931, 959.

9. Ibid., 931.

10. Ibid., 959.

11. Julie E. Miale, "Easing Those Difficult Years," *Today's Health* 34 (February 1956), 30.

12. Ibid., 28.

13. Kenneth C. Hutchin, "The Change and What Husbands Should Know About It," *Today's Health* 44 (September 1966), 54–56, 79–80.

14. Sally Olds, "Menopause—Something to Look Forward To?" *Today's Health* 48 (May 1970), 77.

15. Ibid., 80.

16. In 1986, the earliest year for which I have data, *Health* magazine (the successor to *Family Health*, the successor to *Today's Health*) claimed about five million readers. By comparison, *Reader's Digest* reached 54 million, *Time* reached 25 million, *Newsweek* 20 million, *Ladies' Home Journal* 20 million, *McCall's* 23 million, and *Good Housekeeping* 29 million. *Saturday Evening Post* Complete Demographic Profile as defined by Mediamark Research, Inc. (Spring 1986), 4. I am grateful to Don Sutton of the *Saturday Evening Post* for providing me with a copy of this document.

17. Yankelovich, Skelly and White, "General Mills American Family Study 1978–79: Family Health in an Era of Stress" (October 1978). Roper Center for Public Opinion Research, University of Connecticut, Storrs, CT.

18. Lois Mattox Miller, "Changing Life Sensibly," *Reader's Digest* 35 (October 1939), 101–3.

19. Haberman, "Help for Women over Forty," 67–68.

20. Paul de Kruif, "New Help for Women's Change of Life," *Reader's Digest* 52 (January 1948), 11.

21. Maxine Davis, "The Menopause," *Good Housekeeping* (July 1943), 30.

22. James Scott, "You Need Not Fear the Menopause," *Ladies' Home Journal* (March 1946), 33.

23. Ibid., 33, 190–91.

24. Ruth and Edward Brecher, "The Facts about the Menopause," *Reader's Digest* 73 (July 1958), 79.

25. Lawrence Galton, "What Every Husband Should Know about a Woman's Change of Life," *Better Homes and Gardens* (July 1950), 127.

26. Ibid., 126.

27. "Hormones Rejuvenate Women Past 65, St. Louis Doctor Reports at Conference," *New York Times* (13 July 1950), 27.

28. Waldemar Kaempffert, "Aging Processes Are Arrested by the Injection of Male and Female Sex Hormones," *New York Times* (30 August 1953), E9.

29. "Sex Hormones Fight Age," *Science News Letter* 60 (22 September 1951), 178; "Female Hormone Sparks Minds of Older Women," *Science News Letter* 60 (22 September 1951), 182; "Sex Hormones Make Old People Feel Better," *Science Digest* 30 (December 1951), 73.

30. "Hope for Grandmothers," *Newsweek* 43 (28 June 1954), 82.

31. Anne Fromer, "Can New Drugs Keep You Young?" *Coronet* 37 (February 1955), 35.

32. Sherwin A. Kaufman, M.D., "The Truth about Female Hormones," *Ladies' Home Journal* 82 (January 1965), 22.

33. Ann Walsh, "Pills to Keep Women Young," *McCall's* 93 (October 1965), 104.

34. Houck, "Common Experiences and Changing Meanings," 306.

35. The quotation is taken from the title of chapter 1 of *Feminine Forever*.

36. Houck, "Common Experiences and Changing Meanings," 285–88.

37. *Feminine Forever*, 114.

38. According to Wilson, girls younger than seven demonstrated no estrogen effect, with 0 percent superficial cells, 0 percent intermediate cells, and 100 percent parabasal cells, for a ratio of 0-0-100. After puberty, female vaginal smears showed a ratio of 85-15-0: 85 percent superficial cells, 15 percent intermediate cells, and 0 percent parabasal cells. After menopause, parabasal cells replaced superficial cells, so an untreated postmenopausal woman might score 10-20-70 on the femininity scale. The goal of hormone replacement therapy was to restore the femininity index back to the 85-15-0 ratio.

39. *Feminine Forever*, 173.

40. Bill Davidson, "Menopause: Is There a Cure?" *Saturday Evening Post* 240 (26 August 1967), 71.

41. *Feminine Forever*, 132.

42. Robert A. Wilson, "Which Hormone to Take When," *Vogue* 147 (June 1966), 92–94, 149.

43. "How to Live Young at Any Age," *Vogue* 146 (15 August 1965), 61.

44. "Bazaar's Over-40 Guide on Health, Looks, Sex," *Harper's Bazaar* 106 (August 1973), 87.

45. Alice Lake, "Menopause: Is It Necessary?" *Good Housekeeping* 160 (April 1965), 162.

46. Ibid., 164.

47. Grace Naismith, "Common Sense and the 'Femininity Pill,'" *Reader's Digest* 89 (September 1966), 99.

48. "Estrogens during and after the Menopause," *Medical Letter* 7 (July 1965), 56.

49. Julia Kagan, "Hormone Therapy at Menopause: What Women Doctors Prescribe and Take," *McCall's* 103 (October 1975), 33. The author did not reveal how many questionnaires were initially sent out; thus the survey's response rate is unknown.

50. Ibid., 34.

51. Syntex, "Let's Discuss the Menopause" (September 1969), 6. Records of the American College of Obstetricians and Gynecologists, Washington, D.C. (hereafter, Records—ACOG).

52. Lindsay R. Curtis, "The Menopause: A New Life of Confidence and Contentment" (Bristol, Tennessee: SeMed Pharmaceuticals, 1969), 18–19. Records—ACOG.

53. Ibid., 20.

54. Ibid., 21.

55. Ibid., 31.

56. "Feminine . . . FOR LIFE" (New York: Wilson Research Foundation, 1964). Records—ACOG.

57. "The Complete Woman" (New York: Wilson Research Foundation, 1963). Records—ACOG.

58. "Mistrust without Logic" (New York: Wilson Research Foundation, c. 1964). Records—ACOG.

59. Bernice L. Neugarten, Vivian Wood, Ruth J. Kraines, and Barbara Loomis, "Women's Attitudes toward the Menopause," *Vita Humana* 6 (1963), 140–51.

60. Houck, "Common Experiences and Changing Meanings," 173–77.

61. Neugarten et al., "Women's Attitudes toward the Menopause," 142.

62. Thomas M. Mack et al., "Estrogens and Endometrial Cancer in a Retirement Community," *New England Journal of Medicine* 294 (3 June 1976), 1267. The new prescription estimates were obtained from IMS, Limited, in July 1975.

63. Dianne L. Kennedy et al., "Noncontraceptive Estrogens and Progestins: Use Patterns Over Time," *Obstetrics and Gynecology* 65 (March 1985), 442, 445.

64. Ibid., 442.

65. Mack et al., "Estrogens and Endometrial Cancer in a Retirement Community," 1267.

66. "Let's Discuss the Menopause and Your Need for Estrogens" (Palo Alto, CA: Syntex Laboratories, 1972), n.p. Records—ACOG.

67. U.S. Senate, 94th Congress, 2nd Session, *Oral Contraceptives and Estrogens for Postmenopausal Use,* Joint Hearing before the Subcommittee on Health of the Committee on Labor and Public Welfare and the Subcommittee on Administrative Practice and Procedure of the Committee on the Judiciary (21 January 1976), 68.

68. Kennedy et al., "Noncontraceptive Estrogens and Progestins," 442.

69. B. V. Stadel and N. Weiss, "Characteristics of Menopausal Women: A Survey of King and Pierce Counties in Washington, 1973–1974," *American Journal of Epidemiology* 102 (1975), 209–10.

70. Ibid., 215.

71. Houck, "Common Experiences and Changing Meanings," 328–30.

72. In 1994 the FDA ordered that topically applied hormone drug products be removed from the over-the-counter market.

73. Kathy Peiss, *Hope in a Jar: The Making of America's Beauty Culture* (New York: Owl Books, 1998), 140–41.

74. Ibid., 141.

75. Boncilla ad, *Beauty* (1923). AdAccess, http://scriptorium.lib.duke.edu/adaccess.

76. Edna Wallace Hopper Cosmetics, *Ladies' Home Journal* (1926). AdAccess, http://scriptorium.lib.duke.edu/adaccess.

77. Marie Barlow, *Vogue* (1928). AdAccess, http://scriptorium.lib.duke.edu/adaccess.

78. Dorothy Gray, *Harper's Bazaar* (1936). AdAccess, http://scriptorium.lib.duke.edu/adaccess.

79. Dorothy Gray, *Time* (1937). AdAccess, http://scriptorium.lib.duke.edu/adaccess.

80. Palmolive Company, *Saturday Evening Post* (1933); Proctor and Gamble Company, *Sunday News* (1934). AdAccess, http://scriptorium.lib.duke.edu/adaccess.

81. *New York Times* (12 March 1945), 8.

82. *New York Times* (27 May 1945), 16.

83. Max A. Goldzieher, "The Effects of Estrogens on the Senile Skin," *Journal of Gerontology* 1 (April 1946), 196–200.

84. Howard T. Behrman, "Hormone Creams and the Facial Skin," *Journal of the American Medical Association* 155 (8 May 1954), 119–23.

85. Austin C. Wehrwein, "A.M.A. Hits Drugs as Cosmetic Aids," *New York Times* (7 October 1961), 20; S. Rothman, "Drugs in Cosmetics," *Journal of the American Medical Association* 178 (7 October 1961), 38–42.

86. "Medical Committee Assails Use of Drugs in Cosmetics; Companies Defend Products," *Wall Street Journal* (9 October 1961), 28.

87. Wehrwein, "A.M.A. Hits Drugs as Cosmetic Aids," 20.

88. *New York Times* (29 October 1961), 110.

89. Dorothy Gray, *New York Herald Tribune* (1951). AdAccess, http://scriptorium.lib.duke.edu/adaccess.

90. *New York Times* (6 January 1952), 83; *New York Times* (14 September 1952), SM59; *New York Times* (5 January 1964), 88; *New York Times* (7 June 1959), 118.

91. *New York Times* (4 November 1962), 103.

92. For a thorough discussion of the prescriptive literature for midlife women, see Houck, "Common Experiences and Changing Meanings," chap. 4.

93. See, for example, Bertha Zelda Beck, "Young at Forty-Five," *Today's Health* 29 (May 1950), 44–45.

94. Kenneth L. Woodward, "The Lively Challenge of Middle Life," *McCall's* 99 (October 1971), 81.

95. Ibid.

96. "Sex in the Magazines," *Newsweek* 54 (24 August 1959), 56.

97. Peter Gabriel Filene, *Him/Her/Self: Sex Roles in Modern America* (New York: Mentor Books, 1974), 313.

98. Susan Sontag, "The Double Standard of Aging," *Saturday Review* (23 September 1972), 31, 35.

99. Pauline B. Bart, "Depression in Middle-Aged Women," in Vivian Gornick and Barbara K. Moran, eds., *Woman in Sexist Society: Studies in Power and Powerlessness* (New York: Mentor Books, 1971), 163–86. "Portnoy's Mother's Complaint" appeared in the November–December 1970 issue of *Trans-action*.

100. Bart, "Depression in Middle-Aged Women," 167.

101. Ibid., 169.

102. Ibid., 185–86.

103. "Menopause Speak-Out," *Prime Time* 2 (April 1974), 7–10.

104. See, for example, Lynn Laredo, "Garbage Pail Syndrome," *Prime Time* 2 (May 1974), 5–6.

105. The 1971 edition consisted of chapters on anatomy and physiology, sexuality, "some myths about women," venereal disease, birth control, abortion, pregnancy, prepared

childbirth, postpartum, medical institutions, and "women, medicine, and capitalism." Boston Women's Health Course Collective, *Our Bodies, Ourselves* (Boston: New England Free Press, 1971), in the Records of the National Women's Health Network, Sophia Smith Collection, Smith College, Acc.#96S-47, Box 2.

106. Boston Women's Health Book Collective, *Our Bodies, Ourselves* (New York: Simon and Schuster, 1973), 230.

107. Ibid., 233.

108. The questionnaire and 146 responses can be found in the Records of the Boston Women's Health Book Collective at the Schlesinger Library, Cambridge, MA (hereafter, Records—BWHBC), Accession #99-M147, carton 8.

109. #73, #49, ##444, #401, #455, Records—BWHBC, Accession #99-M147, carton 8.

110. #459, Records—BWHBC, Accession #99-M147, carton 8.

111. #400, Records—BWHBC, Accession #99-M147, carton 8.

112. #433, Records—BWHBC, Accession #99-M147, carton 8.

113. For a discussion of the power of cultural imagery and messages in the construction of the body, see Susan Bordo, *Unbearable Weight: Feminism, Western Culture, and the Body* (Berkeley: University of California Press, 1993).

114. For a discussion of aging women and the youth-centered culture of postwar America, see Elizabeth Haiken, *Venus Envy: A History of Cosmetic Surgery* (Baltimore: Johns Hopkins University Press, 1997), chap. 4.

## Five • From Hero to Villain

1. Donald C. Smith et al., "Association of Exogenous Estrogen and Endometrial Carcinoma," *New England Journal of Medicine* 293 (4 December 1975), 1164–67.

2. Ibid., 1166.

3. Harry K. Ziel and William D. Finkle, "Increased Risk of Endometrial Carcinoma among Users of Conjugated Estrogens," *New England Journal of Medicine* 293 (4 December 1975), 1167–70.

4. The Southern California Permanente Medical Group's department of research and education funded the study, at the modest cost of $1,900, according to the testimony of William D. Finkle in U.S. Senate, 94th Congress, 2nd Session, *Oral Contraceptives and Estrogens for Postmenopausal Use,* Joint Hearing before the Subcommittee on Health of the Committee on Labor and Public Welfare and the Subcommittee on Administrative Practice and Procedure of the Committee on the Judiciary (21 January 1976), 11.

5. Ziel and Finkle, "Increased Risk of Endometrial Carcinoma among Users of Conjugated Estrogens," 1170.

6. Kenneth J. Ryan, "Cancer Risk and Estrogen Use in the Menopause," *New England Journal of Medicine* 293 (4 December 1975), 1200.

7. Ibid.

8. Noel S. Weiss, "Risks and Benefits of Estrogen Use," *New England Journal of Medicine* 293 (4 December 1975), 1201.

9. Richard J. Steckel, Letter to the Editor Re: "Estogens and Endometrial Cancer," *New England Journal of Medicine* 294 (8 April 1976), 847.

10. Leonard B. Goldman, Letter to the Editor Re: "Estogens and Endometrial Cancer," *New England Journal of Medicine* 294 (8 April 1976), 847–48.

11. Douglas R. Shanklin, Letter to the Editor Re: "Estogens and Endometrial Cancer," *New England Journal of Medicine* 294 (8 April 1976), 847.

12. Thomas M. Mack et al., "Estrogens and Endometrial Cancer in a Retirement Community," *New England Journal of Medicine* 294 (3 June 1976), 1262–67.

13. Noel S. Weiss et al., "Increasing Incidence of Endometrial Cancer in the United States," *New England Journal of Medicine* 294 (3 June 1976), 1259–62. The areas selected were Connecticut, Hawaii, Los Angeles County, New Mexico, Oregon, San Francisco Bay Area, Seattle-Tacoma, and Utah.

14. Hysterectomy was the most frequently performed surgery from 1965 (the earliest year for which data were collected) through 1980, when it was surpassed by Cesarean section. Hysterectomy remained the second most common surgical procedure for women of reproductive age into the 1990s. R. Pokras and V. Hufnagel, "Hysterectomies in the United States, 1965–1984," *Vital and Health Statistics* series 13, no. 92 (Washington, DC: Government Printing Office, 1987), 3; Lisa A. Lepine et al., "Hysterectomy Surveillance—United States, 1980–1993," *CDC Surveillance Summaries* Morbidity and Mortality Weekly Report vol. 46, no. SS-4 (8 August 1997), 2.

15. Pokras and Hufnagel, "Hysterectomies in the United States, 1965–1984," 2–3. See also Robert Pokras and Vicki Georges Hufnagel, "Hysterectomy in the United States, 1965–84," *American Journal of Public Health* 78 (July 1988), 852–53.

16. Joseph L. Lyon and John W. Gardner, "The Rising Frequency of Hysterectomy: Its Effect on Uterine Cancer Rates," *American Journal of Epidemiology* 105 (May 1977), 439–43. See also Ziel and Finkle, "Increased Risk of Endometrial Carcinoma among Users of Conjugated Estrogens," 1169.

17. Lynn Rosenberg, Bruce Armstrong, and Hershel Jick, "Myocardial Infarction and Estrogen Therapy in Post-menopausal Women," *New England Journal of Medicine* 294 (3 June 1976), 1256–59.

18. Robert Hoover et al., "Menopausal Estrogens and Breast Cancer," *New England Journal of Medicine* 295 (19 August 1976), 401–5.

19. Judith A. Houck, "Common Experiences and Changing Meanings: Women, Medicine, and Menopause in the United States, 1897–1980" (Ph.D. dissertation, University of Wisconsin-Madison, 1998), 271.

20. Harry M. Marks, *The Progress of Experiment: Science and Therapeutic Reform in the United States, 1900–1990* (Cambridge: Cambridge University Press, 1997), 148.

21. Hugh R. K. Barber, "Estrogen Controversy: A Rational Approach," *The Female Patient* 1 (March 1976), 7.

22. Ibid., 8.

23. Abraham M. Lilienfeld, "Exogenous Estrogens and Endometrial Cancer," *Postgraduate Medicine* 59 (June 1976), 62.

24. Robert W. Kistner, "Estrogens and Endometrial Cancer," *Obstetrics and Gynecology* 48 (October 1976), 479.

25. Ibid.

26. Ibid., 481.

27. "Estrogen Replacement Therapy," *ACOG Technical Bulletin* no. 43 (October 1976), 4. Records—ACOG.

28. *ACOG Newsletter* 21 (January 1977), 5–6.

29. Thomas W. McDonald et al., "Exogenous Estrogen and Endometrial Cancer: Case-control and Incidence Study," *American Journal of Obstetrics and Gynecology* 127 (15 March 1977), 572–80.

30. Laman A. Gray Sr., William M. Christopherson, and Robert N. Hoover, "Estrogens and Endometrial Carcinoma," *Obstetrics and Gynecology* 49 (April 1977), 385–89.

31. Ibid., 388.

32. The number of AMA members was calculated as follows. In 1980 there was a total of 467,679 physicians in the United States, of whom 413,395 were men (88.4%) and 54,284 were women (11.6%). Ellen S. More, *Restoring the Balance: Women Physicians and the Profession of Medicine, 1850–1995* (Cambridge, MA: Harvard University Press, 1999), 225. In 1980, 48% of male physicians and 26.6% of female physicians belonged to the AMA. Paul Starr, *The Social Transformation of American Medicine* (New York: Basic Books, 1982), 427. The resulting estimate (212,870) for 1980 agrees with the figures cited on the AMA Web site: 200,000 members in 1965 and 250,000 in 1983. Allen J. Podraza, "Chronology of AMA History" (updated 7 January 2004). www.ama-assn.org/ama/pub/category/1922.html.

33. E. Stanton Shoemaker, J. Peter Forney, and Paul C. MacDonald, "Estrogen Treatment of Postmenopausal Women: Benefits and Risks," *Journal of the American Medical Association* 238 (3 October 1977), 1524–30.

34. Ibid., 1525.

35. Ibid., 1525–26.

36. *Obstetrics and Gynecology* 46 (August–November 1975).

37. Ibid., 47 (June 1976).

38. *Journal of the American Medical Association* 235 (12 January 1976), 127; *Journal of the American Medical Association* 235 (24 May 1976), 2275.

39. *Obstetrics and Gynecology* 48 (September 1976).

40. Ibid., 53 (April 1979).

41. *Journal of the American Medical Association* 238 (3 October 1977), 1488–91.

42. *Obstetrics and Gynecology* 49 (January 1977).

43. Ibid., 47 (February 1976).

44. *American Journal of Obstetrics and Gynecology* 124 (15 February 1976), 47–50.

45. "Estrogens Hurting, Corticoids Healthy," *Chemical Week* (23 November 1977), 23.

46. Ibid., 23–24.

47. Robert B. Greenblatt, "Estrogens and Endometrial Cancer—Gross Exaggeration or Fact?" *Geriatrics* 32 (November 1977), 60–72; Robert B. Greenblatt and Leland D. Stoddard, "The Estrogen-Cancer Controversy," *Journal of the American Geriatrics Society* 26 (January 1978), 1–8.

48. Greenblatt, "Estrogens and Endometrial Cancer—Gross Exaggeration or Fact?" 62.

49. Ibid., 61, 72.

50. Ralph L. Horwitz and Alvan R. Feinstein, "Alternative Analytic Methods for Case-Control Studies of Estrogens and Endometrial Cancer," *New England Journal of Medicine* 299 (16 November 1978), 1089–94.

51. George B. Hutchison and Kenneth J. Rothman, "Correcting a Bias?" *New England Journal of Medicine* 299 (16 November 1978), 1129–30.

52. Correspondence Re: "Case-Control Studies of Estrogens and Endometrial Cancer," *New England Journal of Medicine* 300 (1 March 1979), 495–97.

53. Frank Cole, Letter to the Editor Re: "Case-Control Studies of Estrogens and Endometrial Cancer," *New England Journal of Medicine* 300 (1 March 1979), 496.

54. Carlos M. F. Antunes et al., "Endometrial Cancer and Estrogen Use: Report of a Large Case-Control Study," *New England Journal of Medicine* 300 (4 January 1979), 9–13.

55. Hershel Jick et al., "Replacement Estrogens and Endometrial Cancer," *New England Journal of Medicine* 300 (1 February 1979), 218–22.

56. Noel S. Weiss et al., "Endometrial Cancer in Relation to Patterns of Menopausal Estrogen Use," *Journal of the American Medical Association* 242 (20 July 1979), 261–64.

57. Ibid., 264.

58. Martin M. Quigley and Charles B. Hammond, "Estrogen-Replacement Therapy—Help or Hazard?" *New England Journal of Medicine* 301 (20 September 1979), 646.

59. Ibid., 647.

60. "NIH Consensus Statements," http://consensus.nih.gov. The Consensus Development Program is still active today.

61. Bernadette M. Eichelberger, "Estrogen Prescribing Practices Scrutinized at NIH Conference," *American Journal of Hospital Pharmacy* 36 (December 1979), 1728.

62. "Estrogen Use and Postmenopausal Women," National Institutes of Health Consensus Development Conference Summary, vol. 2, no. 8 (1979), n.p.

## Six • Enter the Feminists

1. NBC Nightly News (4 December 1975), Vanderbilt Television News Archive, http://tvnews.vanderbilt.edu.

2. ABC Evening News (4 December 1975), Vanderbilt Television News Archive.

3. CBS Evening News (4 December 1975), Vanderbilt Television News Archive.

4. "Estrogen Is Linked to Uterine Cancer," *New York Times* (4 December 1975), 1.

5. Ibid., 55.

6. Jane E. Brody, "Physicians' Views Unchanged on Use of Estrogen Therapy," *New York Times* (5 December 1975), 45.

7. Transcript excerpt from The Six O'Clock Report, WCBS-TV News, Channel 2, New York (3 February 1976), Schlesinger Library, Radcliffe Institute, Cambridge, MA.

8. Display Ad, *New York Times* (3 February 1976), 62.

9. "F.D.A. Panel Gets Data on Estrogen," *New York Times* (17 December 1975), 19; Frances Cerra, "F.D.A. Chief Suspicious of Estrogens," *New York Times* (21 January 1976), 32; "Stronger Warnings on Estrogen Labels Ordered by F.D.A.," *New York Times* (28 September 1976), 21; "Warnings Ordered on Use of Progestin and Estrogen Drugs," *New York Times* (21 July 1977), 18.

10. Albin Krebs, "Notes on People," *New York Times* (7 October 1977), 28. Details of the case were provided by Barbara Seaman in an interview with the author in New York on 27 June 2004.

11. Jane E. Brody, "Why Has Estrogen Fallen on Such Difficult Times?" *New York Times* (23 October 1977), E9.

12. Jane E. Brody, "Personal Health: Menopausal Estrogens—Benefits and Risks of the 'Feminine' Drug," *New York Times* (26 September 1979), C16.

13. "Are Hormones Giving You Cancer?" *Vogue* 166 (February 1976), 114; Marilyn Mercer, "Can Estrogen Therapy Cause Cancer?" *McCall's* 103 (March 1976), 39; Barbara Yuncker, "What's behind the Scary Rise in Uterine Cancer," *Good Housekeeping* 182 (March 1976), 34–39.

14. "Estrogen Therapy: The Dangerous Road to Shangri-La," *Consumer Reports* 41 (November 1976), 642–45.

15. See, for example, Walter S. Ross, "How Safe Are Estrogens?" *Reader's Digest* 113 (November 1978), 261–66; "Perils of Estrogen," *Newsweek* 93 (15 January 1979), 56–57.

16. Paula Weideger, "Estrogen: The Rewards and Risks," *McCall's* 104 (March 1977), 70.

17. "Menopause: The Estrogen Controversy," *Harper's Bazaar* 112 (November 1979), 183.

18. Sara M. Evans, *Tidal Wave: How Women Changed America at Century's End* (New York: Free Press, 2003), 21–32.

19. See Sara Evans, *Personal Politics* (New York: Vintage Books, 1979).

20. Susan Brownmiller, *In Our Time: Memoir of a Revolution* (New York: Delta, 1999), 8. Read together, *In Our Time* and *Tidal Wave* provide a good history of second wave feminism.

21. Sandra Morgen, *Into Our Own Hands: The Women's Health Movement in the United States, 1969–1990* (New Brunswick, NJ: Rutgers University Press, 2002), 4–5. See also Boston Women's Health Book Collective, *Our Bodies, Ourselves* (New York: Simon and Schuster, 1998), 21.

22. Morgen, *Into Our Own Hands*, 7–8.

23. Elizabeth Siegel Watkins, *On the Pill: A Social History of Oral Contraceptives, 1950–1970* (Baltimore: Johns Hopkins University Press, 1998), 108–10, 119–20. See also Morgen, *Into Our Own Hands*, 8–9.

24. Morgen, *Into Our Own Hands*, 5–6. For the complete story of Jane, see Laura Kaplan, *Jane: The Legendary Underground Feminist Abortion Service* (Chicago: University of Chicago Press, 1995).

25. Morgen, *Into Our Own Hands*, 14. See also Wendy Kline, " 'Please Include This in Your Book': Readers Respond to *Our Bodies, Ourselves*," *Bulletin of the History of Medicine* 79 (Spring 2005), 83–88.

26. "Beginnings," *Network News* (October 1976), 1, 6; Morgen, *Into Our Own Hands*, 10; Watkins, *On the Pill*, 130.

27. In 1976, when Reitz had completed her research and was writing her manuscript, two books that included chapters on menopause were published by feminist authors. Both books, *The Curse: A Cultural History of Menstruation*, by Janice Delaney, Mary Jane Lupton, and Emily Toth (New York: Dutton, 1976), and *Menstruation and Menopause: The Physiology and the Psychology, the Myth and the Reality*, by Paula Weideger (New York: Knopf, 1976), took menstruation as their central focus; the biological, psychological, and social effects of menopause, the cessation of menstruation in midlife, were considered as were those of menarche, the onset of menstruation in adolescence, almost as bookends to a woman's

menstrual history. Both books made important contributions to the growing feminist literature on women's bodies and health care, but since menopause was not their primary topic, they are not considered at length here.

28. Rosetta Reitz, *Menopause: A Positive Approach* (New York: Chilton Book Company, 1977), 6–8.

29. The following biographical information comes from an interview conducted by the author with Rosetta Reitz in New York on 26 June 2004.

30. Reitz, *Menopause*, 1.

31. Ibid., 99.

32. Ibid., 104.

33. Reitz interview with the author, 26 June 2004.

34. Barbara Seaman and Gideon Seaman, M.D., *Women and the Crisis in Sex Hormones* (New York: Rawson Associate Publishers, 1977).

35. The following biographical information comes from two interviews conducted by the author with Barbara Seaman in New York on 30 January 1994 and on 27 June 2004.

36. Barbara Seaman, *The Doctors' Case against the Pill* (New York: Peter H. Wyden, Inc., 1969).

37. Barbara Seaman, *Free and Female* (New York: Coward, McCann and Geoghegan, 1972).

38. Ibid., 87–92.

39. Seaman and Seaman, *Women and the Crisis in Sex Hormones*, 511.

40. Sales figures are from *Playbill* (March 1978), n.p.; "Best Sellers," Erie (PA) *Times-News* (1 January 1978), n.p., both in Barbara Seaman Papers, Schlesinger Library, Cambridge, MA, Acc.#82-M33-84-M82, carton 2, folder 91. The best-read nonfiction book in Erie County was *The Beer Can and The Beer Can Collector's Bible;* the best-read fiction book was *Coma*, by Robin Cook.

41. Barbara Seaman and Gideon Seaman, M.D., *Women and the Crisis in Sex Hormones* (New York: Bantam Books, 1982).

42. Barbara Seaman Papers, Schlesinger Library, Cambridge, MA, Acc.#82-M33-84-M82, carton 3, folder 111.

43. *New York Times* (8 May 1976), 13. Louis Parrish, M.D., *No Pause at All* (Binghamton, NY: Reader's Digest Press, 1976).

44. Parrish, *No Pause at All*, 207.

45. Ibid., 11–13.

46. Ibid., 47.

47. Ibid., 39–40.

48. Ibid., 91–92.

49. Ibid., 78.

50. Ibid., 57.

51. Ibid., 207.

52. Ibid., 3.

53. *New York Times* (15 January 1978), BR6. Louisa Rose, ed., *The Menopause Book* (New York: Hawthorn Books, 1977).

54. Rose, ed., *The Menopause Book*, xv–xvi.

55. Ibid., 3–4. For more on this incident, see Christopher Lydon, "Role of Women Sparks Debate by Congresswoman and Doctor," *New York Times* (26 July 1970), 35.

56. Rose, ed., *The Menopause Book*, 44–45.

57. Ibid., 173.

58. Ibid., 65.

59. *New York Times* (27 February 1976), 14. Gloria Heidi, *Winning the Age Game* (New York: Doubleday, 1976).

60. "New Discovery: Public Relations Cures Cancer," *Majority Report* 6 (5–18 February 1977), 1, 10.

61. James T. Patterson, *The Dread Disease: Cancer and Modern American Culture* (Cambridge, MA: Harvard University Press, 1987), 212.

62. Morton Mintz and Victor Cohn, "Hawking the Estrogen Fix," *The Progressive* 41 (September 1977), 24–25.

63. Matt Clark and Mariana Gosnell, "Managing the Menopause," *Newsweek* 97 (9 February 1981), 92; "Menopause the Natural Way," *Good Housekeeping* 193 (October 1981), 275; Jane Porcino, "Menopause Discussion Groups," *Hot Flash* 6 (Fall 1987), 3; Louise Corbett, "Getting Our Bodies Back: Menopausal Self-Help Groups," *Woman Wise* 4 (1981), 2–4.

64. "Menopause Speak-Out," *Prime Time* 2 (April 1974), 7–10.

65. Rosetta Reitz, "Love My Menopause? You Must Be Crazy!" *Prime Time* 4 (April 1976), 17–18.

66. Pauline B. Bart and Marilyn Grossman, "Menopause," *Women and Health* 1 (May–June 1976), 3–11.

67. Anita Johnson, "The Risks of Sex Hormones As Drugs," *Women and Health* 2 (July–August 1977), 8–11.

68. "Editorial," *Women: A Journal of Liberation* 4, no. 4 (1976), 3. Records of the National Women's Health Movement, Sophia Smith Collection, Smith College, Accession #85S-62, box 5.

69. "National Plan of Action, adopted at National Women's Conference, 18–21 November 1977, Houston," Records of the National Women's Health Movement, Sophia Smith Collection, Smith College, Accession #87S-54, box 1.

70. Emily More, "Woman and Health, United States, 1980," *Public Health Reports* 95 (September–October 1980, supplement), 81.

71. Robert N. Butler and Myrna I. Lewis, *Aging and Mental Health: Positive Psychosocial Approaches* (St. Louis: C. V. Mosby Company, 1977), 101; also quoted in More, "Woman and Health, United States," 81.

72. Frances Cerra, "Women and Health: 'False Stereotypes' Challenged for Aged," *New York Times* (12 April 1981), LI1. The first White House Conference on Aging was held in 1961.

73. Tom Ferrell and Margot Slade, "The Rage of Older Women," *New York Times* (19 October 1980), E8.

74. Cerra, "Women and Health," LI10.

75. Dianne L. Kennedy et al., "Noncontraceptive Estrogens and Progestins: Use Patterns over Time," *Obstetrics and Gynecology* 65 (March 1985), 443.

76. Beverly H. Pasley et al., "Prescribing Estrogen during Menopause: Physician Survey of Practices in 1974 and 1981," *Public Health Reports* 99 (July–August 1984), 424–29.

77. Cerra, "Women and Health," LI10.

78. The resulting book, *Lovely Me*, was published by HarperCollins in 1987 and was the basis for a 1998 television movie about the life of Jacqueline Susann.

## Seven • Enter the FDA

1. Paul Starr, *The Social Transformation of American Medicine* (New York: Basic Books, 1982), 379–80.

2. Dennis A. Gilbert, *Compendium of American Public Opinion* (New York: Facts on File, 1988), 25–26.

3. Senator Ted Kennedy, in U.S. Senate, 94th Congress, 2nd Session, *Oral Contraceptives and Estrogens for Postmenopausal Use*, Joint Hearing before the Subcommittee on Health of the Committee on Labor and Public Welfare and the Subcommittee on Administrative Practice and Procedure of the Committee on the Judiciary (21 January 1976), 1.

4. See Elizabeth Siegel Watkins, *On the Pill: A Social History of Oral Contraceptives, 1950–1970* (Baltimore: Johns Hopkins University Press, 1998), chap. 5.

5. U.S. Senate, *Oral Contraceptives and Estrogens for Postmenopausal Use*, 46.

6. Ibid., 48.

7. Ibid., 67–68.

8. Ibid., 5.

9. Ibid., 24.

10. Ibid., 25–26.

11. Ibid., 69.

12. Barbara Resnick Troetel, "Three-Part Disharmony: The Transformation of the Food and Drug Administration in the 1970s" (Ph.D. dissertation, City University of New York, 1996), 2–11.

13. Ibid., 11–14.

14. Ibid., 22.

15. Barbara Seaman, interview by the author, New York, NY, 30 January 1994.

16. Meeting minutes, Obstetrics and Gynecology Advisory Committee, FDA (15–16 December 1975), 17, Records—FDA.

17. Ibid., 15.

18. The only other drug to carry a patient warning label was isoproterenol, an inhalant used by asthmatics, but relatively few people were affected by this 1968 FDA mandate, which thus attracted little attention.

19. For a discussion of the adverse health effects of oral contraceptives and the development of the first patient package insert, see Watkins, *On the Pill*, chaps. 4 and 5.

20. *Federal Register* 35 (10 April 1970), 5962.

21. These letters are on file at the U.S. Food and Drug Administration Dockets Management Branch in Rockville, MD. I am indebted to Suzanne White Junod of the FDA Historians' Office for facilitating my access to these records.

22. *Federal Register* 35 (11 June 1970), 9003.

23. Belita Cowan, executive director of the National Women's Health Network, "Presentation to the Commissioner, FDA" (11 April 1982), Records of the National Women's Health Network, Sophia Smith Collection, Smith College, Acc.#94S-46, box 9.

24. "Estrogens and Endometrial Cancer," *FDA Drug Bulletin* 6 (February–March 1976), 19. "Dear Doctor" letter from John B. Jewell, executive vice president and director of medical affairs, Ayerst Laboratories (December 1975), in U.S. Senate, *Oral Contraceptives and Estrogens for Postmenopausal Use,* 103.

25. *Federal Register* 41 (29 September 1976), 43108–109.

26. These letters are on file at the U.S. Food and Drug Administration Dockets Management Branch in Rockville, MD.

27. Letter 0039, Docket #76N-0381, Records—FDA.

28. Letter 0341, Docket #76N-0381, Records—FDA.

29. Letter 0257, Docket #76N-0381, Records—FDA.

30. Letter 0315, Docket #76N-0381, Records—FDA.

31. Letter 0351, Docket #76N-0381, Records—FDA.

32. Belita Cowan, Public Hearing—Patient Package Inserts, Presentation before the U.S. Food and Drug Administration (30 September 1981). Records of the National Women's Health Network, Sophia Smith Collection, Smith College, Acc#97S-5, box 2.

33. Belita Cowan to Dr. D. T. (16 October 1979). Records of the National Women's Health Network, Sophia Smith Collection, Smith College, Acc#97S-5, box 2.

34. "Move Slowly on Patient Package Insert," *State Medical Society of Wisconsin Medigram* (21 September 1979), 2. Records of the National Women's Health Network, Sophia Smith Collection, Smith College, Acc#97S-5, box 2.

35. Letter 0076, Docket #76N-0381, Records—FDA.

36. Letter 0248, Docket #76N-0381, Records—FDA.

37. Letter 0109, Docket #76N-0381, Records—FDA.

38. Letter 0178, Docket #76N-0381, Records—FDA.

39. Letter 0317, Docket #76N-0381, Records—FDA.

40. Letter 0047, Docket #76N-0381, Records—FDA.

41. Letter 0182, Docket #76N-0381, Records—FDA.

42. *Federal Register* 42 (1977), 37642; *Federal Register* 43 (1978), 4223.

43. *Pharmaceutical Manufacturers Association v. Food and Drug Administration,* Civil No. 77-291, United States District Court (D. Delaware), 484 F. Supp. 1179 (1980).

44. Ibid.

45. *Pharmaceutical Manufacturers Association v. Food and Drug Administration,* Civil No. 77-291, United States Court of Appeals (Third Circuit), 634 F.2d 106 (1980).

46. Troetel, "Three-Part Disharmony," 298.

47. *Federal Register* 40 (7 November 1975), 52075.

48. Troetel, "Three-Part Disharmony," 309.

49. *Federal Register* 45 (12 September 1980), 60754. See also Troetel, "Three-Part Disharmony," 326.

50. Executive Order No. 12291, *Federal Register* 46 (19 February 1981), 13193; *Federal Register* 46 (28 April 1981), 23739; *Federal Register* 47 (17 February 1982), 7200–7201; *Federal Register* 47 (7 September 1982), 39147.

51. Louis A. Morris, Ann Myers, Paul Gibbs, and Chang Lao, "Estrogen PPIs: An FDA Survey," *American Pharmacy* n.s. 20 (June 1980), 22–26.

52. Ibid., 26.

53. Dianne L. Kennedy et al., "Noncontraceptive Estrogens and Progestins: Use Patterns Over Time," *Obstetrics and Gynecology* 65 (March 1985), 443.

54. Fertility and Maternal Health Drugs Advisory Committee, Food and Drug Administration, Transcript, vol. 2 (27 April 1984), 14. Records of the U.S. Food and Drug Administration, Rockville, MD.

55. Victor Cohn, "Patients Pleased but Unpersuaded by More Drug Data," *Washington Post* (27 August 1981).

56. Fertility and Maternal Health Drugs Advisory Committee, Food and Drug Administration, Transcript, vol. 2 (27 April 1984), 178. Records—FDA.

57. For more on the history of consumer information about prescription drugs, see Stuart L. Nightingale, "Written Patient Information on Prescription Drugs," *International Journal of Technology Assessment in Health Care* 11 (1995), 399–409.

58. Troetel, "Three-Part Disharmony," 4.

## Eight • *Resurrecting Estrogen, I*

1. Dianne L. Kennedy, Carlene Baum, and Mary B. Forbes, "Noncontraceptive Estrogens and Progestins: Use Patterns over Time," *Obstetrics and Gynecology* 65 (March 1985), 443; "1975: Top 200 Drugs," *Pharmacy Times* 42 (April 1976), 40.

2. "Top 200 Prescription Drugs of 1980," *American Druggist* 183 (February 1981), 49.

3. Elina Hemminki, Dianne L. Kennedy, Carlene Baum, and Sonja McKinlay, "Prescribing of Noncontraceptive Estrogens and Progestins in the United States, 1974–86," *American Journal of Public Health* 78 (November 1988), 1480; Diane K. Wysowski, Linda Golden, and Laurie Burke, "Use of Menopausal Estrogens and Medroxyprogesterone in the United States, 1982–1992," *Obstetrics and Gynecology* 85 (January 1995), 8.

4. See annual "Top 200 Prescription Drugs" in the February issue of *American Druggist* 1988–2000. The lists from 1995 to the present are available on line at www.rxlist.com.

5. *Federal Register* 37 (25 July 1972), 14826–28. Premarin, for example, was initially approved by the FDA for the treatment of various menopausal symptoms in 1942 because the manufacturer, Ayerst Laboratories, had demonstrated the product's safety, which was the only requirement at the time. After the 1962 amendments to the Food, Drug, and Cosmetic Act of 1938, the FDA required new drug applications to demonstrate the product's efficacy, in addition to its safety. The FDA also undertook a review of all drugs introduced between 1938 and 1962 to assess their efficacy. The National Academy of Sciences (NAS) and the National Research Council (NRC) assisted in this review, known as the Drug Efficiency Study Implementation. In 1972 the NAS-NRC report confirmed Premarin (and several other "estrogen-containing drugs for oral or parenteral use") as "effective" in treating estrogen deficiency associated with menopause and expanded its indications to include "probably effective" in treating selected cases of osteoporosis.

6. "Osteoporosis," NIH Consensus Development Conference Consensus Statement 5 (2–4 April 1984), 1–6.

7. Fertility and Maternal Health Drugs Advisory Committee, Food and Drug Administration, Transcript, vol. 2 (27 April 1984), 168–79. Records—FDA.

8. Maryann Napoli, "Disease of the Week," *New Republic* 195 (1 December 1986), 17–18.

9. A. S. Parkes, "The Rise of Reproductive Endocrinology," *Journal of Endocrinology* 34 (March 1966), xxvi. See also Nicola Perone, "The History of Steroidal Contraceptive Development: The Progestins," *Perspectives in Biology and Medicine* 36 (Spring 1993), 347–48.

10. Perone, "The History of Steroidal Contraceptive Development," 348.

11. Bernard Asbell, *The Pill: A Biography of the Drug That Changed the World* (New York: Random House, 1995), 85–103.

12. Perone, "The History of Steroidal Contraceptive Development," 351–52.

13. Elizabeth Siegel Watkins, *On the Pill: A Social History of Oral Contraceptives, 1950–1970* (Baltimore: Johns Hopkins University Press, 1998), 24.

14. This "off-label" prescribing practice was not unusual. For example, three years before Enovid received FDA approval as the first birth control pill, it was approved for the treatment of gynecological disorders. From 1957 to 1960, physicians who understood that Enovid prevented ovulation sometimes prescribed it as a contraceptive for their patients, even though the FDA had not yet given formal authorization to use the drug for that purpose. Watkins, *On the Pill*, 32.

15. J. W. W. Studd and Margaret Thom, "Estrogen Use and Endometrial Cancer," *New England Journal of Medicine* 300 (19 April 1979), 922.

16. R. Don Gambrell Jr., Tristan A. Castaneda, and Corrine A. Ricci, "Use of the Progestogen Challenge Test to Reduce the Risk of Endometrial Cancer," *Obstetrics and Gynecology* 55 (June 1980), 732–38; R. Don Gambrell Jr., "Estrogens, Progestogens and Endometrial Cancer," *Journal of Reproductive Medicine* 18 (June 1977), 301–6; R. Don Gambrell Jr. et al., "Reduced Incidence of Endometrial Cancer among Postmenopausal Women Treated with Progestogens," *Journal of the American Geriatrics Society* 27 (September 1979), 389–94; R. Don Gambrell Jr., "The Prevention of Endometrial Cancer in Postmenopausal Women with Progestogens," *Maturitas* 1 (1978), 107–12.

17. M. I. Whitehead et al., "Endometrial Histology and Biochemistry in Climacteric Women during Oestrogen and Oestrogen/progestogen Therapy," *Journal of the Royal Society of Medicine* 72 (May 1979), 322–27; M. I. Whitehead et al., "Effects of Estrogens and Progestins on the Biochemistry and Morphology of the Postmenopausal Endometrium," *New England Journal of Medicine* 305 (31 December 1981), 1599–1605.

18. Lombardo F. Palma, "Postmenopausal Osteoporosis and Estrogen Therapy: Who Should Be Treated?" *Journal of Family Practice* 14 (February 1982), 355.

19. Martin B. Wingate, "Postmenopausal Osteoporosis: Concerns and Costs in Clinical Management," *Journal of Medicine* 15 (1984), 324.

20. "Osteoporosis: Looking at the Whole Picture," *Medical World News* (14 January 1985), 38.

21. Lila E. Nachtigall, Richard H. Nachtigall, Robert D. Nachtigall, and E. Mark Beckman, "Estrogen Replacement Therapy I: A 10-Year Prospective Study in the Relationship to Osteoporosis," *Obstetrics and Gynecology* 53 (March 1979), 277–81.

22. Information about the background, design, and funding of this study comes from an interview conducted by the author with Lila Nachtigall in New York on 5 January 2005.

23. The technique of single photon absorptiometry was first described by J. R. Cameron and J. Sorenson ("Measurement of Bone Mineral *In Vivo*. An Improved Method," *Science* 142 [1963], 230–36). Dual photon absorptiometry was first described in 1966 by G. W. Reed ("The Assessment of Bone Mineralization from the Relative Transmission of Am-241 and Cs-137 Radiations," *Physics in Medicine and Biology* 11 [1966], 174).

24. Nachtigall et al., "Estrogen Replacement Therapy I," 280.

25. Lila E. Nachtigall, Richard H. Nachtigall, Robert D. Nachtigall, and E. Mark Beckman, "Estrogen Replacement Therapy II: A Prospective Study in the Relationship to Carcinoma and Cardiovascular and Metabolic Problems," *Obstetrics and Gynecology* 54 (July 1979), 74–79.

26. Ibid., 74.

27. R. Lindsay, D. M. Hart, Colin Forest, and Clive Baird, "Prevention of Spinal Osteoporosis in Oophorectomised Women," *The Lancet* (29 November 1980), 1151–54.

28. Tom A. Hutchinson, Stanley M. Polansky, and Alvan R. Feinstein, "Post-menopausal Oestrogens Protect against Fractures of Hip and Distal Radius," *The Lancet* (6 October 1979), 705–9; Noel S. Weiss et al., "Decreased Risk of Fractures of the Hip and Lower Forearm with Postmenopausal Use of Estrogen," *New England Journal of Medicine* 303 (20 November 1980), 1195–98; Richard E. Johnson and Elmer E. Specht, "The Risk of Hip Fracture in Postmenopausal Females with and without Estrogen Drug Exposure," *American Journal of Public Health* 71 (February 1981), 138–44; Annlia Paganini-Hill et al., "Menopausal Estrogen Therapy and Hip Fractures," *Annals of Internal Medicine* 95 (July 1981), 28–31; Nancy Krieger et al., "An Epidemiologic Study of Hip Fracture in Postmenopausal Women," *American Journal of Epidemiology* 116 (1982), 141–48; Bruce Ettinger, Harry K. Genant, and Christopher E. Cann, "Long-Term Estrogen Replacement Therapy Prevents Bone Loss and Fractures," *Annals of Internal Medicine* 102 (March 1985), 319–24.

29. Fred W. Lafferty and Dennis O. Helmuth, "Post-menopausal Estrogen Replacement: The Prevention of Osteoporosis and Systemic Effects," *Maturitas* 7 (1985), 147–59.

30. Lombardo F. Palma, "Postmenopausal Osteoporosis and Estrogen Therapy: Who Should Be Treated?" *Journal of Family Practice* 14 (February 1982), 355–59.

31. Paul D. Saville, "Post-menopausal Osteoporosis and Estrogens: Who Should Be Treated and Why," *Postgraduate Medicine* 75 (1 February 1984), 135–43.

32. Council on Scientific Affairs, "Estrogen Replacement in the Menopause," *Journal of the American Medical Association* 249 (21 January 1983), 359–61.

33. NIH Consensus Development Conference Statement Online, "Osteoporosis" 5, no. 3 (2–4 April 1984), 1–6.

34. See, for example, C. M. Wylie, "Hospitalization for Fractures and Bone Loss in Adults. Why Do We Regard These Phenomena as Dull?" *Public Health Reports* 92 (January–February 1977), 33–38; L. Solomon, "Bone Density in Ageing Caucasian and African Populations," *The Lancet* (22–29 December 1979), 1326–30; J. Mangaroo et al., "Prevalence of Bone Demineralization in the United States," *Bone* 6 (1985), 135–39.

35. "Estrogen Use and Postmenopausal Women," National Institutes of Health Consensus Development Conference Summary, vol. 2, no. 8 (1979), n.p.

36. NIH Consensus Development Conference Statement Online, "Osteoporosis" 5, no. 3 (2–4 April 1984), 1–6.

37. See, for example, A. Horsman et al., "Prospective Trial of Oestrogen and Calcium in Postmenopausal Women," *British Medical Journal* (24 September 1977), 789–92; R. R. Recker, P. D. Saville, and R. P. Heaney, "Effects of Estrogens and Calcium Carbonate on Bone Loss in Postmenopausal Women," *Annals of Internal Medicine* 87 (December 1977), 649–55; J. F. Aloia et al., "Prevention of Involutional Bone Loss by Exercise," *Annals of Internal Medicine* 89 (September 1978), 356–58; M. K. White et al., "The Effects of Exercise on the Bones of Postmenopausal Women," *International Orthopaedics* 7 (1984), 209–14.

38. "Calcium for Postmenopausal Osteoporosis," *Medical Letter on Drugs and Therapeutics* 24 (26 November 1982), 105, A41–44.

39. *Obstetrics and Gynecology* 52 (August 1978), A41–44.

40. *Journal of the American Medical Association* 245 (16 January 1981), 272–74.

41. Ibid., 251 (22–29 June 1984), 372–74.

42. Ibid., 252 (27 July 1984), 477–79.

43. NIH Consensus Development Conference Statement Online, "Osteoporosis" 5, no. 3 (2–4 April 1984), 1–6.

44. "Menopausal and Postmenopausal Women's Survey," *Saturday Evening Post* 258 (December 1986).

45. NIH Consensus Development Conference Statement Online, "Osteoporosis" 5, no. 3 (2–4 April 1984), 1–6.

46. Fertility and Maternal Health Drugs Advisory Committee, Food and Drug Administration, Transcript, vol. 2 (16 December 1983), 102–7. Records—FDA.

47. Endocrinologic and Metabolic Drugs Advisory Committee, Food and Drug Administration, Transcript (18 February 1977), see esp. pages 114–51. Records—FDA.

48. Fertility and Maternal Health Drugs Advisory Committee, Food and Drug Administration, Transcript, vol. 2 (27 April 1984), 168–78. Records—FDA.

49. Ibid., 221.

50. Ibid., 130.

51. Ibid., 95.

52. Ibid.

53. Fertility and Maternal Health Drugs Advisory Committee, Food and Drug Administration, Transcript, vol. 1 (26 April 1984), 189. Records—FDA.

54. Fertility and Maternal Health Drugs Advisory Committee, Food and Drug Administration, Transcript, vol. 2 (27 April 1984), 6.

55. Endocrinologic and Metabolic Drugs Advisory Committee, Food and Drug Administration, Transcript, 24 June 1985, 33. Records—FDA.

56. Ibid., 38.

57. Ibid., 44.

58. *Federal Register* 51 (11 April 1986), 12568–69.

59. Jack S. Gruber and Charlene T. Luciani, "Physicians Changing Postmenopausal Sex Hormone Prescribing Regimens," *Progress in Clinical and Biological Research* 216 (1986), 325–35.

60. Elizabeth Barrett-Connor, "Postmenopausal Estrogens—Current Prescribing Patterns of San Diego Gynecologists," *Western Journal of Medicine* 144 (May 1986), 620–21.

61. Ibid.

## Nine • Resurrecting Estrogen, II

1. Sheila Raviv, quoted in Tacie Dejanikus, "Major Drug Manufacturer Funds Osteo-porosis Public Education Campaign," *Network News* (May–June 1985), 1. Source material for this paragraph comes from this article, continued on pages 3 and 8 of this issue of *Network News*.

2. Testimony of Gloria Kinney, vice president for nutrition education, National Dairy Council, in U.S. Senate, 99th Congress, 1st Session, *Reviewing the Treatment and Diagnosis of Osteoporosis,* Hearing before the Subcommittee on Aging of the Committee on Labor and Human Resources (20 June 1985), 145–62.

3. Statement submitted by Joseph J. Westwater, Chief Executive Officer, National Dairy Promotion and Research Board, in ibid., 170–75.

4. Ibid., 137–41.

5. Ibid., 141.

6. H.J.Res. 610, 98th Congress.

7. H.J.Res. 46 and S.J.Res. 61, 99th Congress; Public Law No. 99-42.

8. U.S. Senate, *Reviewing the Treatment and Diagnosis of Osteoporosis,* 1–2.

9. Ibid., 4.

10. Ibid., 124.

11. Ibid., 183.

12. Ibid., 93–94.

13. Ibid., 167.

14. Dejanikus, "Major Drug Manufacturer Funds Osteoporosis Public Education Cam-paign," 1, 3, 8.

15. Ibid., 1.

16. See, for example, Maryann Napoli, "Disease of the Week," *New Republic* 195 (1 December 1986), 17–18, and the testimony of the National Women's Health Network on Osteoporosis in U.S. Senate, *Reviewing the Treatment and Diagnosis of Osteoporosis,* 166–69.

17. Testimony of Richard B. Mazess, in U.S. Senate, *Reviewing the Treatment and Diag-nosis of Osteoporosis,* 114.

18. Ibid.

19. Napoli, "Disease of the Week," 17.

20. Mariamne H. Whatley and Nancy Worcester, "The Role of Technology in the Co-optation of the Women's Health Movement: The Cases of Osteoporosis and Breast Cancer Screening," in Kathryn Strother Ratcliff et al., eds., *Healing Technology: Feminist Perspectives* (Ann Arbor: University of Michigan Press, 1989), 201.

21. Robert B. Mazess, in U.S. Senate, *Reviewing the Treatment and Diagnosis of Os-teoporosis,* 246.

22. Whatley and Worcester, "The Role of Technology in the Co-optation of the Wom-en's Health Movement," 207–8.

23. Stanley J. Reiser, "The Emergence of the Concept of Screening for Disease," *Health and Society* 56 (1978), 403–4.

24. Ambiguities surrounding the Pap smear are discussed in Monica J. Casper and

Adele E. Clarke, "Making the Pap Smear into the 'Right Tool' for the Job: Cervical Cancer Screening in the USA, circa 1940–95," *Social Studies of Science* 28 (April 1998), 255–90.

25. Michael S. Goldstein, *The Health Movement: Promoting Fitness in America* (New York: Twayne, 1992), 4, 10.

26. Dava Sobel, "A Decade of Planets and DNA and Bottom Quarks," *New York Times* (1 January 1980), 14.

27. Robert Crawford, "Healthism and the Medicalization of Everyday Life," *International Journal of Health Services* 10 (1980), 370. For an interesting discussion of the rise of the relationship between preventive medicine and the individualization of health care responsibility in the postwar years, see Dorothy Porter, *Health, Civilization and the State* (London: Routledge, 1999), chap. 13.

28. Floris W. Wood, ed., *An American Profile—Opinions and Behavior, 1972–1989* (Detroit: Gale Research, 1990), 676.

29. Dennis A. Gilbert, *Compendium of American Public Opinion* (New York: Facts on File, 1988), 16–17.

30. Renee C. Fox, "The Medicalization and Demedicalization of American Society," *Daedalus* 106 (1977), 21.

31. For example, the *New York Times* launched a special weekly section on science and medicine in 1979.

32. Napoli, "Disease of the Week," 17.

33. William A. Nolen, M.D., "Estrogen Therapy at Menopause: Weighing the Risks," *McCall's* 108 (May 1981), 59.

34. Dr. Saul B. Gusberg, quoted in Matt Clark and Mariana Gosnell, "Managing the Menopause," *Newsweek* 97 (9 February 1981), 93.

35. Clark and Gosnell, "Managing the Menopause," 93. See also "Menopause the Natural Way," *Good Housekeeping* 193 (October 1981), 275, and Louise Corbett, "Getting Our Bodies Back: Menopausal Self-Help Groups," *Woman Wise* 4 (1981), 2–4.

36. Margaret Markham, "Breezing through Menopause," *Harper's Bazaar* 115 (September 1982), 260.

37. Ibid., 206, 260–62; Penny Wise Budoff, "Hot Flash: Good News about Estrogen," *Harper's Bazaar* 116 (September 1983), 70, 80; "Good News on Estrogen for Older Women," *Newsweek* 101 (28 February 1983), 74.

38. "Osteoporosis, the 'silent' disease, strikes one in four women over 65, an expert warns," *People* (1 April 1985), 23.

39. "The calcium controversy: an expert warns that supplements are not the cure-all for dowager's hump," *People* (13 April 1987), 27.

40. Riggs, in ibid., 70.

41. Ibid.

42. Jonathan Probber, "An Explosion of Calcium-Fortified Foods," *New York Times* (3 December 1986), C7.

43. "Calcium 'Fad' Boon to Jersey Concerns," *New York Times* (16 February 1986), NJ23.

44. "Theory on Calcium and Bone Loss Is Disputed," *New York Times* (22 January 1987), B18.

45. Jane E. Brody, "Dozens of Factors Critical in Bone Loss among Elderly," *New York Times* (17 February 1987), C1.

46. Jane E. Brody, "Personal Health," *New York Times* (25 February 1987), C10.

47. Robin Marantz Henig, "Estrogen, in and out of Favor," *Washington Post* (8 May 1985), 12.

48. Lila Nachtigall and Joan Rattner Heilman, "Estrogen: The Facts Can Change Your Life," *Redbook* 167 (September 1986), 22.

49. Mary Powers, "Estrogen Therapy Is Revived," *[Memphis]Commercial Appeal* (13 November 1988).

50. Sylvie Reice, "The Anti-Aging Lifestyle," *Ladies' Home Journal* 101 (September 1984), 115.

51. Porter, *Health, Civilization and the State,* 297.

52. See Susan Faludi, *Backlash: The Undeclared War against American Women* (New York: Crown, 1991).

53. *MIDLIFE—NO CRISIS* (Summit, NJ: Ciba, 1987), 7.

54. *For the Woman Approaching Menopause* (Evansville, IN: Mead Johnson Laboratories, 1985), 54.

55. Other examples (and this list is by no means exhaustive) include: Penny Wise Budoff, M.D., *No More Hot Flashes and Other Good News* (New York: Putnam, 1983); Winnifred Berg Cutler, *Menopause: A Guide for Women and the Men Who Love Them* (New York: Norton, 1983); Sadja Greenwood, *Menopause, Naturally: Preparing for the Second Half of Life* (Volcano, CA: Volcano Press, 1984); Wulf H. Utian, M.D., Ph.D., and Ruth S. Jacobowitz, *Managing Your Menopause* (New York: Prentice Hall, 1990).

56. Morris Notelovitz and Marsha Ware, *Stand Tall! Every Woman's Guide to Preventing Osteoporosis* (Gainesville, FL: Triad Publishing, 1982).

57. Ibid., 4.

58. Ibid., 1.

59. Ibid., 15.

60. Ibid., 40.

61. Ibid., 15.

62. Ibid., 112.

63. Ibid., 113.

64. Ibid., 120.

65. Ibid., 121.

66. Cader Books, "Bestseller Lists, 1900–1995," www.caderbooks.com/best80.html. Sales of *Jane Fonda's Workout Book* taken from dust jacket of Jane Fonda with Mignon McCarthy, *Women Coming of Age* (New York: Simon and Schuster, 1984).

67. Fonda, *Women Coming of Age,* dust jacket.

68. Ibid., 24.

69. Ibid., 19, 22.

70. Ibid., 46–49.

71. Ibid., 51.

72. Ibid., 175.

73. Cader Books, "Bestseller Lists, 1900–1995," www.caderbooks.com/best80.html.

74. Lila Nachtigall and Joan Ratner Heilman, *Estrogen: The Facts Can Change Your Life* (New York: Harper and Row, 1986), 19.

75. Ibid., 114.

76. Ibid., 84.

77. Ibid., 11–12.

78. Ibid., 138.

79. Ibid., 194–95.

80. Ibid., 199.

81. Shere Hite, *The Hite Report: A Nationwide Study of Female Sexuality* (New York: Macmillan, 1976). Cader Books, "Bestseller Lists, 1900–1965," www.caderbooks.com/best70.html.

82. Lila Nachtigall with Joan Heilman, *The Lila Nachtigall Report* (New York: Putnam's, 1977), 232.

83. Ibid., 87.

84. Ibid., 67.

85. Ibid., 191.

86. Lila Nachtigall, interview with the author, New York, 5 January 2005.

87. The following biographical information comes from an interview conducted by the author with Lila Nachtigall in New York, 5 January 2005.

88. Opening statement of Washington Women's Liberation at Women's Hearings on the Birth Control Pill, 7 March 1970, in U.S. Congress, Senate Subcommittee on Monopoly, *Competitive Problems in the Drug Industry* (Washington, DC: Government Printing Office, 1967), 7283.

## Ten • Skeptics and Believers

1. *WIN (Women's International Network) News* (22 March 1979), 19, in Records of the Boston Women's Health Book Collective at the Schlesinger Library, Cambridge, Mass. (hereafter, Records—BWHBC), Accession #99-M125, carton 1, folder "Minutes: (July–Dec) 1980 [2]."

2. Paula Brown Doress and Diana Laskin Siegal, "Project Coordinators' Preface," *Ourselves, Growing Older* (New York: Simon and Schuster, 1987), xv.

3. Tish Sommers, "Foreword," *Ourselves, Growing Older*, xiv.

4. Ibid., xiii.

5. *Ourselves, Growing Older*, 117.

6. Ibid., 123.

7. Ibid., 80.

8. Ibid., 88.

9. Ibid., 260, 269–71.

10. In addition to the scholarship discussed in the text, see also: Frances B. McCrea, "The Politics of Menopause: The "Discovery" of a Deficiency Disease," *Social Problems* 31 (October 1983): 111–23; Margaret Lock, "Models and Practice in Medicine: Menopause as Syndrome or Life Transition," in Robert A. Hahn and Atwood D. Gaines, eds., *Physicians of Western Medicine: Anthropological Approaches to Theory and Practice* (Dordrecht: D. Reidel,

1985), 115–39; Kathleen I. MacPherson, "Menopause as Disease: The Social Construction of a Metaphor," *Advances in Nursing Science* 3 (January 1981), 95–113; Kathleen I. MacPherson, "Osteoporosis and Menopause: A Feminist Analysis of the Social Construction of a Syndrome," *Advances in Nursing Science* 7 (July 1985), 11–22.

11. Judith Posner, "It's All in Your Head: Feminist and Medical Models of Menopause (Strange Bedfellows)," *Sex Roles* 5 (1979), 179–90.

12. Patricia A. Kaufert, "Myth and the Menopause," *Sociology of Health and Illness* 4 (July 1982), 141–66.

13. Ibid., 159.

14. "Our 9th Anniversary," *Hot Flash* 9 (Winter 1990), 2.

15. "The History of AFI," www.afriendindeed.ca/about_afi.htm.

16. Delores Hemphill and Yvonne Kimber, *A Positive Look at Menopause: A Teacher Training Manual* (Columbia, MO: Planned Parenthood of Missouri, 1982). Copy located in Sophia Smith Collection, Smith College, Northampton, MA.

17. Matt Clark and Mariana Gosnell, "Managing the Menopause," *Newsweek* 97 (9 February 1981), 92; "Menopause the Natural Way," *Good Housekeeping* 193 (October 1981), 275; Jane Porcino, "Menopause Discussion Groups," *Hot Flash* 6 (Fall 1987), 3; Louise Corbett, "Getting Our Bodies Back: Menopausal Self-Help Groups," *Woman Wise* 4 (1981), 2–4.

18. Louise Corbett, "Getting Our Bodies Back," 4.

19. "Treatment Resources: Medical Clinics," *McCall's* 115 (November 1987), 94. For more on the establishment and practices of menopause clinics, especially in countries outside the United States, see Marilys Guillemin, "Working Practices of the Menopause Clinic," *Science, Technology, and Human Values* 25 (Autumn 2000), 449–71. The first menopause clinic in the world was opened in 1967 in Cape Town, South Africa, by Wulf Utian. A decade later he moved to Cleveland, where he began a large clinical research center. Utian went on to found several scientific organizations for the study of menopause (the International Menopause Society plus national menopause societies in North America, Europe, and Asia); he has remained in the forefront of menopause research for forty years. Germaine Greer sarcastically named Utian "the Grand Master in Menopause" (*The Change* [New York: Knopf, 1992], 13).

20. Paula Doress-Worters, personal communication (13 April 2005); Lila Nachtigall, interview (5 January 2005).

21. "Does Sex End at Menopause?" *McCall's* 115 (November 1987), 92. The circulation figure comes from "*The Saturday Evening Post* Complete Demographic Profile as defined by Mediamark Research, Inc., Fall 1986," 4, which lists the number of readers for *McCall's* as 21,621,000. Thanks to Don Sutton of the *Saturday Evening Post* for providing me with this document.

22. Circulation figures from ibid.

23. Demographic information from ibid.

24. Howard M. Fillit, M.D., "Might Estrogen Prevent Memory Loss?" *Saturday Evening Post* 258 (December 1986), 50.

25. Ibid., 50.

26. Ibid., 51.

27. Ibid.

28. Ibid., 52, 110.

29. Editor's Note in ibid., 50.

30. Cory SerVaas, M.D., "Is an Estrogen Skin Patch for You?" *Saturday Evening Post* 258 (December 1986), 53.

31. Cory SerVaas, M.D., "More about Estrogen Skin Patches," *Saturday Evening Post* 259 (January–February 1987), 52–55.

32. SerVaas, "Is an Estrogen Skin Patch for You?" 56.

33. I am grateful to Wendy Braun and Georgia Ratliff at the *Saturday Evening Post* for facilitating my access to these surveys in August 2003.

34. #4199.

35. #1448.

36. #3929.

37. #643.

38. #3094.

39. #1440.

40. #3021.

41. #4289.

42. #3043.

43. #3476.

44. #3071.

45. #4517.

46. #4573.

47. #41, #112, #281, #136.

48. #3364.

49. #3075.

50. #4167.

51. #143.

52. #3728.

53. #1618.

54. #2062.

55. #3131.

56. #3680.

57. #1136.

58. #4471.

59. #717.

60. #2620.

61. #2538.

62. Dianne L. Kennedy, Carlene Baum, and Mary B. Forbes, "Noncontraceptive Estrogens and Progestins: Use Patterns over Time," *Obstetrics and Gynecology* 65 (March 1985), 441–46; Elina Hemminki, Dianne L. Kennedy, Carlene Baum, and Sonja M. McKinlay, "Prescribing of Noncontraceptive Estrogens and Progestins in the United States, 1974–86," *American Journal of Public Health* 78 (1988), 1479–81; Diane K. Wysowski, Linda Golden, and Laurie Burke, "Use of Menopausal Estrogens and Medroxyprogesterone in the United States, 1982–1992," *Obstetrics and Gynecology* 85 (January 1995), 6–10.

63. Wysowski, Golden, and Burke, "Use of Menopausal Estrogens and Medroxyprogesterone," 8.

64. Kennedy, Baum, and Forbes, "Noncontraceptive Estrogens and Progestins," 441–46; Hemminki, Kennedy, Baum, and McKinlay, "Prescribing of Noncontraceptive Estrogens and Progestins," 1479–81; Wysowski, Golden, and Burke, "Use of Menopausal Estrogens and Medroxyprogesterone," 6–10.

65. Hemminki, Kennedy, Baum, and McKinlay, "Prescribing of Noncontraceptive Estrogens and Progestins," 1480.

66. Kennedy, Baum, and Forbes, "Noncontraceptive Estrogens and Progestins," 445.

67. Hemminki, Kennedy, Baum, and McKinlay, "Prescribing of Noncontraceptive Estrogens and Progestins," 1480; Wysowski, Golden, and Burke, "Use of Menopausal Estrogens and Medroxyprogesterone," 9.

68. Wysowski, Golden, and Burke, "Use of Menopausal Estrogens and Medroxyprogesterone," 9.

69. Hemminki, Kennedy, Baum, and McKinlay, "Prescribing of Noncontraceptive Estrogens and Progestins," 1480; Wysowski, Golden, and Burke, "Use of Menopausal Estrogens and Medroxyprogesterone," 9.

70. Hemminki, Kennedy, Baum, and McKinlay, "Prescribing of Noncontraceptive Estrogens and Progestins," 1480.

71. Wysowski, Golden, and Burke, "Use of Menopausal Estrogens and Medroxyprogesterone," 9.

72. Hemminki, Kennedy, Baum, and McKinlay, "Prescribing of Noncontraceptive Estrogens and Progestins," 1480.

73. Veronica A. Raviknar, "Compliance with Hormone Therapy," *American Journal of Obstetrics and Gynecology* 156 (May 1987), 1332.

74. Wysowski, Golden, and Burke, "Use of Menopausal Estrogens and Medroxyprogesterone," 9. The one in six figure was derived as follows: "Given that there were 36 million prescriptions for oral and transdermal estrogen products in 1992 and that there were about 36 million women 50 years old or older in the United States in 1991, and considering that most oral prescriptions are for a 2-month supply, an estimated six million women, or roughly one in six women 50 years or older, would be exposed."

75. Renate Klein and Lynette J. Dumble, "Disempowering Women: The Science and Politics of Hormone Replacement Therapy (HRT)," *Women's Studies International Forum* 17 (1994), 331.

76. Frances B. McCrea and Gerald E. Markle, "The Estrogen Replacement Controversy in the USA and UK: Different Answers to the Same Question?" *Social Studies of Science* 14 (1984), 7–9.

### Eleven • *Weighing the Benefits and Risks of HRT*

1. Meeting Minutes, Fertility and Maternal Health Drugs Advisory Committee Workshop on Current Status of Combined Hormone Replacement Therapy, vol. 1 (20 June 1991), 50.

2. Barron H. Lerner, *The Breast Cancer Wars: Fear, Hope, and the Pursuit of a Cure in*

*Twentieth-Century America* (New York: Oxford University Press, 2001), 172–74, 258. See also James T. Patterson, *The Dread Disease: Cancer and Modern American Culture* (Cambridge, MA: Harvard University Press, 1987).

3. Meeting Minutes, Fertility and Maternal Health Drugs Advisory Committee Workshop on Current Status of Combined Hormone Replacement Therapy, vol. 2 (21 June 1991), 249–50.

4. Ibid., 212.

5. William N. Spellacy, "A Perspective on Progestogens in Oral Contraceptives," *American Journal of Obstetrics and Gynecology* 142 (15 March 1982), 717.

6. Valerie Beral, "Cardiovascular-Disease Mortality Trends and Oral-Contraceptive Use in Young Women," *The Lancet* (13 November 1976), 1047–51.

7. T. W. Meade, "Effects of Progestogens on the Cardiovascular System," *American Journal of Obstetrics and Gynecology* 142 (15 March 1982), 776–80; Philip E. Sartwell and Paul D. Stolley, "Oral Contraceptives and Vascular Disease," *Epidemiologic Reviews* 4 (1982), 95–109.

8. Trudy L. Bush, "Letter to the Editor: Postmenopausal Estrogen Use and Heart Disease," *New England Journal of Medicine* 315 (10 July 1986), 134.

9. Interview conducted by the author with Lila Nachtigall in New York, 5 January 2005.

10. Lila E. Nachtigall, Richard H. Nachtigall, Robert D. Nachtigall, and E. Mark Beckman, "Estrogen Replacement Therapy, II: A Prospective Study in the Relationship to Carcinoma and Cardiovascular and Metabolic Problems," *Obstetrics and Gynecology* 54 (July 1979), 74–79, quote on 78.

11. Peter W. F. Wilson, Robert J. Garrison, and William P. Castelli, "Postmenopausal Estrogen Use, Cigarette Smoking, and Cardiovascular Morbidity in Women over 50: The Framingham Study," *New England Journal of Medicine* 313 (24 October 1985), 1038–43.

12. Meir J. Stampfer et al., "A Prospective Study of Postmenopausal Estrogen Therapy and Coronary Heart Disease," *New England Journal of Medicine* 313 (24 October 1985), 1044–49.

13. John C. Bailar III, "When Research Results Are in Conflict," *New England Journal of Medicine* 313 (24 October 1985), 1080–81.

14. Ibid., 1081.

15. Ibid.

16. Warren G. Thompson, "Letter to the Editor: Postmenopausal Estrogen Use and Heart Disease," *New England Journal of Medicine* 315 (10 July 1986), 133–34.

17. Philip R. J. Burch, "Letter to the Editor: Postmenopausal Estrogen Use and Heart Disease," *New England Journal of Medicine* 315 (10 July 1986), 134.

18. Albert M. Van Hemert, "Letter to the Editor: Postmenopausal Estrogen Use and Heart Disease," *New England Journal of Medicine* 315 (10 July 1986), 133.

19. John C. Bailar III, "Letter to the Editor: Postmenopausal Estrogen Use and Heart Disease," *New England Journal of Medicine* 315 (10 July 1986), 136.

20. For a good history of the clinical trial, see Abraham M. Lilienfeld, "*Ceteris Paribus:* The Evolution of the Clinical Trial," *Bulletin of the History of Medicine* 56 (Spring 1982), 1–18. See also J. Rosser Matthews, *Quantification and the Quest for Medical Certainty* (Princeton, NJ: Princeton University Press, 1995), and Harry M. Marks, *The Progress of Experiment:*

*Science and Therapeutic Reform in the United States, 1900–1990* (Cambridge: Cambridge University Press, 1997).

21. Graham A. Colditz et al., "Menopause and the Risk of Coronary Heart Disease in Women," *New England Journal of Medicine* 316 (30 April 1987), 1105–10; Karen A. Matthews et al., "Menopause and Risk Factors for Coronary Heart Disease," *New England Journal of Medicine* 321 (7 September 1989), 641–46; Jay M. Sullivan et al., "Postmenopausal Estrogen Use and Coronary Atherosclerosis," *Annals of Internal Medicine* 108 (March 1988), 358–63; Jay M. Sullivan et al., "Estrogen Replacement and Coronary Artery Disease," *Archives of Internal Medicine* 150 (December 1990), 2557–62.

22. Elizabeth Barrett-Connor and Trudy L. Bush, "Estrogen Replacement and Coronary Heart Disease," *Cardiovascular Clinics* 19 (1989), 170.

23. Bruce Ettinger, "Hormone Replacement Therapy and Coronary Heart Disease," *Obstetrics and Gynecology Clinics of North America* 17 (December 1990), 753–54.

24. See, for example, Meir J. Stampfer and Graham A. Colditz, "Estrogen Replacement Therapy and Coronary Heart Disease: A Quantitative Assessment of the Epidemiologic Evidence," *Preventive Medicine* 20 (January 1991), 47–63, and Bruce M. Psaty et al., "A Review of the Association of Estrogens and Progestins with Cardiovascular Disease in Postmenopausal Women," *Archives of Internal Medicine* 153 (28 June 1993), 1421–27.

25. Antoine Lacassagne, "Endocrine Factors Concerned in the Genesis of Experimental Mammary Carcinoma," *Journal of Endocrinology* 13 (October 1955), ix.

26. Robert A. Wilson, "The Roles of Estrogen and Progesterone in Breast and Genital Cancer," *Journal of the American Medical Association* 182 (27 October 1962), 327–31.

27. Roy Hertz, "An Appraisal of Certain Problems in the Use of Steroid Compounds for Contraception," in Advisory Committee on Obstetrics and Gynecology, Food and Drug Administration, *Report on the Oral Contraceptives* (Washington, DC: Government Printing Office, 1 August 1966), 51.

28. Barbara S. Hulka, "Hormone-Replacement Therapy and the Risk of Breast Cancer," *CA-Cancer Journal* 40 (September–October 1990), 289.

29. Ibid.

30. Ibid., 290.

31. Ibid., 295.

32. William D. Dupont and David L. Page, "Menopausal Estrogen Replacement Therapy and Breast Cancer," *Archives of Internal Medicine* 151 (January 1991), 67–72; Karen K. Steinberg et al., "A Meta-analysis of the Effect of Estrogen Replacement Therapy on the Risk of Breast Cancer," *Journal of the American Medical Association* 265 (17 April 1991), 1985–90.

33. Dupont and Page, "Menopausal Estrogen Replacement Therapy and Breast Cancer," 71.

34. Brian E. Henderson, Annlia Paganini-Hill, and Ronald K. Ross, "Decreased Mortality in Users of Estrogen Replacement Therapy," *Archives of Internal Medicine* 151 (January 1991), 75–78.

35. Thomas E. Moon, "Estrogens and Disease Prevention," *Archives of Internal Medicine* 151 (January 1991), 17.

36. Ibid., 18.

37. Steinberg et al., "A Meta-analysis of the Effect of Estrogen Replacement Therapy on the Risk of Breast Cancer," 1989.

38. Graham A. Colditz et al., "Prospective Study of Estrogen Replacement Therapy and Risk of Breast Cancer in Postmenopausal Women," *Journal of the American Medical Association* 264 (28 November 1990), 2648–53; David W. Kaufman et al., "Estrogen Replacement Therapy and the Risk of Breast Cancer: Results from the Case-Control Surveillance Study," *American Journal of Epidemiology* 134 (15 December 1991), 1375–85; Julie R. Palmer et al., "Breast Cancer Risk after Estrogen Replacement Therapy: Results from the Toronto Breast Cancer Study," *American Journal of Epidemiology* 134 (15 December 1991), 1386–95; Deborah Grady and Virginia Ernster, "Invited Commentary: Does Postmenopausal Hormone Therapy Cause Breast Cancer?" *American Journal of Epidemiology* 134 (15 December 1991), 1396–1400; David W. Kaufman, Julie R. Palmer, Lynn Rosenberg, and Samuel Shapiro, "Authors' Response to 'Invited Commentary: Does Postmenopausal Hormone Therapy Cause Breast Cancer?'" *American Journal of Epidemiology* 134 (15 December 1991), 1401.

39. Janet B. Henrich, "The Postmenopausal Estrogen/Breast Cancer Controversy," *Journal of the American Medical Association* 268 (14 October 1992), 1901.

40. Ibid., 1902.

41. Regine Sitruk-Ware, "Estrogens, Progestins and Breast Cancer Risk in Post-menopausal Women: State of the Ongoing Controversy in 1992," *Maturitas* 15 (October 1992), 132.

42. Ibid., 133.

43. Margaret J. Nachtigall et al., "Incidence of Breast Cancer in a 22-Year Study of Women Receiving Estrogen-Progestin Replacement Therapy," *Obstetrics and Gynecology* 80 (November 1992), 827–30.

44. Sitruk-Ware, "Estrogens, Progestins and Breast Cancer Risk in Post-menopausal Women," 137.

45. Roger Lobo, in Meeting Minutes, Fertility and Maternal Health Drugs Advisory Committee (14 June 1990), 28–29.

46. Ibid., 32–33.

47. Ibid., 38. Although earlier Lobo had noted that women accounted for 58 percent of the $40 to $60 billion cost of cardiovascular disease, he did not repeat this qualification when recapping these figures in his closing remarks to the committee.

48. Cynthia Pearson, in Meeting Minutes, Fertility and Maternal Health Drugs Advisory Committee (14 June 1990), 22–23. The Fertility and Maternal Health Drugs Advisory Committee debated the bioequivalence of conjugated estrogens on 6 January 1989. Two generic formulations came onto the market in the mid-1970s and had captured a significant part of the estrogen market by the mid-1980s. See "Synthetic Generic Conjugated Estrogens: Timeline" at www.fda.gov/cder/news/cetimeline.htm.

49. Lee Goldman and Anna N. A. Tosteson, "Uncertainty about Postmenopausal Estrogen," *New England Journal of Medicine* 325 (12 September 1991), 800–802.

50. Nancy E. Davidson, "Hormone-Replacement Therapy—Breast versus Heart versus Bone," *New England Journal of Medicine* 332 (15 June 1995), 1638–39.

## Twelve • 1992: The Year of the Menopause

1. January 3, 1993, "Year of the Woman," U.S. Senate Historical Minute Essays, www .senate.gov/artandhistory/history/minute/year_of_the_woman.htm.

2. Bills proposed in House and Senate, 96th Congress (1979–1980) through 106th Congress (1999–2000), identified by keywords "women's health." Thomas: Legislative Information on the Internet, Library of Congress, http://thomas.loc.gov.

3. The proposed ORWH budget figure came from Sec. 486H of the "Women's Health Research Act, H. R. 5290, 101st Congress (17 July 1990), http://thomas.loc.gov. The total NIH budget came from "The NIH Almanac—Appropriations," www.nih.gov/about/alma nac/appropriations/part2.htm.

4. For an interesting discussion, see Sheryl Burt Ruzek and Julie Becker, "The Women's Health Movement in the United States: From Grass-Roots Activism to Professional Agendas," *Journal of the American Medical Women's Association* 54 (Winter 1999), 4–8.

5. U.S. Senate, 102nd Congress, 1st Session, *The Role of Menopause and Gender Difference in Aging on the Development of Disease in Mid-life and Older Women*, Hearing before the Subcommittee on Aging of the Committee on Labor and Human Resources (19 April 1991).

6. Ibid, 1.

7. Ibid, 25.

8. Ibid, 3.

9. Ibid, 32–33.

10. Ibid, 52.

11. Interview conducted by the author with Florence Haseltine in San Francisco, 19 November 2005.

12. U.S. Senate, 102nd Congress, 1st Session, *The Role of Menopause and Gender Difference in Aging*, 8.

13. Ibid, 24.

14. Ibid, 11.

15. U.S. House of Representatives, 102nd Congress, 1st Session, *Women at Midlife: Consumers of Second-Rate Health Care?*, Hearing before the Subcommittee on Housing and Consumer Interests of the Select Committee on Aging (30 May 1991).

16. Ibid., 76.

17. Ibid., 141.

18. Bernadine Healy, "Women's Health, Public Welfare," *Journal of the American Medical Association* 266 (24–31 July 1991), 566.

19. Bernadine Healy, "The Yentl Syndrome," *New England Journal of Medicine* 325 (25 July 1991), 274–76.

20. Ibid., 275.

21. Healy, "Women's Health, Public Welfare," 567.

22. Ibid., 566; Fawn Vrazo, "Menopause Becoming 'Au Courant' as It Hits Women of Baby Boom," *Philadelphia Inquirer* (31 May 1991), A1.

23. David Perlman, "Estrogen Therapy Helps Prolong Life, Study Says," *San Francisco Chronicle* (5 January 1991), A2; "Estrogen's Health Benefits Exceed Risk of Cancer, Re-

searchers Find," *New York Times* (5 January 1991), 8; "Estrogen Therapy Prolongs Lives," *St. Louis Post-Dispatch* (16 January 1991), 4E; Sandy Rovner, "Estrogen Use May Lower Death Rates of Older Women," *Washington Post* (15 January 1991), Z5.

24. Judy Foreman, "US Planning $500m Study on Women's Health," *Boston Globe* (20 April 1991), 1; Judy Foreman, "Studies Point to Bias in Care; Women Seen Treated Less Aggressively for Heart Disease," *Boston Globe* (25 July 1991), 1.

25. Gina Kolata, "Estrogen after Menopause Cuts Heart Attack Risk, Study Finds," *New York Times* (12 September 1991), 1; David Brown, "Estrogen's Benefits; Hormone Found to Cut Heart Risks in Women," *Washington Post* (12 September 1991), A1; "Major Study: Estrogen Reduces Risk of Heart Disease in Women," *St. Petersburg Times* (12 September 1991), 1A.

26. Jean Seligman et al., "Not Past Their Prime," *Newsweek* 116 (6 August 1990), 66.

27. Ibid., 67.

28. Ibid., 68.

29. Vrazo, "Menopause Becoming 'Au Courant,'" A1.

30. Julia Kagan and Jo David, "What Every Woman over 35 Needs to Know about Her Body," *McCall's* 118 (June 1991), 60.

31. Martha King, "The Baby Boom Meets Menopause," *Good Housekeeping* 214 (January 1992), 46.

32. Joan Lippert, "Women's Health Update," *Ladies' Home Journal* 109 (January 1992), 68.

33. Renee Asher, "Estrogen: Deciding If It's Right for You," *McCall's* 119 (February 1992), 28; Madeline Chinnici, "The Estrogen Question," *Good Housekeeping* 213 (October 1991), 91.

34. Ronni Sandroff, "Menopause: Is It a Medical Problem?" *New York Times Magazine* (26 April 1992), 22, 24.

35. Jane Gross, "Aging Baby Boomers Take Fresh Look at a Milestone," *New York Times* (17 May 1992), 1.

36. Ibid., 30.

37. Jane E. Brody, "Can Drugs 'Treat' Menopause? Amid Doubt, Women Must Decide," *New York Times* (19 May 1992), C1.

38. Ibid.

39. Ibid, C8.

40. Melinda Beck et al., "Menopause," *Newsweek* 119 (25 May 1992), 71–79.

41. Ibid., 71.

42. Ibid., 79.

43. Nina Darnton, "The Battle of the Bulges," *Newsweek* (2 March 1992), 70.

44. Bickley Townsend, "Boomers Facing 50," *Marketing Research* 4 (Spring 1992), 48.

45. "CBS Evening News" (27 May 1992), CBS News Transcripts, Lexis-Nexis Academic Universe.

46. Gail Sheehy, *The Silent Passage: Menopause* (New York: Random House, 1992). The citations that follow in this section are taken from the paperback edition, published by Pocket Books in 1993.

47. Brenda Warner Rotzoll, "'Silent Passage' Reveals Little on Menopause," *Chicago Sun-Times* (13 May 1992), 51.

48. Diana Morgan, "Gail Sheehy's Lukewarm Flash," *Washington Post* (11 May 1992), B1.

49. Sheehy, *The Silent Passage*, 14.

50. Ibid., 18.

51. Ibid., 198–99.

52. Ibid., 179, 208.

53. Ibid., 182, 189.

54. Ibid., 49.

55. Ibid., 198–99.

56. Ibid., 208.

57. Sherwin A. Kaufman, M.D., "The Truth about Female Hormones," *Ladies' Home Journal* 82 (January 1965), 22.

58. Sheehy, *The Silent Passage*, 221.

59. Germaine Greer, *The Change: Women, Aging and the Menopause* (New York: Alfred A. Knopf, 1992). The citations that follow in this section are taken from the paperback edition, published by Ballantine Books in 1993. *The Change* was first published in Great Britain in 1991 by Hamish Hamilton Ltd., London.

60. Ibid., 378.

61. Ibid., 295.

62. Ibid., 359.

63. See, for example, Natalie Angier, "The Transit of Woman," *New York Times* (11 October 1992), BR1; Sharon Curtin, "Greer: The Change and the Choices," *Washington Post* (20 October 1992), E2; Natalie Danford, "Greer Hails the Aging of Women," *Chicago Sun-Times* (25 October 1992), 12. *The Change* was also reviewed in magazines such as *Time*, *Newsweek*, and the *New Yorker*.

64. Cader Books, "Bestseller Lists 1900–1995," www.caderbooks.com/best90.html.

## Thirteen • Meno-Boomers

1. American College of Physicians, "Guidelines for Counseling Postmenopausal Women about Preventive Hormone Therapy," *Annals of Internal Medicine* 117 (1992), 1038–41.

2. Colin P. Kerr, "Eight Underused Prescriptions," *American Family Physician* 50 (15 November 1994), 1497–1504.

3. The Writing Group for the PEPI Trial, "Effects of Estrogen or Estrogen/Progestin Regimens on Heart Disease Risk Factors in Postmenopausal Women: The Postmenopausal Estrogen/Progestin Interventions (PEPI) Trial," *Journal of the American Medical Association* 273 (18 January 1995), 199–208.

4. Jane E. Brody, "New Therapy for Menopause Reduces Risks," *New York Times* (18 November 1994), A1.

5. P. A. Newcomb and B. E. Storer, "Postmenopausal Hormone Use and Risk of Large-bowel Cancer," *Journal of the National Cancer Institute* 87 (19 July 1995), 1067–71; "Estrogen vs. a Cancer," *New York Times* (19 July 1995), C8; F. Grodstein, G. A. Colditz, and M. J. Stampfer, "Post-menopausal Hormone Use and Tooth Loss: A Prospective Study," *Journal of the American Dental Association* 127 (March 1996), 370–77; "Estrogen Supplements

for Women Could Help to Prevent Tooth Loss," *New York Times* (20 March 1996), C13; M. C. Nevitt et al., "Association of Estrogen Replacement Therapy with the Risk of Osteoarthritis of the Hip in Elderly Women," *Archives of Internal Medicine* 156 (14 October 1996), 2073–80; Susan Gilbert, "Estrogen May Help Prevent Osteoarthritis," *New York Times* (2 October 1996), C11; The Writing Group for the PEPI Trial, "Effects of Hormone Therapy on Bone Mineral Density: Results from the Postmenopausal Estrogen/Progestin Interventions (PEPI) Trial," *Journal of the American Medical Association* 276 (6 November 1996), 1389–96; Jane E. Brody, "Hormone Therapy Can Increase Bone Mass, New Study Says," *New York Times* (6 November 1996), A19; M. X. Tang et al., "Effect of Oestrogen during Menopause on Risk and Age at Onset of Alzheimer's Disease," *Lancet* 348 (17 August 1995), 429–32; "Estrogen May Reduce Alzheimer's Risk in Women, Study Says," *New York Times* (16 August 1996), A18; B. Ettinger et al., "Reduced Mortality Associated with Long-term Postmenopausal Estrogen Therapy," *Obstetrics and Gynecology* 87 (January 1996), 6–12; "Study Finds Estrogen Hormone Helps Postmenopausal Women," *New York Times* (1 January 1996), 12.

6. Jane E. Brody, "New Clues in Balancing the Risks of Hormones after Menopause," *New York Times* (15 June 1995), A1; G. A. Colditz et al., "The Use of Estrogens and Progestins and the Risk of Breast Cancer in Postmenopausal Women," *New England Journal of Medicine* 332 (15 June 1995), 1589–93.

7. Gina Kolata, "Cancer Link Contradicted by New Hormone Study," *New York Times* (12 July 1995), A1; J. L. Stanford et al., "Combined Estrogen and Progestin Hormone Replacement Therapy in Relation to Risk of Breast Cancer in Middle-aged Women," *Journal of the American Medical Association* 274 (12 July 1995), 137–42.

8. Jeremy Greene, "The Abnormal and the Pathological: Cholesterol, Statins, and the Threshold of Disease," in Andrea Tone and Elizabeth Siegel Watkins, eds., *Medicating Modern America: Prescription Drugs in History* (New York: New York University Press, 2007), 216.

9. Larry J. Copeland, ed., *Textbook of Gynecology* (Philadelphia: W. B. Saunders Company, 1993), 620.

10. Ibid., 631.

11. J. Claude Bennett and Fred Plum, *Cecil Textbook of Medicine*, 20th ed. (Philadelphia: W. B. Saunders Company, 1996), 1311–12.

12. Kurt J. Isselbacher et al., eds., *Harrison's Principles of Internal Medicine*, 13th ed. (New York: McGraw-Hill, 1994), 2033.

13. Mark C. Fishman, Andrew R. Hoffman, Richard D. Klausner, and Malcolm S. Thaler, eds., *Medicine*, 4th ed. (Philadelphia: Lippincott-Raven, 1996), 226.

14. Eugene Braunwald, ed., *Heart Disease: A Textbook of Cardiovascular Medicine*, 5th ed. (Philadelphia: W. B. Saunders Company, 1997), 1709.

15. *Obstetrics and Gynecology* 77 (February 1991), A52–54.

16. *Journal of the American Medical Association* 261 (13 January 1989), 221–22.

17. *Obstetrics and Gynecology* 79 (February 1992), A49–51.

18. Ibid., 87 (January 1996).

19. Ibid., 75 (February 1990), inside cover.

20. Ibid., 85 (February 1995).

21. Boris Kaplan et al., "Gynecologists' Trends and Attitudes toward Prescribing Hormone Replacement Therapy during Menopause," *Menopause* 9 (September–October 2002), 354.

22. Jacques E. Roussouw, "Estrogens for Prevention of Coronary Heart Disease: Putting the Brakes on the Bandwagon," *Circulation* 94 (1996), 2982–85.

23. Katherine M. Newton et al., "What Factors Account for Hormone Replacement Therapy Prescribing Frequency?" *Maturitas* 39 (25 July 2001), 1–10.

24. Todd B. Seto et al., "Effect of Physician Gender on the Prescription of Estrogen Replacement Therapy," *Journal of General Internal Medicine* 11 (April 1996), 199.

25. Ibid., 201.

26. Sally E. McNagny, Nanette Kass Wenger, and Erica Frank, "Personal Use of Postmenopausal Hormone Replacement Therapy by Women Physicians in the United States," *Annals of Internal Medicine* 127 (15 December 1997), 1093–96.

27. Ibid., 1096.

28. Ibid., 1095.

29. *Journal of the American Medical Association* 275 (7 February 1996), 352a; *Obstetrics and Gynecology* 87 (February 1996).

30. "Top 200 Prescriptions of 1996," *American Druggist* 214 (February 1997), 30.

31. Ibid., 30–37. Trimox, a generic penicillin, edged into the number one spot, with 40,370,000 prescriptions sold.

32. Diane K. Wysowski, Linda Golden, and Laurie Burke, "Use of Menopausal Estrogens and Medroxyprogesterone in the United States, 1982–1992," *Obstetrics and Gynecology* 85 (January 1995), 6–10; Adam L. Hersh, Marcia L. Stefanick, and Randall S. Stafford, "National Use of Postmenopausal Hormone Therapy: Annual Trends and Response to Recent Evidence," *Journal of the American Medical Association* 291 (7 January 2004), 47–53.

33. Wayne L. Pines, "A History and Perspective on Direct-to-Consumer Promotion," *Food and Drug Law Journal* 54 (1999), 493, 496–97.

34. Francis B. Palumbo and C. Daniel Mullins, "The Development of Direct-to-Consumer Prescription Drug Advertising Regulation," *Food and Drug Law Journal* 57 (2002), 423.

35. Richard Frank et al., "Trends in Direct-to-Consumer Advertising for Prescription Drugs," Report prepared for the Kaiser Family Foundation (February 2002), 10; see www .kff.org/rxdrugs/loader.cfm?url=/commonspot/security/getfile.cfm&PageID=14881. "Top 200 Prescriptions of 1998," *American Druggist* 216 (February 1999), 41.

36. Frank et al., "Trends in Direct-to-Consumer Advertising for Prescription Drugs," 10. Claritin topped the list in both 1998 and 1999, with $139.1 million spent in 1998 and $126.4 million spent in 1999.

37. *Better Homes and Gardens* (March 1999), 66–67.

38. *Prevention* (June 1999), following 120.

39. Veronica A. Ravnikar, "Compliance with Hormone Replacement Therapy: Are Women Receiving the Full Impact of Hormone Replacement Therapy Preventive Health Benefits?" *Women's Health Issues* 2 (Summer 1992), 80; Charles B. Hammond, "Women's Concerns with Hormone Replacement Therapy—Compliance Issues," *Fertility and Sterility* 62 (December 1994), 157S; Dorothy L. Faulkner et al., "Patient Noncompliance with Hor-

mone Replacement Therapy: A Nationwide Estimate Using a Large Prescription Claims Database," *Menopause* 5 (1998), 226.

40. Bruce Ettinger, Alice Pressman, and Paula Silver, "Effect of Age on Reasons for Initiation and Discontinuation of Hormone Replacement Therapy," *Menopause* 6 (1999), 282–89.

41. Hersh, Stefanick, and Stafford, "National Use of Postmenopausal Hormone Therapy," 52. A telephone survey of 495 postmenopausal women between the ages of 50 and 74 found that almost 38 percent were currently using HRT. See Nancy L. Keating et al., "Use of Hormone Replacement Therapy by Postmenopausal Women in the United States," *Annals of Internal Medicine* 130 (6 April 1999), 545–53.

42. Usha Sambamoorthi et al., "Estrogen Replacement Therapy among Elderly Women: Results from the 1995 Medicare Current Beneficiary Survey," *Women's Health Issues* 9 (November–December 1999), 286.

43. Kate M. Brett and Jennifer H. Madans, "Differences in Use of Postmenopausal Hormone Replacement Therapy by Black and White Women," *Menopause* 4 (1997), 66–70; Randall S. Stafford et al., "The Declining Impact of Race and Insurance Status on Hormone Replacement Therapy," *Menopause* 5 (1998), 140–44. See also Keating et al., "Use of Hormone Replacement Therapy by Postmenopausal Women in the United States," 545–53.

44. Nancy E. Avis and Catherine Johannes, "Socioeconomic Status and HRT Use," *Menopause* 5 (1998), 138.

45. Charles W. Schauberger et al., "A Quality Improvement Project to Increase the Use of Postmenopausal Hormone Replacement Therapy (HRT)," *Wisconsin Medical Journal* 95 (October 1996), 697–701.

46. Laurie A. MacDougall, Joshua I. Barzilay, and Charles G. Helmick, "Hormone Replacement Therapy: Awareness in a Biracial Cohort of Women Aged 50–54 Years," *Menopause* 6 (1999), 251–56.

47. Wanda K. Nicholson et al., "Hormone Replacement Therapy for African American Women: Missed Opportunities for Effective Intervention," *Menopause* 6 (1999), 147–55.

48. Louise Pilote and Mark A. Hlatky, "Attitudes of Women toward Hormone Therapy and Prevention of Heart Disease," *American Heart Journal* 129 (June 1995), 1237–38; Donna B. Jeffe, Michael Freiman, and Edwin B. Fisher Jr., "Women's Reasons for Using Postmenopausal Hormone Replacement Therapy: Preventive Medicine of Therapeutic Aid?" *Menopause* 3 (1996), 106–16; Katherine M. Newton et al., "Women's Beliefs and Decisions about Hormone Replacement Therapy," *Journal of Women's Health* 6 (1997), 459–65.

49. Jeffe, Freiman, and Fisher, "Women's Reasons for Using Postmenopausal Hormone Replacement Therapy," 106. See also Patricia Kaufert et al., "Women and Menopause: Beliefs, Attitudes, and Behaviors: The North American Menopause Society 1997 Menopause Survey," *Menopause* 5 (1998), 197–202.

50. Loran M. Salamone et al., "Estrogen Replacement Therapy: A Survey of Older Women's Attitudes," *Archives of Internal Medicine* 156 (24 June 1996), 1293.

51. Douglas S. Rabin et al., "Why Menopausal Women Do Not Want to Take Hormone Replacement Therapy," *Menopause* 6 (1999), 66.

52. Jeffe, Freiman, and Fisher, "Women's Reasons for Using Postmenopausal Hormone Replacement Therapy," 115.

53. Nancy Fugate Woods and Ellen Sullivan Mitchell, "Anticipating Menopause: Observations from the Seattle Midlife Women's Health Study," *Menopause* 6 (1999), 170.

54. L. A. Bastian et al., "Attitudes and Knowledge Associated with Being Undecided about Hormone Replacement Therapy: Results from a Community Sample," *Women's Health Issues* 9 (November–December 1999), 336.

55. Sandra Morgen, *Into Our Own Hands: The Women's Health Movement in the United States, 1969–1990* (New Brunswick, NJ: Rutgers University Press, 2002), 149.

56. Ibid., 147–48.

57. Ibid., 235.

58. Kaufert et al., "Women and Menopause," 197–202.

59. Wulf H. Utian and Isaac Schiff, "NAMS—Gallup Survey on Women's Knowledge, Information Sources, and Attitudes to Menopause and Hormone Replacement Therapy," *Menopause* 1 (1994), 39–48.

60. Julie Kagan and Jo David, "The Facts of Life: What Every Woman over 35 Needs to Know about Her Body," *McCall's* 118 (June 1991), 71.

61. Books in Print search (20 December 2001).

62. Faye Rice, "Menopause and the Working Boomer," *Fortune* 130 (14 November 1994), 203–12; Martha King, "The Baby Boom Meets Menopause," *Good Housekeeping* 214 (January 1992), 46–50.

63. Kathleen Kehoe, "Summary of ERT/HRT Coverage in Lay Press" (May 1991), unpublished paper, National Women's Health Network, Washington, D.C.

64. Patricia Lopez Baden, "Estrogen: Friend or Foe?" *Better Homes and Gardens* 74 (March 1996), 94–95.

65. A search of the Lexis-Nexis database (23 August 2001) of television and radio mentions of menopause and estrogen yielded the following number of "hits" per year: 1990—11, 1991—19, 1992—44, 1993—66, 1994—217, 1995—261, 1996—441, 1997—1000, 1998—806, 1999—439, 2000—937.

66. "CBS This Morning" (23 July 1990), CBS News Transcripts, Lexis-Nexis Academic Universe.

67. "CBS Evening News" (27 May 1992), CBS News Transcripts, Lexis-Nexis Academic Universe.

68. Melinda Beck, "Menopause," *Newsweek* 119 (25 May 1992), 71–72; Julia Kagan and Jo David, "Cosby Copes with Menopause," *McCall's* 118 (June 1991), 66.

69. The commercial was paid for by the Wyeth-Ayerst Women's Health Research Institute (Wyeth-Ayerst, now simply Wyeth, is the manufacturer of the Premarin brand of estrogen replacement therapy and the Prempro brand of combined estrogen-progestin replacement therapy). The commercial was viewed during the ABC Evening News on 21 January 2002. For a discussion of the implications of direct-to-consumer pharmaceutical advertising, see Adele E. Clarke, Laura Mamo, Jennifer R. Fishman, Janet K. Shim, and Jennifer Ruth Fosket, "Biomedicalization: Technoscientific Transformations of Health, Illness, and U.S. Biomedicine," *American Sociological Review* 68 (2003), 178.

70. "Power Surge," www.power-surge.com/educate.htm.

71. Margaret Morganroth Gullette, "What, Menopause Again?" *Ms.* 4 (July–August 1993), 34.

72. Katha Pollitt, "Hot Flash," *The Nation* 254 (15 June 1992), 808.

73. Melinda Beck, "Menopause," *Newsweek* 119 (25 May 1992), cover; Claudia Wallis, "The Estrogen Dilemma," *Time* 145 (26 June 1995), 47.

74. Erica Jong, *Fear of Fifty: A Midlife Memoir* (New York: HarperCollins, 1994), xxii, xxix.

75. Helen Gurley Brown, *The Late Show: A Semiwild but Practical Survival Plan for Women over 50* (New York: Avon Books, 1994).

76. Sara Rimer, "Kicking and Screaming, Baby Boomers Begin to Talk about Aging," *New York Times* (30 March 1998), A10.

77. Sue Hubbell, "A Gift Decade," *New York Times* (19 February 1995), SM25.

78. Lynn Darling, "Age, Beauty and Truth," *New York Times* (23 January 1994), sec. 9, 1, 5.

79. In the same vein as Germaine Greer's *The Change,* see Francine Du Plessix Gray, "The Third Age," *New Yorker* 72 (26 February–4 March 1996), 186–92.

80. Laurence Zuckerman, "A Guru for Women over 40," *Time* 131 (29 February 1988), 67.

81. Caroline Wang, "*LEAR'S* Magazine 'For the Woman Who Wasn't Born Yesterday': A Critical Review," *Gerontologist* 28 (October 1988), 600.

82. *More* 1 (Fall 1997), 101.

83. David M. Eisenberg et al., "Unconventional Medicine in the United States: Prevalence, Costs, and Patterns of Use," *New England Journal of Medicine* 328 (28 January 1993), 246–52.

84. David M. Eisenberg et al., "Trends in Alternative Medicine Use in the United States, 1990–1997," *Journal of the American Medical Association* 280 (11 November 1998), 1569–75.

85. For a comprehensive history of alternative medicine, see James C. Whorton, *Nature Cures: The History of Alternative Medicine in America* (New York: Oxford University Press, 2002).

86. For recent scholarship on the practices of alternative medicine in the twentieth century, see the collection of articles in Robert D. Johnston, ed., *The Politics of Healing: Histories of Alternative Medicines in Twentieth-Century North America* (New York: Routledge, 2004).

87. James C. Whorton, "From Cultism to CAM: Alternative Medicine in the Twentieth Century," in Johnston, ed., *The Politics of Healing,* 303.

88. Susan M. Love with Karen Lindsey, *Dr. Susan Love's Hormone Book: Making Informed Choices about Menopause* (New York: Random House, 1997), quotes on xviii and xvi.

89. Christiane Northrup, *The Wisdom of Menopause* (New York: Bantam Books, 2001), 190.

90. Malcolm Gladwell, "The Estrogen Question: How Wrong Is Dr. Susan Love?" *New Yorker* 73 (9 June 1997), 54–61. For an interesting discussion of women and alternative medicine, see Amy Sue Bix, "Engendering Alternatives: Women's Health Care Choices and Feminist Medical Rebellions," in Johnston, ed., *The Politics of Healing,* 153–80.

91. Kaufert et al., "Women and Menopause," 201.

92. Jane L. Levere, "Advertising: Campaigns for Supplements for That Midlife Event Contend That Mother Nature Knows Best," *New York Times* (18 August 1998), D5.

93. Sheryl Gay Stolberg, "The Estrogen Alternative," *New York Times Sunday Magazine* (6 May 2001), 108.

94. Laurie Tarkan, "Natural Remedies for Menopause Gain Popularity," *New York Times* (20 June 2000), F12, F7.

95. For surveys of women's attitudes on menopause and hormone replacement therapy, see Utian and Schiff, "NAMS—Gallup Survey," 39–48, and Kaufert et al., "Women and Menopause," 197–202. For a discussion of the disconnection between increased media coverage and better-informed consumers on the subject of women's health risks, more broadly, see Cristine Russell, "Hype, Hysteria, and Women's Health Risks: The Role of the Media," *Women's Health Issues* 3 (Winter 1993), 191–97.

96. Deborah Lupton explores the parallels and tensions between mainstream and feminist arguments about the use of hormone replacement therapy by menopausal women in her article "Constructing the Menopausal Body: The Discourses on Hormone Replacement Therapy," *Body and Society* 2 (1996), 91–97.

97. For an interesting discussion of aging women and the youth-centered culture of postwar America, see Elizabeth Haiken, *Venus Envy: A History of Cosmetic Surgery* (Baltimore: Johns Hopkins University Press, 1997), chap. 4. See also Lois Banner, *In Full Flower: Aging Women, Power, and Sexuality* (New York: Knopf, 1992).

## Fourteen • The "Gold Standard"

1. See, for example, Rita Rubin, "U.S. Halts Study on Hormone Therapy," *USA Today* (9 July 2002), 1A; Gina Kolata, "Study Is Halted over Rise Seen in Cancer Risk," *New York Times* (9 July 2002), 1; Patricia Guthrie, "Study Says Halt Hormone Therapy," *Pittsburgh Post-Gazette* (9 July 2002), 1; Cheryl Clark, "Risks Cited in Hormone Therapy for Women; Harm Reportedly Greater Than Benefit; Trial Halted," *San Diego Union-Tribune* (9 July 2002), A-1.

2. Melody Petersen, "Wyeth Stock Falls 24% after Report," *New York Times* (10 July 2002), A16.

3. Melody Petersen, "Survey Halted, Drug Makers Seek to Protect Hormone Sales," *New York Times* (17 July 2002), C1.

4. The figures on the decrease in Premarin and Prempro sales come from NDC-Health's Web site (www.ndchealth.com/index.asp). See also Adam L. Hersh, Marcia L. Stefanick, and Randall S. Stafford, "National Use of Postmenopausal Hormone Therapy: Annual Trends and Response to Recent Evidence," *Journal of the American Medical Association* 291 (7 January 2004), 47–53.

5. The information in this section is based on interviews conducted by the author with Deborah Grady and Stephen Hulley in San Francisco on 21 November 2005.

6. Stephen Hulley et al., "Randomized Trial of Estrogen Plus Progestin for Secondary Prevention of Coronary Heart Disease in Postmenopausal Women," *Journal of the American Medical Association* 280 (19 August 1998), 605.

7. Ibid., 605–13.

8. Diana B. Petitti, "Hormone Replacement Therapy and Heart Disease Prevention:

Experimentation Trumps Observation," *Journal of the American Medical Association* 280 (19 August 1998), 651.

9. Leon Speroff, "The Heart and Estrogen/Progestin Replacement Study," *Maturitas* 31 (1998), 13.

10. Trudy L. Bush, "Lessons from HERS: The Null and Beyond," *Journal of Women's Health* 7 (September 1998), 781–83.

11. For a rejection of the HERS findings, see C. Lauritzen, "A Critical European View of the HERS Trial," *Maturitas* 31 (1998), 15–19.

12. CBS Evening News (10 November 1998), abstract from Vanderbilt Television News Archive, http://tvnews.vanderbilt.edu.

13. Serge Rozenberg, Caroline Fellemans, and Hamphrey Ham, "Opinion Study towards Hormone Replacement Therapy in the Prevention of Coronary Heart Disease," *Maturitas* 38 (2001), 273–77.

14. "Study Casts Doubt That Estrogen Cuts Women's Heart Disease," *New York Times* (19 August 1998), A20.

15. Ron Winslow, "Study Raises New Questions about Estrogen," *Wall Street Journal* (19 August 1998), B1; "Study Raises Doubt about an Estrogen Benefit," *Washington Post* (19 August 1998), A9; Brenda C. Coleman, "Estrogen May Be Overrated; Study Doubts Heart Benefit," *Chicago Sun-Times* (19 August 1998), 30.

16. ABC Evening News, CBS Evening News, CNN Evening News (18 August 1998), abstracts from Vanderbilt Television News Archive, http://tvnews.vanderbilt.edu.

17. "Menopause: A Guide to Smart Choices," *Consumer Reports* 64 (January 1999), 50.

18. Adam L. Hersh, Marcia L. Stefanick, and Randall S. Stafford, "National Use of Postmenopausal Hormone Therapy," *Journal of the American Medical Association* 291 (7 January 2004), 47–53.

19. Susan Thaul and Dana Hotra, eds., *An Assessment of the NIH Women's Health Initiative* (Washington, DC: National Academy Press, 1993), 19. I am grateful to Florence Haseltine for alerting me to this report.

20. Ibid., 37–38.

21. Ibid., 1.

22. Gina Kolata, "Estrogen Use Tied to Slight Increase in Risks to Heart," *New York Times* 149 (April 5, 2000), 1.

23. Hersh, Stefanick, and Stafford, "National Use of Postmenopausal Hormone Therapy," 47–53.

24. Guthrie, "Study Says Halt Hormone Therapy," 1; Kolata, "Study Is Halted over Rise Seen in Cancer Risk," A1; Patricia Guthrie, "Health Risk to Women Halts Hormone Study," *Atlanta Journal-Constitution* (9 July 2002), 1A.

25. Writing Group for the Women's Health Initiative, "Risks and Benefits of Estrogen Plus Progestin in Health Postmenopausal Women: Principal Results from the Women's Health Initiative Randomized Controlled Trial," *Journal of the American Medical Association* 288 (17 July 2002), 321–33.

26. Ibid., 333.

27. Leigh Hopper, "Menopause Drug Warning Issued," *Houston Chronicle* (9 July 2002), A1.

28. Guthrie, "Health Risk to Women Halts Hormone Study," 1A.

29. Suzanne W. Fletcher and Graham A. Colditz, "Failure of Estrogen Plus Progestin Therapy for Prevention," *Journal of the American Medical Association* 288 (17 July 2002), 366.

30. Ibid., 367.

31. Mary Duenwald, "Patients Weigh Quitting Drug after Research Indicates Risk," *New York Times* (10 July 2002), 16.

32. Gina Kolata with Melody Petersen, "Hormone Replacement Study a Shock to the Medical System," *New York Times* (10 July 2002), A1.

33. Lila E. Nachtigall, unpublished correspondence (2002).

34. "Amended Report from the NAMS Advisory Panel on Postmenopausal Hormone Therapy" (6 October 2002).

35. "Quality of Life Improved with Postmenopausal Hormone Therapy," NAMS Press Release (2 October 2002).

36. "The Menopause Experience Swings from Mild to Awful," *Pittsburgh Post-Gazette* (23 July 2002), F-4.

37. Alice E. Bisk, "Letter to the Editor," *New York Times* (22 July 2002), 16.

38. John C. Stevenson and Malcolm I. Whitehead, "Hormone Replacement Therapy: Findings of Women's Health Initiative Trial Need Not Alarm Users," *British Medical Journal* 325 (20 July 2002), 113–14.

39. Gina Kolata, "Replacing Replacement Therapy," *New York Times* (27 October 2002), IV-2.

40. Gina Kolata, "F.D.A. Orders Warning on All Estrogen Labels," *New York Times* (9 January 2003), A18.

41. Jennifer Hays et al., "Effects of Estrogen Plus Progestin on Health-Related Quality of Life," *New England Journal of Medicine* 348 (8 May 2003), 1839–54.

42. Sally A. Shumaker et al., "Estrogen Plus Progestin and the Incidence of Dementia and Mild Cognitive Impairment in Postmenopausal Women," *Journal of the American Medical Association* 289 (28 May 2003), 2651–62.

43. Stephen R. Rapp, "Effect of Estrogen Plus Progestin on Global Cognitive Function in Postmenopausal Women," *Journal of the American Medical Association* 289 (28 May 2003), 2663–72.

44. Sylvia Wassertheil-Smoller et al., "Effect of Estrogen Plus Progestin on Stroke in Postmenopausal Women," *Journal of the American Medical Association* 289 (28 May 2003), 2673–84.

45. Rowan T. Chlebowski et al., "Influence of Estrogen Plus Progestin on Breast Cancer and Mammography in Healthy Postmenopausal Women," *Journal of the American Medical Association* 289 (25 June 2003), 3243–53.

46. Christopher I. Li et al., "Relationship between Long Durations and Different Regimens of Hormone Therapy and Risk of Breast Cancer," *Journal of the American Medical Association* 289 (25 June 2003), 3254–63.

47. Deborah Grady, "Postmenopausal Hormones—Therapy for Symptoms Only," *New England Journal of Medicine* 348 (8 May 2003), 1835–37; Kristine Yaffe, "Hormone Therapy and the Brain: Déjà vu All over Again?" *Journal of the American Medical Association* 289 (28

May 2003), 2717–19; Peter H. Gann and Monica Morrow, "Combined Hormone Therapy and Breast Cancer: A Single-Edged Sword," *Journal of the American Medical Association* 289 (25 June 2003), 3304–6.

48. Gina Kolata, "Hormone Therapy, Already Found to Have Risks, Is Now Said to Lack Benefits," *New York Times* (18 March 2003), A26.

49. "Delusions of Feeling Better," *New York Times* (19 March 2003), A26.

50. The Women's Health Initiative Steering Committee, "Effects of Conjugated Equine Estrogen in Postmenopausal Women with Hysterectomy," *Journal of the American Medical Association* 291 (14 April 2004), 1701–12.

51. Ibid.

52. Ibid., 1710.

53. Sally A. Shumaker et al., "Conjugated Equine Estrogens and Incidence of Probable Dementia and Mild Cognitive Impairment in Postmenopausal Women," *Journal of the American Medical Association* 291 (23–30 June 2004), 2947–58.

54. Mark A. Espeland et al., "Conjugated Equine Estrogens and Global Cognitive Function in Postmenopausal Women," *Journal of the American Medical Association* 291 (23–30 June 2004), 2959–68.

55. Robert L. Brunner et al., "Effects of Conjugated Equine Estrogen on Health-Related Quality of Life in Postmenopausal Women with Hysterectomy," *Archives of Internal Medicine* 165 (26 September 2005), 1976–86.

56. Ibid., 1978.

57. Interview with Wulf Utian by the author in San Diego on 29 September 2005.

58. Sumit R. Majumdar, Elizabeth A. Almasi, and Randall S. Stafford, "Promotion and Prescribing of Hormone Therapy after Report of Harm by the Women's Health Initiative," *Journal of the American Medical Association* 292 (27 October 2004), 1983–88.

59. Judith Parsells Kelly et al., "Use of Postmenopausal Hormone Therapy since the Women's Health Initiative Findings," *Pharmacoepidemiology and Drug Safety* 14 (5 April 2005), 837–42.

60. Jennifer S. Haas et al., "Changes in the Use of Postmenopausal Hormone Therapy after the Publication of Clinical Trial Results," *Annals of Internal Medicine* 140 (3 February 2004), 184–89.

61. Rob Stein, "Hormone Therapy Proves Tough to Quit," reprinted in *Pittsburgh Post-Gazette* (7 August 2003), A-1.

62. Ibid., A-12.

63. Virginia Linn, "HRT Update: Many Women go Back On," *Pittsburgh Post-Gazette* (16 December 2003), E-1.

64. Priscilla Grant, "HRT: Ten Women, Ten Choices," *More* (July–August 2003), 102–6, 146.

65. NDCHealth (www.ndchealth.com/index.asp); Hersh, Stefanick, and Stafford, "National Use of Postmenopausal Hormone Therapy," 47–53.

66. Majumdar, Almasi, and Stafford, "Promotion and Prescribing of Hormone Therapy after Report of Harm," 1983–88.

67. Grant, "HRT: Ten Women, Ten Choices," 102.

68. Nancy Gibbs, "Midlife Crisis? Bring It On!" *Time* 165 (16 May 2005), 53.

69. Ibid., 54.

70. Christine Gorman, "Menopause: A Healthy View," *Time* 165 (16 May 2005), 57.

71. NIH State-of-the-Science Panel, "National Institutes of Health State-of-the-Science Conference Statement: Management of Menopause-Related Symptoms," *Annals of Internal Medicine* 142 (21 June 2005), 1003–13.

72. S. M. Harman, "KEEPS: The Kronos Early Estrogen Prevention Study," *Climacteric* 8 (Match 2005), 3–12. See also www.kronosinstitute.org.

73. Suzanne Somers, *The Sexy Years: Discover the Hormone Connection* (New York: Crown, 2004).

74. *The Sexy Years* debuted at number three on the *New York Times Book Review*'s "Advice, How-to and Miscellaneous: Hardcover Best Sellers List" on 28 March 2004. See also the full-page ad in the *New York Times* (25 March 2004), B9.

75. Celeste Fremon, "Suzanne Somers: Still Sexy (After All These Years)," *Good Housekeeping* 238 (March 2004), 128–31.

76. *New York Times* (25 March 2004), B9.